Urban Geography

Urban Geography is a science of comparatively recent origin. It is concerned with the nature and function of towns, with their situations and sites, their relations with each other and with the countryside, their plans and planning problems, and the people who live in them. Its development as a branch of Human Geography has taken place mainly in the post-war period, and although there is already a vast amount of written material on the subject, very few treatises have been produced in recent years. As new cities are now being built in every part of the world and old ones extended and transformed, there is a need for a comprehensive survey of the whole field, and this book accordingly provides an up-to-date summary of the main facts and theories of Urban Geography.

GEOGRAPHIES FOR ADVANCED STUDY
EDITED BY PROFESSOR S. H. BEAVER M.A., F.G.S.

Urban Geography

GEOGRAPHIES FOR ADVANCED STUDY
Edited by Professor Stanley H. Beaver, M.A., F.G.S.

The Tropical World
The Soviet Union
Malaya, Indonesia, Borneo and the Philippines
West Africa
The Scandinavian World
A Regional Geography of Western Europe
The British Isles—A Geographic and Economic Survey
Central Europe
Geomorphology
Statistical Methods and the Geographer
The Polar World
An Historical Geography of South Africa
North America
Land, People and Economy in Malaya
The Western Mediterranean World
Human Geography
Geography of Population
Historical Geography of Western Europe before 1800

Urban Geography

J. Beaujeu-Garnier
PROFESSOR OF GEOGRAPHY AT THE SORBONNE

and

G. Chabot
PROFESSOR OF GEOGRAPHY AT THE SORBONNE

Translated by G. M. Yglesias
and S. H. Beaver

JOHN WILEY & SONS INC
New York N.Y.

Published throughout the world except the United States by Longmans, Green & Co. Ltd.

English translation © Longmans, Green and Co. Ltd., 1967

FIRST PUBLISHED 1967
SECOND IMPRESSION 1969

Printed in Great Britain by
Spottiswoode, Ballantyne & Co. Ltd., London and Colchester

FOREWORD

For a long time the town as a concept remained to some extent a matter of speculation. People spoke of the town, but its affairs were left in the hands of architects and mayors, and geographers found no place for it except in their regional surveys.

It was not until the end of the nineteenth century that urban geography made its appearance, notably in Germany, with Friedrich Ratzel. In 1907 K. Hassert devoted a special work to it: *Die Städte Geographisch Betrachtet*.

Fifteen years ago, a book by one of us appeared which, for the first time in France, was devoted to urban geography. Since that time further books and articles have appeared on the same subject.[1] A place is found for it in all geographical congresses. Studies have been devoted to tropical cities, hitherto practically neglected. Man has been put into his environment and the cities have sprung to life: urban populations have been the subject of widespread research.

At the same time new cities have sprung up, old ones have been extended and transformed and their buildings have escalated skywards. Urban problems have not only multiplied, they have also taken a new form. Urban geography needs rethinking.

It is true that statistical records are far from perfect; we cannot draw final conclusions from them. But at least they provide a basis for hypothesis and thus suggest themes for further research.

We are aware that geography and cartography form a unity, and we try to give a cartographical interpretation of cities.

This exposition can therefore be considered as a reappraisal. It is designed for students of urban geography and for all who are concerned with the problems of cities and who see in them the main element in all regional organisation. It endeavours to make available the ideas and methods which have come to light in various countries.

---

[1] For example, Pierre George, *La Ville. Le fait urban*, Paris, 1952, and *Précis de géographie urbaine*, Paris, 1961, Pierre Lavedan, *La géographie des villes*, 2nd edn., Paris, 1959.

CONTENTS

Foreword   v

Chapter 1   The Urban Phenomenon: Urban Concentration   1

Development of urban concentration   2
The attraction of cities and the rural exodus   21

Chapter 2   *Definition of the Town*   23

Chapter 3   *The Cartographic Representation of Towns*   32

Part One   The world's towns   39

Chapter 4   *European Towns*   41

The towns of Norden   41
         West-Central Europe   48
         Mediterranean Europe   54
         Eastern Europe   59

Chapter 5   *The U.S.S.R.*   61

Old Russia and Eastern Europe   61
Soviet Asia   65

Chapter 6   *Australasia and the Americas*   69

Australia and New Zealand   70
America North of the Rio Grande   72
America South of the Rio Grande   78

Chapter 7   *North Africa and non-Soviet Asia*   83

North Africa   83
The 'Middle East'   86
South and South-East Asia   88
The Far East   90

Chapter 8   *Africa South of the Sahara*   94

Part Two Urban Functions — 102

Chapter 9 The Origin of Towns — 103

Chapter 10 Urban Functions: General Principles — 106

Definition and classification — 106
Determination of urban functions — 108
Function and geographical situation — 113
Representation of urban functions — 114

Chapter 11 The Military Function — 119

Chapter 12 The Commercial Function — 123

Primitive forms of trade — 123
Continental trade — 125
Commercial centres — 129
Sea ports — 133
Classification of ports — 137
The towns of ports — 142
Airports — 146
The Emporium — 147

Chapter 13 The Industrial Function — 148

Mining towns — 149
Manufacturing industries — 154

Chapter 14 The Cultural Function — 168

Religious associations — 168
University towns — 171
Literary and artistic centres — 175
Museum towns — 176
Festival and congress towns — 177

Chapter 15 The Town as a Resort — 179

Hospital towns and spas — 179
Holiday resorts — 182
Towns for the retired — 187

CONTENTS

Chapter 16 The Administrative and Political Function	189
The administrative rôle	189
National Capital	190
Supercapitals	196
Chapter 17 Variations and Plurality of Functions	197
How functions change	197
The Metropolis	200

Part Three The Plan and Extent of Towns 205

Chapter 18 The Part Played by the Site	207
Chapter 19 The Town Plan	211
The Gridiron Plan	212
The radioconcentric plan	214
The linear town	216
Adaptation to obstacles	217
Remodelling the plan	218
Chapter 20 The Expansion of Towns	221
Accretion	221
The star	225
Absorption of villages	226
Shanty towns	227
Polynuclear expansion	228
Planned expansion	233
Chapter 21 The Suburbs	238
The various types	239
The development of the suburbs	240
Chapter 22 Town Clusters and Conurbations	243
Conurbations: definition	243
'Interurbations'	246
The development of a conurbation	247
'Megalopolis'	249
Chapter 23 Satellite Towns	252

Chapter 24 Agglomerations	256

Part Four   Life in Towns 262

Chapter 25 The Concept of Urban Concentration	263
The extent of urban concentration	263
Comparison of urban densities	267
Degrees of concentration	270
Factors in concentration	274
Chapter 26 Problems of Space	277
Difficulties of expansion and administration	277
The need for space	278
The price of land	281
Upwards and outwards	284
Land use within towns	288
Industrial zones	292
The central areas	297
Residential areas	304
Movement and transport	313
Supplies	321
The problem of water	331
Disposal of waste	337
Urban saturation	340
Urban renewal	343
Chapter 27 The Urban Population	355
The urban melting pot	355
Sex composition	358
Age composition	362
Families in towns	369
The health of the citizens	377
Social behaviour	382
Socio-professional classes	385
Social mobility	391

Part Five   The Town and its Region 393

Chapter 28 The Nature of Urban Influence	395
Urban influence on agriculture	395
Supplying the town with food	397
Supplying the town with raw materials	400

The town as supplier of the region	401
Industry in the countryside	405
Administration	407
Medical services	407
Cultural influences	407
Consciousness of urban influence	410
Recreational links	411

Chapter 29   Spheres of Influence   412

Methods of investigation	412
The town in regional demography	414
Zones of Influence	425

Chapter 30   Conclusion   449

Bibliography	453
Index	467

ACKNOWLEDGEMENTS

The photographs are from the author's own collections with the exception of the following:

Plate 5, Fox Photos Ltd; Plates 13,14, Dupaquier; Plate 15, Pictorial Press; Plate 21, Barnaby's Picture Library; Plate 22, Mike Andrews.

PLATES

PLATE		FACING PAGE
1	Reykjavik: quarter near the port	48
2	Trondheim: warehouses in the port	48
3	Leeds: working-class housing and industry	49
4	Vällingby: shopping centre in the new town	49
5	London: the city seen from the dome of St Paul's. Bomb-damaged areas partly replaced by new buildings, partly remaining as car parks (1960)	80
6	Frankfurt-am-Main: reconstruction (1953)	81
7	Warsaw: postwar reconstruction of old market square	81
8	Venice: the city rising from the waters; Lido in distance	176
9	Monte-Carlo and the rocky promontory of Monaco	176
10	Rome: the Forum and Trajan's market; on the right, the Loggia of the Knights of Rhodes	177
11	Palermo: narrow street near the port	177
12	Rome: Piazza del Popolo from the Pincio gardens; in the distance the dome of St Peter's	177
13	Moscow: the Moskva river from the Gorki park	208
14	Tashkent: a high school amidst mud huts	208
15	New York by night	209
16	Sydney: the geometry of suburbia	240
17	Noumea (New Caledonia): gridiron pattern of town centre, with dispersed housing on the hills	240
18	Salvador (Bahia): the core of the old city	241
19	Salvador (Bahia): housing on stilts in the lagoon	241
20	São Paulo: the transformation of the city centre	272
21	Rio de Janeiro: fragmented sited around the bay (Sugar-Loaf in centre); sky-scrapers rising amongst the old villas	272
22	Brasilia: city of the year 2,000	273
23	Brazil: a present day shanty-town	273
24	Ghardaia: market place	336
25	Moulay Idris: ancient Moroccan town	336
26	Istanbul: old houses and the Blue Mosque	337
27	Cairo: citadel, Mehemet mosque and south-eastern quarter seen from the minaret of Ibn Tulan mosque	337
28	Bombay: the English quarter	368
29	Bombay: old Indian houses	368
30	Bangkok: the floating quarter	369
31	Kankan): hutments	369
32	Labé (Guinea): market place	369

MAPS AND DIAGRAMS

FIGURE PAGE

1 Towns with over 100,000 population 6–7
2 Major towns in the USSR: centres and satellites 10
3 Evolution of towns and built-up areas in the United States and India (1940–1960) 12
4 Various methods of cartographic representation of towns and urbanisation 33
5 Diagram representing the same urban function in different towns 111
6 Diagram showing the distribution of towns according to their functions 114
7 The development of town functions in Rumania from 1930 to 1956 115
8 Cartographic representation of urban functions 117
9 Various types of development of port towns 144-5
10 Plan of gridiron town: Kuopio, in Finland 211
11 San Francisco: Gridiron plan to the west of the Bay 213
12 An example of a radioconcentric plan: Milan 214
13 An example of a radioconcentric plan: Beaune 215
14 Plan of a linear town 216
15 The expansion of Vienna 232
16 Megalopolis 248
17 Population densities of six districts of Paris in relation to the total surface area 264
18 Population densities of six central districts of Paris 265
19 Variations in density from the centre to the perimeter in London 272
20 Densities of São Paulo (different administrative areas) and the large towns of Brazil 273
21 Evolution of population profile across Tokyo 274
22 Three types of urban 'zoning' 290
23 Urban structure of Chicago 291
24 Location of factories in Roubaix–Tourcoing 293
25 Commercial structure of Paris 294
26 The urban structure of Calcutta 296
27 The business centre of São Paulo 301
28 Diagram of population distribution and commuting in the New York region 302
29 Zones of the New York district 303
30 Annual variation in treatment of domestic refuse in Paris 339
31 Development of Soviet towns 346
32 The new Japanese town of Senri 353
33 Age pyramids of Lens, Cannes, Tulle and Dakar 366
34 Dakar. Age pyramid for European population 367
35 Triangular diagram for certain large towns 368

36 Urban vegetable supplies in southern Finland	403
37 The zone of influence of Växjö as a shopping centre	404
38 Newspapers 'hinterlands' in the Rhine–Main region	409
39 An example of isochrones	413
40 Commuting to New York	418
41 Incomes and dominant occupations in the New York region	419
42 Commuting into Lille	422
43 Commuting outwards from Lille	423
44 Zones of influence of French towns with more than 50,000 inhabitants	430
45 Respective zones of influence of three centres	433
46 Theoretical scheme for urban influence	434
47 Theoretical scheme of urban hierarchy	437
48 The urban mesh of Mediterranean Languedoc	441
49 The urban mesh of the Rhenish Palatinate	443

CHAPTER 1

THE URBAN PHENOMENON: URBAN CONCENTRATION

The urban phenomenon[1] is, without doubt, one of the most striking features of contemporary civilisation. Even the use of the expression 'phenomenon', signifying an astonishing appearance, demonstrates something paradoxical in the development of cities. We do not, after all, speak of a rural phenomenon, as it seems to us perfectly natural that men should scratch the earth and sow seed in order to gain their bread. Perhaps there was a time when even that seemed extraordinary, when men, ceasing to drive their flocks before them, stayed in their own fields. But since prehistoric times there has never been any question of finding anything astonishing in rural life.

It is true to say that the existence of cities did not raise great problems in bygone days. It seemed normal that men should congregate at certain points. Mediterranean civilisation embraced a large number of cities. When the site of Troy was excavated, several cities were discovered to be superimposed on the original and the Troy of Homer's day was by no means the most ancient. We have even formed the habit of distinguishing periods of ancient history by the names of cities: Thebes, Memphis, Babylon, Athens, Sparta; all these are used as labels of glory. Cities were born later in western Europe, later still in the north. They developed gradually and this growth seemed to belong to the natural order of things. Everything was changed in the second half of the nineteenth century. Towns arose almost overnight, notably in America, and the expression 'mushroom growth' became familiar. New cities and new extensions to old ones increased at a raging speed. It was then that we became aware of the urban phenomenon. Historians are readily able to distinguish between the pre-industrial and the industrial age into which different countries have entered at varying speeds for the last century. These are economic terms. The most obvious, the most spectacular and the most geographical aspect of the industrial phase is precisely the development of cities, the human stampede towards ill-defined agglomerations.

The pace has increased in recent times. Formerly there were peaks and periods of growth and decline which were largely dependent on the

[1] Cf. Georges Chabot, 'À propos du phénomène urbain' in *Mélanges géographiques offerts au Doyen E. Bénévent*, 1954; Philippe Pinchemel, 'Le phénomène urbain', *Revue de l'Action populaire*, 1963.

flow of trade. The Hellenic period and the Roman Empire witnessed a flowering of cities. The Middle Ages became urban from the moment that travel became easier. The ninth century, predominantly rural, is contrasted by the historian Henri Pirenne with the eleventh, when towns multiplied. By contrast the barbarian invasion of Europe initiated a period of decline.

DEVELOPMENT OF URBAN CONCENTRATION

It was, however, in the nineteenth century that the great movement of urban concentration, the effects of which are felt today, was born. From the beginning of the first half of the century, urban population increased relatively far more rapidly than total population and this tendency continues.[1]

	Growth of world population %	Growth of urban population (1) %
1800–1850	29·2	175·4
1850–1900	37·3	192
1900–1950	49·2	228

(1) Places of over 5,000 inhabitants.

The economic revolution, by concentrating the means of production through the transition from cottage industry to factories, brought the workers together. Capitalism promotes trade and concentrates it; Socialism attracts populations to the centres of industry; Colonialism creates cities of whites surrounded by native populations; transfers of population, often accentuated by political partition (for example, the Indian Federation and Pakistan), in uprooting men contribute to the flow towards the cities. The development of transport has been the determining factor. For a long time it was possible to transport foodstuffs and goods in reasonably large quantities by waterways but, at least until the arrival of steam navigation, this traffic was slow and hazardous. As for land traffic, this had hardly progressed since ancient times until the period when the railways became the fundamental instrument of change and stimulated the growth of the urban network. More recently the progress of road transport has accentuated certain characteristics of this development.

[1] K. Davis and H. Hertz, 'Patterns of World Urbanisation', *Bull. Int. Stat. Inst.*, 1954; Ph.-M Hauser, *Le Phénomène de l'Urbanisation en Asie et en Extrême-Orient*, Calcutta, 1959, Introduction, pp. 53 ff.

Even the demand for the production of food supplies is no longer enough to limit urbanisation. For a long time cities developed according to the provision that the neighbouring countryside could afford; this restriction disappeared with the development of transport which allowed them to obtain provisions from all over the world. It is significant that even catastrophes have not succeeded in interrupting the growth of cities. Wars, which destroyed so many European cities, were only the starting point, after a brief pause, for new urban growth. And an urban mentality developed which was itself a factor of urbanisation. The Canadian west took on straight away an urban character because the immigrants brought with them European customs very different from those brought by the French-Canadians.

There were about 492,000 inhabitants of Paris at the time of Louis XIV and 548,000 on the eve of the First Empire, in 1801: progress was thus almost imperceptible; with the start of the railway network there was a considerable acceleration, and the city registered its first million inhabitants in 1851. The effects of railways have also proved significant for towns of lesser importance. About 1830 Alençon and Le Mans played practically identical roles in the heart of rich agricultural country and had approximately equal populations. In the 1850s Le Mans was clever enough to secure the route of the main line from Paris to Brittany (the station was opened in 1853)[1] and now has 132,000 inhabitants; Alençon has barely 20,000. Springfield and Northampton in the north-east of the United States had each, about 1,800, a little over 2,000 inhabitants. The former secured the route of the east–west railway from Boston, in 1839, and is an active centre of 163,000 people while the latter vegetates with fewer than 30,000.[2]

The results of this dynamic urbanisation are shown by the statistics. Since the beginning of the nineteenth century up to the end of the first half of the twentieth the population of the world has multiplied by 2·64; during this period the population living in localities of 5,000 inhabitants or more has increased exactly ten times more rapidly (multiplied by 26·3).[3] (See table overleaf.)

An increasing part of the world's population occupies larger and larger cities. About 1800 it was estimated that there were 750 towns with a population of more than 5,000, 200 towns with more than 20,000 and 45 with more than 100,000. In 1950 the number of these urban centres rose to 27,600, 5,500 and 875 respectively. At the present time nearly one-third of the world's population is concentrated in urban areas with populations of more than 500,000; more than a

[1] J. Garnier, 'Le Mans', *La vie urbaine*, 1939, pp. 135–59.
[2] G. Cestre, *Evolution urbaine de Northampton (Massachusetts)*, Paris, 1963, pp. 52, 326.
[3] Davis and Hertz, *op. cit.*

Relative Growth of Different Categories of Towns

	1800–1950 Multiplied by
World population	2·64
Localities of over 5,000 inhabitants	26·3
Localities of over 100,000 inhabitants	20·1

fifth in localities with more than 20,000 inhabitants and 13 per cent in towns with over 100,000 inhabitants.

In 1800 there was no city in the world whose population reached or even approached a million inhabitants. In 1950, 133 cities boasted a population of more than 500,000 and forty-nine cities had more than a million.

The rate of growth has been estimated in the following manner:[1]

Comparison between Urban Population of the World and Total Population (1800–1950)

		Urban localities				
		5,000 and over		20,000 and over		100,000 and over
Year	World population (millions)	Population (millions)	Percentage of world population	Population (millions)	Percentage of world population	Population (millions)	Percentage of world population
1800	0·906	27·2	3·0	21·7	2·4	15·6	1·7
1850	1·171	74·9	6·4	50·4	4·3	27·5	2·3
1900	1·608	218·7	13·6	147·9	9·2	88·6	5·5
1950	2·400	716·7	29·8	502·2	20·9	313·7	13·1

The rate continues to accelerate and shows no signs of slowing down. For cities of more than 100,000 the population growth was 75 per cent between 1800 and 1850, 220 per cent between 1850 and 1900 and 250 per cent between 1900 and 1950. The rate continues and as more precise statistics become available, its natural development is reinforced by new statistical accuracy. In 1960 it was estimated that there were 1,413 agglomerations of more than 100,000 and seventy-one of more than a million.[2] This wave of urbanisation did not occur

[1] Hauser, *op. cit.*, p. 56.
[2] *Demographic Year Book of the United Nations*, 1960, Tables 7 and 8.

everywhere at a uniform speed and does not show the same pattern of distribution throughout the world.

In Europe, where the modern economic revolution was born and developed, urban expansion was very strong from the beginning of the nineteenth century, the populations of towns containing more than 100,000 inhabitants increased by 150 per cent between 1800 and 1850 and by 250 per cent from 1850 to 1900. On the other hand since 1900 a distinct slowing up of the movement is noticeable and the rate of growth rose to only 97 per cent. Certain countries such as the United Kingdom seem already to have uprooted all the available rural population and the movement is now from one town to another rather than from the countryside towards the towns. In England and Wales the percentage of urban population was already 77 per cent in 1901 and had only risen to 80·7 per cent by 1951.[1]

In North America and Australia, countries where the population was inflated in the nineteenth century and the beginning of the twentieth by a considerable influx of Europeans, the rate of urbanisation began later and developed on a much larger scale. It also continued longer than in Europe. In Australia the population in towns of more than 100,000 quadrupled between 1900 and 1950 while in Europe the figure only doubled. In the United States the number of urban centres increased sixfold between 1800 and 1850, multiplied by twelve between 1850 and 1900 and again by three between 1900 and 1950.[2]

In the countries which are now in the middle of recent rapid development, the growth of the number of cities which began relatively recently is going on at greater speed. Japan and above all the U.S.S.R. can be taken as examples. The population of Russian towns of more than 100,000 inhabitants doubled between 1800 and 1850, quintupled between 1850 and 1900 and has been multiplied by seven between 1900 and 1950. The rate of change has also accelerated in the underdeveloped countries, but for quite other reasons. In Asia in so far as we can trust the available statistics, the population of towns of more than 100,000 only increased by 24 per cent between 1800 and 1850 and by 50·2 per cent between 1850 and 1900 but it has more than quintupled between 1900 and 1950. It is estimated that the speed of urban growth in India was the same between 1910 and 1950 as that of the United States between 1810 and 1850. In Japan and in India the more important the agglomeration, the stronger is the growth of population, but this does not seem to be the case in all regions of Asia: Malaya provides an exception. However, Asiatic urbanisation,

[1] A. Guilcher and J. Beaujeu-Garnier, *Les Iles britanniques*, Paris, 1963, pp. 163 ff.
[2] J. Bogue, *The Population of the United States*, New York, 1959, pp. 19 ff.

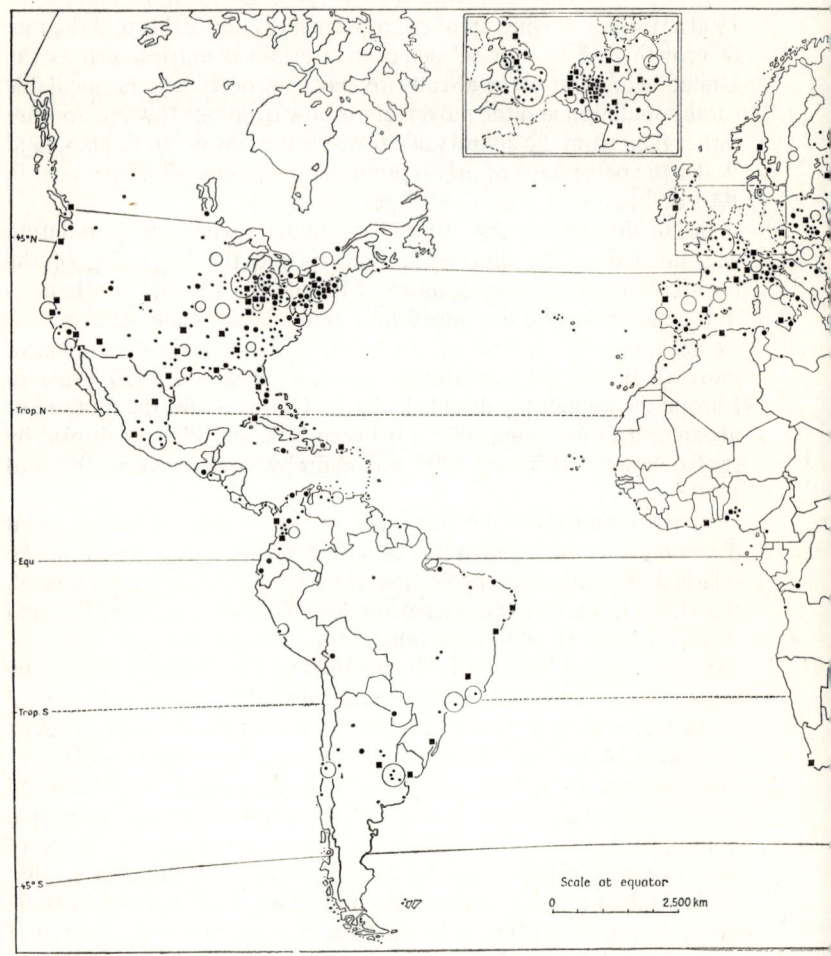

URBAN CONCENTRATION

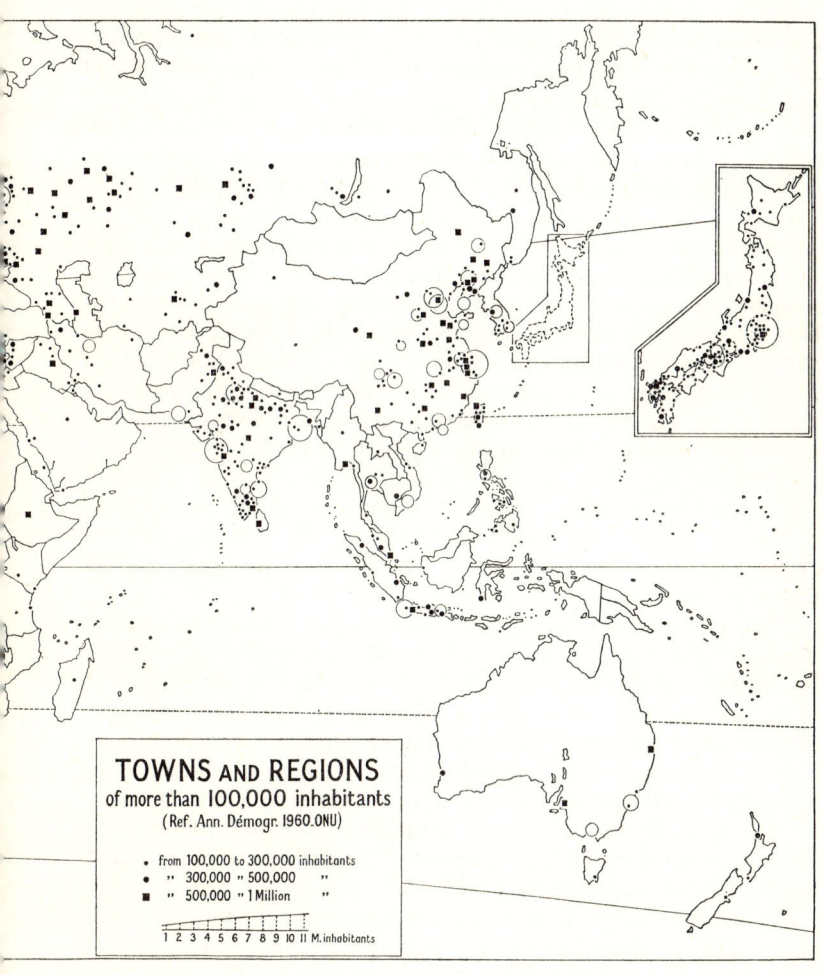

TOWNS AND REGIONS
of more than 100,000 inhabitants
(Ref. Ann. Démogr. 1960.ONU)

- from 100,000 to 300,000 inhabitants
- " 300,000 " 500,000 "
- " 500,000 " 1 Million "

1 2 3 4 5 6 7 8 9 10 11 M. inhabitants

in spite of its rapid progress, does not possess the same characteristics as those of highly urbanised countries. The proportion in the agglomerations of less than 20,000 is relatively greater, while that in the most important is appreciably less.[1]

The development of population in Africa and Latin America is, to a certain extent, comparable with what has happened in Asia. After stagnation or slow development until the beginning of the twentieth century, a rapid impetus occurred. In Chile between 1920 and 1960, the proportion of inhabitants living in districts of more than 20,000 rose from 28 per cent to 47 per cent. In Brazil the number of towns with populations of more than 100,000 rose from 6 per cent in 1920 to 31 per cent in 1960 and the proportion of the total population contained in them rose from 8·7 per cent to 18·6 per cent.[2]

It is possible, therefore, to distinguish four types in the development of urban concentration. First, in Europe, where urbanisation has been linked with the general development of the population and now has a tendency to slow down; secondly in countries outside Europe which have been populated by Europeans, where an enormous impetus is still strongly felt, although declining; thirdly in the U.S.S.R. where moderate expansion corresponds with economic development; and lastly in the underdeveloped countries where, after long stagnation, there is a sudden explosion of urban population.

The ever-widening urban sprawl. The convergence of hundreds of millions of people towards the central points which we call cities is expressed both by the multiplicity of the points of assembly, and by the growth and sprawl of its density.

Built-up areas extend almost indefinitely and teem with the invasion of urban life. This life overflows the traditional administrative boundaries and spreads out into the country. An ever increasing part of the population moves further away from the city centre and, with the passage of time, the agglomerations disperse at a greater rate. In 1801 a zone containing 2,500 inhabitants to the square mile existed 3·7 miles from the centre of London; in 1921 it was 13 miles from the capital. In Paris the distance was 4·3 miles in 1856 and 10·5 miles in 1931. In spite of the lack of complete accord, the parallel growth of the two agglomerations is very marked.[3]

This is expressed in demographic statistics which demonstrate that, alongside a vast increasing urban population, a certain development of rural populations is taking place, quite unconnected with work on the

[1] Hauser, *op. cit.*
[2] Brasil, VII Recenseamento geral, 1960.
[3] S. Korzybski, 'Le profil de densité de population dans l'étude des zones urbaines de Londres et de Paris', *La vie urbaine*, 1952, no. 2, pp. 113–56.

land. These people ('commuters') live in the country but work in the town, and they come and go mornings and evenings, according to their working hours. Thus in the United States it has been necessary to distinguish, in the censuses of 1950 and 1960, between agricultural and non-agricultural rural populations. Populations are classed therefore in three large categories.[1]

Population of the United States of America

	1950	1960
Urban	64	69·9
Non-agricultural rural	20·7	21·4
Agricultural	15·3	8·7

There is a definite tendency towards the phenomenon we call 'suburbanisation'. Between 1940 and 1950 it could be estimated roughly that the population in the urban fringes had increased two and a half times more rapidly than in the central areas and five and a half times more rapidly between 1950 and 1960. 'Residence in single storey houses, spread over scores of square miles around the urban nucleus, represents a waste of space which only nations the size of continents can allow themselves. And such development is unthinkable without the automobile.'[2]

The example of the United States is not an isolated one, but the wide range of the tendency there is specially remarkable because the high standard of living allows the inhabitants to choose their place of residence, while the improvement of transport and technical achievements and the vast areas of space available make all sorts of dreams possible. In this respect California beats all records. Not only do the built-up areas stretch indefinitely, but at the same time the utilisation of ground in the City Centre is inadequate, especially at Los Angeles. This is the result of the part played by the automobile, the preference of the early immigrants from the Middle West for separate houses, earthquakes which have made a low urban skyline necessary, resulting in the typical Californian house, and the Californian taste, in their enchanting climate, for the pleasures of suburban life.

In 1950 more than half the population of the Standard Metropolitan Areas of Los Angeles and San Francisco lived outside the city centre

[1] Censuses of 1950 and 1960. The figures given by the Department of Agriculture show 11·1 per cent rural agricultural in 1960.
[2] H. von Borch, *U.S.A., Société inachevée*, Paris, 1962, p. 77.

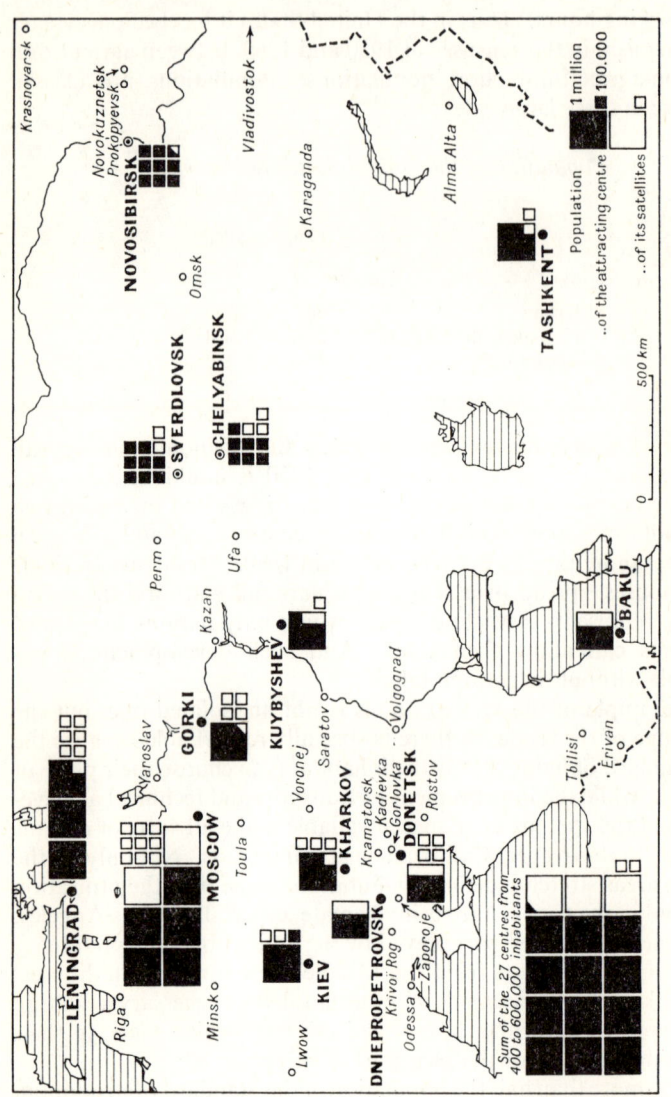

Fig. 2. *Major towns in the USSR: centres and satellites (proportion of the respective populations)*

as compared with 49·9 per cent for the other Standard Metropolitan Areas of the United States.[1]

Suburbanisation occurs all over the world but is particularly concentrated in highly developed regions: in Europe, in Australia, in New Zealand. In Australia a city such as Sydney, where the municipal territory was increased in 1949 by the joining together of eight lesser units, covers only 10·6 square miles whereas the suburban zone stretches over more than 230 square miles and houses seven times more people than the city proper.

In the U.S.S.R. progress is equally rapid in the ten biggest agglomerations which contain more than one million inhabitants; 31 per cent of them live in 'satellite towns'. In Moscow betewen 1939 and 1959 the city grew by 20 per cent and the suburbs by 90 per cent.[2]

By contrast, in regions where urbanisation is retarded, even if growth is extremely rapid, we do not yet observe a comparable suburban development.

Towns of over 10,000 Inhabitants and their Agglomerations

	U.S.A.	India	Brazil
Percentage of growth of population of towns, 1940–60	31·9	162	200
Percentage of growth of the population of agglomerations, 1940–60		54	140
Percentage of population of towns in relation to agglomerations:			
in 1940		55	87·9
in 1960		47·3	94·8

Source: *U.N. Demographic Yearbook*, 1960, and censuses of the countries in question.

Comparison between India and the United States underlines the two types of evolution. In the Indian peninsula the present development of urban population is extremely rapid but the proportion of inhabitants living in the great cities as compared with the suburbs is very high and has a tendency to increase; it is the reverse, as we have seen, in the United States.

Those who populate the urban fringes may be of two origins: they

[1] H. F. Gregor, 'Spatial disharmonies in California population growth', *Geog. Rev.*, 1963, pp. 100–22.

[2] Symposium, *Goroda Sputniki*, Moscow, 1961, pp. 16, 34.

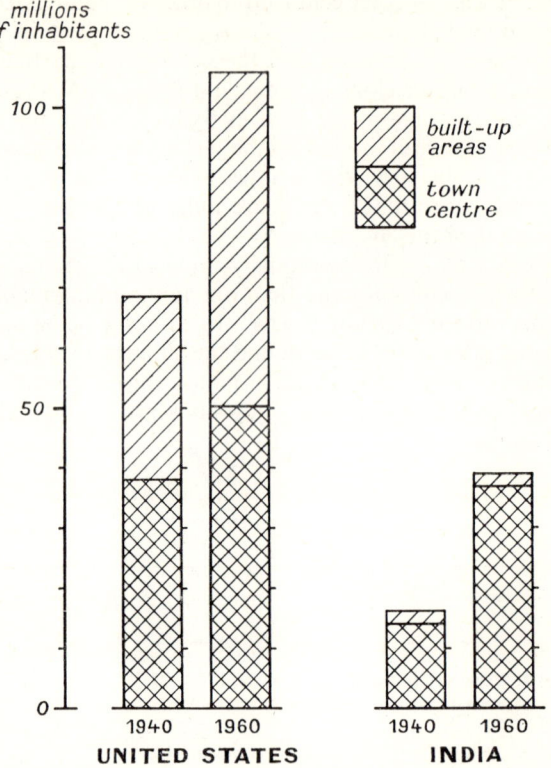

Fig. 3. Evolution of towns and built-up areas in the United States and India. (1940–1960)

Note the small part of the population in Indian towns outside the main urban area; the contrary is the case in the United States.

may be new arrivals who have settled in peripheral districts where they feel closer to the countryside or which are more like their own small native towns, and where living is cheaper; or they may be town dwellers of long standing who have had enough of the overcrowding, noise and difficult conditions of life in the central areas, and who decide to leave often unhealthy buildings for the more salubrious air of the suburbs. In the United States out of 6.9 per cent of white migrants who poured into the suburbs between 1940 and 1950, 3·6 per cent came from the countryside and 3·3 per cent from dense populated areas.[1]

[1] Bogue, *op. cit.* p. 407.

Patterns of urban growth. Where do the multitudes come from who swell the cities of the earth? The statistics all agree: there are two great phenonema, natural growth and migration. The relative importance of these two factors varies according to region, period, and type of city.

In general, as already noted, in those regions which have been urbanised for a long time and now include a high proportion of town dwellers, natural growth plays the main part.

In the United States between 1940 and 1950 it is estimated that about 70 per cent of the increase of town dwellers was due to natural growth and about 30 per cent to the migration of country dwellers to the towns. The birth rate in urban populations was about 22·9 per thousand while the mortality rate fluctuated around 10 per thousand which reflects an actual annual increase of 12·9 for every thousand inhabitants. In spite of all this the part played by the countryside was far from negligible.[1]

In countries in the course of full demographic growth the proportion is almost exactly reversed. In Brazil between 1940 and 1950, 29 per cent of the development of the eight largest agglomerations was due to natural increase and 71 per cent to migratory movement. The figures are reasonably homogeneous, with a maximum of 76 per cent for Recife, and excepting only Belem where the proportion is exactly reversed and where natural surplus produced 71 per cent of urban growth. In Brazil, as in the United States, if the statistics are to be believed, the rate of natural growth appears to be about the same, that is 11·46 per thousand per annum (birth rate for the eight agglomerations under consideration, 28·16 per thousand; mortality 16·7 per thousand).[2]

In the great cities of developing countries where the rhythm of increase is greater, the rôle played by migrations appears to be the determining factor. This factor is most significant in the case of major cities which possess a great power of attraction, if not always a great power of absorption. The capitals of the impoverished states of Brazil seem full of promise to the country people of the north-east, cheated of their harvest by drought or thrown out of work by tyrannical estate owners. In the medium-sized towns of 5,000 inhabitants or more, on the other hand, there is a balance between the two contributing factors, with a very slight predominance in favour of natural growth.

The part played by migrations in heavily urbanised countries which are pledged to economic change and a fresh expansion is also of great

[1] *Ibid.*, pp. 39–40.
[2] *Contribuçoes para o estudo da demografia do Brasil*, Rio de Janeiro, 1961, pp. 316–20.

importance. Thus, in the U.S.S.R. it is estimated that up to 74 per cent of the population in the new mining centres of the Donetz basin have come from comparatively distant areas, either rural or industrial, while in the towns of the Far East, which include 80 per cent of the urban population, 99 per cent are Russians from the towns and countryside of Europe.[1]

The universal law of demographic urban growth is that of movement.
In nineteenth-century Europe it was the peasant communities which poured into the expanding cities while male European emigrants went off to found innumerable homes across the ocean. At the present time other countries are undergoing similar changes, but it would perhaps be an over-simplification to put forward a classification based only on a stage of general development reached by the country under consideration. Indeed, in a country like the United States at least three types of city can be discerned: the stagnant organisms of the industrial cities of New England where urban growth hardly matches the natural population growth; cities of moderate growth in the centre which correspond more or less to the general average; and vigorous expanding cities like those of the Pacific states. Los Angeles in particular, where the population has increased by 50 per cent in the course of the last ten years, owes more than three-quarters of its growth to the multitudes who have poured in from other parts of the United States.

We are therefore able to consider on the one hand cities with a limited development and little appeal to immigrants, and on the other hand those of rapid growth which attract people born elsewhere. The first group includes great cities which suffered an inflated population in the last century and now continue to grow at a more moderate pace, like the great cities of Europe or those of the eastern seaboard of North America. In this group are also found more modest towns not yet fully developed and also certain ancient cities of rural European regions. Within the second group of dynamic urban organisms it is necessary to effect a subdivision based on the degree of economic evolution, and to distinguish harmonious growth where economic and demographic rhythms run parallel, from haphazard growth in which they are disconnected. Examples of the first type are São Paulo in Brazil, the new industrial cities of the Soviet Union and the colonial centres in Asiatic Russia; of the second type, the cities of northeastern Brazil, southern Italy and a great part of Africa and Asia.

In certain extreme cases there is a lack of connexion between the

[1] O. A. Konstantinov, 'Les villes de la République Socialiste Soviétique de l'Ukraine', *Bull. de la Soc. de Géog. de l'U.R.S.S.*, 1954, no. 2, pp. 215–28, and census of 1959.

natural demographic evolution of the city and its local development; the balance of births and deaths is negative. Thus it was in certain ancient cities of Europe (e.g. in Vienna before the *Anschluss*), and in certain country towns: in Toulouse the ratio was only positive on two occasions between 1856 and 1946.[1] It was the same in towns in tropical countries ravaged by epidemics, such as the famous influenza epidemic in India at the time of the first World War. Yet these cities continue to grow. Turin would have decreased from 336,000 to 330,000 inhabitants between 1901 and 1951 if its population had only been renewed at the 1901 rate; on the contrary there was a population of 711,000 in 1951 and almost a million at the present time. Many of its new inhabitants came from the south of Italy. The situation is similar in Milan, where a quarter of the recent migrants came from southern Italy, with many others from the agricultural and mountainous regions closer at hand.[2]

The magnetic attraction of towns. This attraction rests as much on a psychological idea as on economic fact and the appalling poverty of the countryside plays a part at least as important as the advantages, real or imaginary, of the town. The spectacle of thousands of people uprooting themselves in order to grasp at the prizes offered by the great metropolitan cities also embraces others headed towards urban agglomerations which are incapable of providing them with improved conditions.

The area of urban attraction varies perceptibly according to divergent economic and political conditions. The partition of India and the exodus of the Muslims built up the population of Karachi from 386,655 in 1941 to 916,000 in 1961. The creation of Israel and the influx of Jews from all quarters of the world made Tel Aviv, the capital, which began in 1909 as a few villas scattered in the sand dunes to the north of Jaffa, into an important industrial centre of more than 600,000 people. The development of the wealth of Asiatic Russia has caused the population of all European Russia to converge on the urban centres situated beyond the Urals just as, a century earlier, the attraction of gold in California or Australia produced mass migration from Europe and from the already crowded areas of the continent where these discoveries were made. But this movement of populations across frontiers and oceans towards some more privileged part of the world is bound up with exceptional circumstances. Within these general movements it is possible to distinguish certain broad characteristics.

[1] J. Coppolani, *Toulouse*, 1960, p. 284.
[2] E. Migliorini, 'Migrazioni interne e spostamenti territoriali della popolazione italiana', *Actes du 18e Congrès géographique italien*, April 1961.

Certain large cities hold a powerful attraction for people all over the world. There are the great capitals of the West or certain commercial centres. New York has its Chinese and Italian quarters and its Greek and Portorican districts. Paris includes 11 per cent foreign-born people.

In Hong Kong there may be only 1 per cent of foreigners, but these embrace more than fifty nationalities. In São Paulo 15 per cent of the population were born in other countries and 13·7 per cent retain their original nationality.

Capital cities and great centres of economic activity also exercise a powerful attraction on the outlying districts of their own countries. Their attraction may be due to the network of communication, or even to their topographical setting. Thus in Paris only 45 per cent of the inhabitants are natives of the city and 44 per cent came from the provinces. The département of Seine-et-Oise ranks highest with 76·2 per cent, but all northern France sends some of its sons to Paris, while hardly a soul comes from the more distant south-western and south-eastern provinces, which are separated by the mountainous mass of the Central Plateau.[1] In the same way, in England, the district which has sent most migrants to London is contained in a triangle between Wales and the Pennines. In the city of São Paulo, 15 per cent of the Brazilian population was born outside the state of São Paulo, and of these half came from the neighbouring state of Minas Gerais, 10 per cent from Bahia, with all the other states and territories of Brazil represented in the list. The same diversity of catchment can be observed in the new capital of Brasilia, created in 1956; out of 141,742 inhabitants in the 1960 census, 23 per cent came from the state of Goias in which the capital lies, a little more than 20 per cent from neighbouring Minas Gerais and 13·6 per cent from Bahia. About a quarter came from the north-east, and the new capital even attracted some 3,000 former inhabitants of São Paulo and over 1,000 from Rio.[2] These movements have been facilitated by the convergence of new roads built towards the new capital from all the far-flung cities of Brazil, and also by what can be called the mystique of the conquest of the interior.

The same facts are observed in the urban civilisation of other continents. In Hong Kong in 1931, only one-third of the inhabitants were native to the city.[3] During the last few decades Calcutta and Howrah have expanded with the influx of population from those regions of Bengal ravaged by flood and famine, with refugees from East Pakistan and with immigrants from more distant regions. Among

[1] *Paris, 1960;* cf. M. Croze, *L'Origine des Parisiens,* pp. 47–69.
[2] Brasilia; special census, 1960.
[3] R. H. Hughes, 'Hong-Kong', *Geog. Journal,* 1951, pp. 1–23.

the latter group are those from Bihar, Uttar Pradesh, Orissa, Nepal, the Punjab, Assam and even from the south of India.[1] At Dakar the proportion of adults born in the city represents only a quarter of the total. The composition of the remaining three-quarters is dominated by those from Senegal (34 per cent); while the other African territories (Guinea, the Sudan, Mauritania) provide 11 per cent and France 10 per cent.[2]

There are many further examples which all confirm this trend and may be used to illustrate two additional observations. In highly developed countries the migrations to the great cities generally arise from other smaller towns. On the other hand in regions in the course of urban development, it is often the peasants who migrate direct to the city. Between 1940 and 1950 the volume of migration from town to town in the United States represented 32·5 per cent of all population movements.[3]

The regional market towns provide a third type of centre of attraction, but save in exceptional instances, these have a much more limited power of attraction and are dependent upon their surrounding regions. Records in France show that many provincial towns with a population of 50,000 to 100,000 draw more than 90 per cent from their own departments and the adjacent districts, that is to say from those regions within a radius of 60 to 120 miles. At Bahia, the majority of immigrants came from Reconcavo and its border, that is within a radius of 75 miles.[4]

In considering the very busy industrial regions, even those of limited size, we observe that the conditions of immigration can vary either because of the great numbers of people needed for industry, for example in the mining regions or because of the need for certain specialists. Thus in the coal districts of north-west Europe there are clusters of small mining towns and, as the French and Belgian populations are attracted elsewhere by higher salaries and better conditions, it has been found necessary to appeal to immigrants from less privileged regions. The Poles came in force to France between 1920 and 1930 and were replaced by southern Italians, or Algerians after the Second World War.

The result of these migrations can be the transformation of the urban network; in the older countries it is a common phenomenon to find a great number of small agricultural market towns situated close to each other with larger regional market towns at intervals.

[1] A. B. Chatterjee, 'Demographic features of Howrah City', *Geog. Rev. of India*, 1958, vol. 20, pp. 150–69.
[2] Dakar census.   [3] J. Bogue, *op. cit.*
[4] J. Beaujeu-Garnier, 'Les migrations vers Salvador (Bahia)', *Cahiers d'Outre-mer*, 1962, pp. 291–300.

Changes in transportation and in the commercialisation of agricultural products resulted in the decline of many towns and the concentration of activities and population in a small number of centres more favourably placed and better equipped. At the same time industrial centres and a number of tertiary towns appear and the whole structure of the urban network undergoes a profound change.

The localisation of great urban centres. A glance at a map which shows the great cities of the world demonstrates that four areas stand out in the general picture of the continents: the central Far East, which is particularly dense between the 22nd and the 48th parallels, embracing Japan, Korea, Formosa, Manchuria and continental China and penetrating into the interior of the continent of Asia to a depth of about 1,000 miles; the peninsula of India (*sensu lato*) extending from the 8th to the 35th parallel; Europe, (including Russia as far as the Urals) stretching from the 40th to the 60th parallels and to which can be added the northern fringe of North Africa; finally the centre of North America between the 36th and 47th parallels, that is the region that stretches from the estuary of the St Lawrence to the western extremity of the Great Lakes and the centre of the east coast. Elsewhere in the world we find clusters of big cities, always situated along the coasts except in the highlands of Mexico and Colombia, at certain points in the Middle East, in the Soviet part of Central Asia and in the north east of the Republic of South Africa.

Localisation by Continents of Agglomerations of over 100,000 Inhabitants (1960)

	Number of agglomerations	Percentage of total
Asia	442	32·6
of which: Far East (1)	238	27·8
Europe	327	24·8
America	323	24·4
of which: North (2)	217	16·2
Central	35	
South	71	
U.S.S.R.	147	11
Africa	84	6·4
Australia	11	0·8
	1334	

Source: U.N. *Demographic Yearbook*, 1960. Tables 7 and 8.
(1) Far East: China, Japan, Korea, Formosa.
(2) U.S.A. and Canada.

The United Nations *Demographic Year Book* gives a list of 1,334 agglomerations in which the population exceeds 100,000. These great human groups appear irregularly throughout the different continents (see the Table above). This picture is based on the last census or on estimates, but it is possible that a certain number of big towns are omitted, especially in China where there is only the census of 1953 to go on, plus a few sporadic figures for 1957.

By studying the census figures for the period nearest to 1950 it has been possible to calculate the percentage of the world population living in cities of 20,000 inhabitants and more, in cities of a population of more than 100,000 and finally the percentage of the urban population living in cities of more than 100,000.

Percentage of Urban Population in the Principal Regions of the World

	Percentage of total of population living in towns of 20,000 inhabitants and over	Percentage of population living in towns of 100,000 inhabitants and over	Percentage of urban population living in towns of 100,000 inhabitants and over
World	21	13	62
Oceania	47	41	87
North America (1)	42	29	69
Europe (2)	35	21	60
U.S.S.R.	31	18	58
South America	26	18	69
Central America	21	12	57
Asia	13	8	63
Africa	9	5	51

Source: Hauser, *op. cit.*, p. 107.
(1) Including U.S.A. and Canada.
(2) Excluding U.S.S.R.

These statistics emphasise the paradoxical position of areas where urbanisation is weak; Latin America, Asia and Africa have a small percentage of urban populations, although relatively a greater proportion of large towns.

The case of Asia is particularly striking: this continent contains more than half the world's population but only a little more than one-third of those who live in towns larger than 20,000, and a similar proportion of those who live in towns larger than 100,000. In all those Asian countries which supply statistics, the proportion of populations living in towns of more than 100,000 is always equal to or higher than two-fifths of the total of urban populations with the exception of North Borneo. The proportion rises to 70 per cent and more in Malaya, South Korea, China, Indonesia, the former Indo-China and Thailand.

On the other hand, Europe, North America and the U.S.S.R. contain a much higher proportion of urban populations in towns with populations of more than 20,000 and 100,000 inhabitants than their relative share of the world's population. For Europe the figures are respectively 16·4 per cent of world population, 27·5 per cent of those living in towns of more than 20,000 inhabitants and 26·5 per cent of those in towns of more than 100,000.

It would be interesting to calculate different rates of urbanisation; particularly the proportion of populations in each country contained in towns of more than 20,000 and of more than 100,000. The United Nations *Year Book* 1960 gives these percentages which have been calculated for a certain number of countries. We have extracted some of these, characteristic of countries containing the largest number of agglomerations:

	Number of towns with over 100,000 inhabitants	Percentage of population in towns of over 100,000 inhabitants	over 20,000 inhabitants
U.S.S.R.	148	23·5	35·5
Japan	115	41·2	
U.S.A.	107	29·5	41·2
China	102	8·3	
India	97	8·6	
West Germany	51	30·7	
United Kingdom	45	36·9	

Source: U.N. *Demographic Yearbook*, 1960, Table 8.

The Table substantiates the remarks already made concerning the special case of the very large Asiatic cities, apart from those in Japan. We must not, however, draw too precise conclusions, as the documents supplied by the various countries are far from being comparable. First from the point of view of date; the figures used by the United Nations *Year Book* are, for example, those of 1959 for the U.S.S.R. and of 1950 for the United States. Then from the standpoint of classification of urban types: the U.S.S.R. includes towns and only two agglomerations, Moscow and Leningrad; whereas in the United States in 1960, 124 towns were given which exceeded a population of 100,000 but 203 agglomerations; for the United Kingdom 45 and 71 respectively.

We must wait for better information before being able to give a more precise picture of the great laws of world urbanisation but the fact of the rural exodus towards the cities remains.

THE ATTRACTION OF THE CITIES AND THE RURAL EXODUS

The causes of the movement of rural populations towards the city have often been analysed and here it suffices to recall them. There are economic causes such as the concentration of industry and commerce and the disparity of wages. There are social reasons, such as the desire for educational opportunities or to enjoy more varied leisure time pursuits. There are other reasons, which are not always sufficiently emphasised. The peasant is no longer a serf as in the Middle Ages, but he remains a slave to his cattle. However small the farm there will be at least a minimum of livestock; the beasts have to be fed and groomed, the cows have to be milked. In the Jura Mountains there are peasant families who spend the winter with nothing whatever to do for half or three-quarters of the time. Snow covers the woods and the paths, it is too cold to cut wood; all day long they play cards or listen to the radio, but a holiday is impossible as they have to be in the cowshed early in the morning and at the end of the day to see to the livestock. And the lot of the woman is the hardest. There are peasant women who work in the fields, look after the farmyard, care for the children, do the cooking and the mending and fetch water from the well, which may be a couple of hundred yards away. It is the woman who urges her husband to migrate so that she may have an easier life.

In underdeveloped countries the situation is not quite the same although the result may be similar. In Italy the census of 1961 shows that the rate of urbanisation had grown in the south in the same way as in the north. The city is often the outlet for miserable rural conditions and towards it rush the starving and the unemployed.

No one knows how to protect the towns in north-east Brazil or in Africa from this desperate immigration. And yet here the woman does not benefit as she does in Europe. In the towns of black Africa the woman's task is not easier; the mens' wages are wretched, hardly reaching half of what is needed as the bare minimum for the family. The woman has to go out and work in the plantation several miles from the town and returns in the evening bent under the weight of wood for the kitchen or clusters of bananas brought in from the forest. Pierre

Lasserre[1] analyses in detail the reasons for the rural exodus in Gabon and he shows that the Africans come into the town without any clearly conceived motives. 'They have,' he says 'an idealised notion of the city as a whole, without stopping to analyse its advantages and disadvantages.' Probably our peasants are more knowledgeable than the Negroes of a primitive society, but one cannot be certain that to some extent all over the world there is not some of the same inconsequential movement as one finds amongst the natives of Gabon.

Thus the movement of urbanisation continues. The whole earth is covered with cities; their influence penetrates more and more into the countryside. New districts are added to old cities and completely transform their appearance. Great buildings with fifteen to twenty storeys, modern frontages and new windows, characteristic of an architecture previously found only in Manhattan, illustrate astonishingly the twentieth-century rush towards the cities.

[1] *Libreville*, Paris, 1958.

CHAPTER 2

DEFINITION OF THE TOWN

In the preceding chapter towns were described as those places in the world where ever increasing groups of mankind congregate more and more rapidly. At first sight this idea seems clear enough since each one of us can readily distinguish between town and country. But the criteria which we apply, whether consciously or unconsciously, vary according to individuals. The concept of a town changes according to historical and geographical setting even though the distinction between town and country may remain constant. Each civilisation has its own conception of a town, as Knut Hassert[1] observed in his pioneer work on the subject. Indeed some of the most celebrated cities of ancient times, inhabited mostly by farmers, would certainly not qualify as such in terms of twentieth-century standards. It is surprising how, in these days, the label of 'urban' is given to people who would certainly not qualify as town dwellers. This is the case in certain countries with very dispersed population where any agglomeration has the appearance of a town and is counted as such in the statistics.[2]

In spite of this we can enquire into the characteristics which, at least in our own day, make up our idea of a town. We must not gloss over the difficulties. These are reflected in an imprecise vocabulary. In German, *Stadt* is the only word used, while in English *town* and *city* are more or less synonymous. In French the term *cité* represents the town considered from a social and political point of view. In Swedish *Staden* means a class of towns below which are two other classes: market towns (*Köping*) and municipal agglomerations (*Municipalsamhällen*). Swedish statisticians finally adopted a more general term, *tätort* which includes all agglomerations in which the houses are not more than 200 metres apart and which contain a population of more than 200. These districts are classified as *urban areas*. On the other hand the Swedish word *by* which signifies village, is the same word used in Norwegian to mean town.

These linguistic complications do not make the task an easy one, but they do demonstrate the effort made in each language to contrast the idea of the town with that of the country. The Americans are even more precise in that they distinguish between rural agricultural

[1] *Die Städte geographisch betrachtet*, Leipzig, 1907.
[2] Sven Godlund, 'Urbaniseringen i Sverige', *Medd. från Göteborgs Universitets geografiska Institutionen*, 1961.

population (*rural farm*) and non-agricultural rural population (*rural non-farm*).

Many geographers have been at pains to attempt a definition of the town. A preliminary definition is indeed the first scientific step in any study. But a rapid survey of such definitions shows how difficult it is for an author to be precise.[1]

When F. Ratzel outlined the framework of human geography at the end of the nineteenth century, he gave a definition of which several elements remain of value: A city is 'a continuous and dense agglomeration of people and dwellings occupying a large area of ground and lying at the focus of great trade routes'. This definition needs to be amplified by the author's examples contained in his *Anthropogeographie*. Here he contrasts the differences between the way of life in villages based on arable and livestock farming and in towns based on commerce and industry. He also contrasts the less crowded dwellings in villages with the denser concentrations in towns. Elsewhere he specifies that below a population of 2,000 the group loses its urban character. As Hallstein Myklebost quite rightly observes, Ratzel's analysis contains three conditions which form the basis of all other definitions: (1) some sort of professional activity (2) a concentration of housing (3) a minimum number of inhabitants. However, Ratzel was writing at a time not far removed from the pre-industrial epoch and he stressed particularly the importance of trade (*Handel und Verkehr*). This is a point of view very similar to that of H. Wagner for whom towns were 'concentration points of human commerce'.[2]

All writers have since agreed with Ratzel that the activities within the town must form an essential part of its definition; but these are so numerous that it is difficult to summarise them. Indeed it is often only in a negative sense, by contrast with the countryside, that such activities can be recognised. According to Fr. von Richthofen[3] the town consists 'of an organised group in which normally the main occupations are concerned with commerce and industry as opposed to agricultural pursuits'. Similarly W. Christaller emphasises commerce, administration, and small businesses; but he relegates industry to second place. M. Aurousseau characterises a town as a place where transport, manufacture, the import and export of raw materials, education, the administration of national affairs are carried on; or, quite simply, a place of residence.

---

[1] For a full analysis of these definitions, see Hallstein Myklebost, *Norges Tettbygde steder*, Oslo, 1960.

[2] *Lehrbuch*, 1923, p. 844.

[3] *Vorlesungen über allgemeine Siedlungs und Verkehrsgeographie*, Berlin, 1908, p. 262.

To make such a list is so complicated and it is so likely to be incomplete that often it is preferable to follow Fr. von Richthofen and to exclude agricultural occupations. This is what R. E. Dickinson does. Certainly in the world we live in today this is the clearest definition although it cannot be applied, as we have seen, in ancient times or in the Middle Ages. Nevertheless, even this definition may prove false, for Pierre George emphasises that, in a socialist economy, the city may be 'a residential centre bound up almost exclusively with agriculture'.

However, if all the attempted definitions are combined, certain common features emerge. A town must form a dense agglomeration of sufficiently large dimensions. At the same time, certain features spring from subjective appreciation rather than objective criteria. We visualise the urban scene in order to distinguish the town from the village, but this is not enough. Something else must be added to distinguish the townsman from the countryman. Here again it is the old idea of 'a way of life', made familiar to geographers by Vidal de la Blache, to which we turn most readily because its very imprecision gives it a greater elasticity.

The way of life must be considered as a whole. J. Brunhes and P. Deffontaines consider only one angle of it: 'A town can be said to exist if the majority of the population spends the greater part of its time within the bounds of the agglomeration.' This definition, attractive in itself, cannot be applied either to dormitory towns or to agricultural urban areas and it takes no account of the size of the agglomeration which could be reduced to a mere factory surrounded by a few houses.

Several Swedish writers insist and not without reason, on the occupational aspect. H. W. Ahlmann and later W. William-Olsson defined the town as an agglomeration characterised by a certain variety of occupation.[1] Towns no longer exist where all the population does the same work. This means that some large Mediterranean and Hungarian villages must be excluded from the classification. It means that today the large agricultural centres would be excluded. Also around the cities there are dormitory towns where everyone does similar work but which are still towns, and the people in them live like the inhabitants of the nearby city. They have the same preoccupations, the same amusements and often work in the same undertakings.

Without detracting from the value of other definitions, let us keep this concept of occupation as an essential factor. Our definition will then approach that of H. Bobek, who adds to the qualifications accepted by all (continuity and agglomeration, sufficiently large

[1] J. Frodin, 'Staden som geografisk företeelse', *Ymer*, 1946.

dimensions) another idea somewhat difficult to define: *städtisches Leben*. Now we can hardly define urban life except by contrasting it with rural life and this is almost a tautology. It clearly illustrates the stumbling block which has faced all writers who have attempted a definition. Something always remains which is impossible to pinpoint precisely.

There are cases, however, where for practical purposes or statistical classification it has been necessary to determine what, in fact, constitutes a town. In military terms a town, as a political or administrative centre, with its museums and works of art and its eminent citizens, requires special protection. It is necessary to lay down boundaries for such protection. This was achieved by a walled circle of ramparts and moats. The town lay entrenched behind these fortifications with access to the world outside only through well guarded gateways and drawbridges. The district beyond these precincts formed no part of the town in those days. When the city spread, its boundaries were displaced at those points where concentric rings of roads marked the ancient limits. The siege and capitulation of a town had a very precise meaning which, in the case of defeat, was symbolised by the surrender of the keys to the conqueror.

In fiscal affairs also it was important to define a town's precise limits. The supply of provisions posed problems which could only be resolved by strict regimentation, and a tax on urban supplies was one of the easiest to collect. All that needed to be done was to place an official at the gates.

This is not all. A town cannot be administered like a rural parish; powerful citizens make their interests felt and demand a right to play their part in the administration. Problems of municipal organisation arise which are unknown in country districts. Towns have to be granted certain privileges, such as the right to their own administration or the right to administer certain services. One of the most precious of these was the right to hold a market; these rights were seized by the crown, the Church and the nobility. It was important to know to whom such rights were granted. In all countries in the Middle Ages the name of a town was a title under which solemn legislation was enacted, and in certain countries this custom has persisted. In Scandinavia, Great Britain and Japan there are numerous towns created under royal charter. From time to time new towns are created under such charters, but these hardly affect the rest. In Sweden the ancient fortress of Marstrand (outside Göteborg), with 1,213 inhabitants, remains a town while, apart from the communities in the suburbs of Stockholm, Finspångs in Östergotland, with a population of 15,196, is still denied this status.

A precise definition is not absolutely necessary until we approach statistical classification. Generally speaking urban populations can be distinguished from rural ones. As a rule this is a definition based on numerical considerations but even this raises numerous problems. In France an agglomeration of 2,000 is considered a town. Thus it becomes necessary to define 'agglomeration':

An agglomeration may be considered as a population which dwells in adjoining houses and which makes use of communal facilities such as parks and gardens or places of work or other similar centres of this kind, even if the houses are separated by a road, path, river, canal, railway or city wall. Suburbs and other parts of the city must also be considered as belonging to the agglomeration as well as those zones which, as a result of war, are separated from the centre and not yet reconstructed. No interruption of the agglomeration can be said to exist, even if unenclosed, on account of un-built-up spaces between buildings (as for example, in new districts in the course of construction). But the agglomeration may be said to be interrupted by undefined stretches of land or by cultivated land (fields, meadows, vineyards, whether or not enclosed by fences, hedges or wire) if this ground is of a size clearly larger than a normal size of a group of buildings.'[1]

The minimum population figure of 2,000 adopted by France has been the one recommended since 1887. Without actually adopting the same figure, a good many countries have accepted the principle of a minimum size. In the Argentine and in Portugal, as in France all agglomerations with a population of 2,000 and over are classed as towns.[2] The United States and Thailand have raised this figure to 2,500, Spain and Greece to 10,000, Canada, Malaysia and Scotland on the other hand, have reduced it to 1,000, Iceland to 300 and Denmark to 250.[3] Certain countries also demand a minimum size, not only for agglomerations but also for territorial districts (communes): the Netherlands (20,000), Switzerland (10,000), Belgium and Austria (5,000), Germany and Czechoslovakia (2,000).

Other countries add to the numerical qualification other criteria: Guatemala considers as towns population groups of 2,000, but the limit is lowered to 1,500 if running water is laid on. The Republic of Panama defines as towns centres with a minimum population of 1,500 which display 'urban characteristics'. Some countries abandon

[1] 'Instruction aux maires relative à la décomposition de la population des communes', *Recensement de 1934, Population de la France*, Paris, 1954.
[2] No account is taken here of the slight variation by which in France this refers specifically to the administrative centre of the district.
[3] The very low figures for the Scandinavian countries are explained by the general dispersion of their population: all agglomerations fulfil urban functions; G. Enequist, 'Sveriges mindre Tätorter', *Ymer*, 1947, and 'Vad är en Tatort' in *Tatorter ach Omland*, Lund, 1951.

the numerical criterion altogether. Chile considers as urban those centres which are concerned with certain public and municipal services; Israel those centres of a non-agricultural nature. In the same way all the chief administrative centres are considered as towns in the following countries: Bolivia, Brazil, Costa-Rica, Equador, Salvador, Haïti, Honduras, Nicaragua.

In the light of these difficulties many countries cling to lists approved by the government: Bulgaria, Ceylon, England and Wales, Finland, Japan, New Zealand, Norway, Paraguay, Poland, Rumania, Sweden, The Republic of South Africa, Hungary. Among these some consider that the designation 'urban' demands a minimum size of population: South Korea (4,000), Eire (1,500), the Republic of India (5,000 as well as the presence of urban characteristics), Finland (less than 50 per cent of the population to be agricultural workers); while in Turkey, to agglomerations given the official title of towns are added those with a population exceeding 5,000. In the U.S.S.R. the official title of town is granted to towns with a population of more than 12,000 (10,000 in the Ukraine, Moldavia and Tadjikistan, 5,000 in Georgia and Turkmenistan) provided that the agricultural population does not exceed a certain percentage. (This must be less than 15 per cent in the Russian Soviet Federal Socialist Republic, 25 per cent in Georgia, 33 per cent in Turkmenistan, and less than half in the Ukraine and Moldavia.)[1]

All these statistical definitions are very varied and demonstrate the difficulties and discrepancies within a single country. As has been shown the numerical factor is present in almost all the definitions. We need not labour the objections that can be made to any sort of classification which is subject to a precise figure. But this figure is difficult to determine and is not laid down everywhere according to the same criteria. We have seen the definition of agglomeration as given by the French statistical services and this has the advantage of being unequivocal. In other countries the authorities are less precise. Gerd Enequist estimates 70 metres as being the maximum distance which should separate two houses, though in large towns this distance may be greater if not given up to fields or forests.[2] Hallstein Myklebost reduces this distance to 50 metres, though he avoids too precise a ruling.[3] It is easy to contrast those countries where the houses are crowded together either by tradition or by military necessity, with those where they are more spread out in order to enjoy more sunshine. These differences, hardly noticeable in the centre of the town, appear more sharply when it is necessary to define the boundaries of

[1] P. George, *Précis de géographie urbaine*, Paris, 1961.
[2] Enequist, *op. cit.*   [3] Myklebost, *op. cit.*

the suburbs. Indeed the latter are closely linked to the town and form an integral part of it.[1]

The differences of principle which arise from a consideration of the examples cited above are of a more serious nature. According to the country under consideration, more or less importance is attached to the occupations of the inhabitants or to the siting of the buildings. It seems to be impossible to lay down an international rule[2] and the solution can only be a national one 'when it deals with a comparatively small and homogeneous country' (Pierre George). The European Conference of statistics in Prague has recommended that a nucleus of people living in compact groups of a minimum size of 2,000 should be regarded as an urban population on condition that, in the case of nuclei of less than 10,000, agricultural workers do not exceed 25 per cent of the total. Above 10,000 the group can be classified automatically as urban.[3]

Fishing ports pose a special problem. The inhabitants do not make their living from the land but their work is certainly a 'primary' activity. However, even the development of fishing, as Gerd Enequist observes,[4] tends to give an increasingly urban aspect to fishing ports. This is the case of Norwegian fishing villages concerned with whaling.

So we are bound to include as towns groups which run contrary to our previous definitions and which were rejected in early studies of urban geography. We have included in our calculations a 'normal' town in the same way that we have accepted 'normal' erosion. We rejected as 'large villages' certain Asiatic or even Mediterranean groupings such as Barletta in Italy, cited by H. Myklebost, which has a population of 64,000 but which includes 40,000 peasants. In fact in other equally large agglomerations an urban life has been established as of necessity. Urban functions which elsewhere are concentrated in special centres are here bound up with agricultural pursuits. If there is a difference, this is felt right at the heart of the agglomeration. In this way we are often able to distinguish, in the case of large agglomerations, a central nucleus where commercial activities are concentrated and an outer fringe inhabited by agricultural workers, from which they can easily reach their fields.

It was the same in France in the Middle Ages, when in the autumn the vine-growers brought their presses into the streets of Clermont-Ferrand. It is not surprising that agricultural towns are even more

[1] G. Chabot, 'Faubourgs, banlieues, zones d' influence', *Urbanisme et habitation*, 1954.
[2] George, *Précis de géographie urbaine*, p. 9.
[3] 'Méthodes de recensement de la population', *Nations Unies, Etudes démographiques*, New York, 1949.
[4] G. Enequist, *Geographica*, 1948.

numerous in Asiatic countries. In Indo-China Pierre Gourou observes such towns with populations of more than 10,000; and the same is true of China.

This does not mean that we should necessarily exclude more sparsely populated towns, comprising perhaps only a few hundred inhabitants, which may fully deserve the title of town, since they exercise the functions of one. These small urban nuclei play the same rôle vis-à-vis the big agglomerations that the Scandinavian *tätort* plays in relation to the scattered farms. If the agricultural townships appear to us as a feature of less developed countries, we may also note that the Soviet agricultural townships have taken an ultramodern form.

More powerful than any statistical or administrative definition, stronger than any professional formula, is the conception of the town that characterises different regions. In each country a town can be said to exist if the people of the district feel themselves to be in a town.

But the idea of the town in itself seems exaggerated. Another and broader concept is appearing which goes beyond the town and includes neighbouring localities which depend on the life of the town, even if they themselves are not completely continuous, such as dormitory towns and industrial centres; to this group we often give the name of agglomeration; we speak of the Parisian agglomeration which stretches out over three or four départements.

The publications of the United Nations, following the example of the American statistics, describe an 'urbanised zone' stretching beyond the town proper and adding to it a continuous built-up area, provided that this conforms to certain criteria, notably in regard to demographic density.

The metropolitan area, more spread out, is the zone that surrounds a big town and where the activities of the population are concentrated towards the latter. This definition is vague enough. But in spite of this the expression 'metropolitan area' which has been adopted by American authors, appears to correspond well enough with the definition of agglomeration. The centre of urban research at the University of California gives it a very precise meaning: 'This is a population group of at least 100,000, containing at least one town (continuous urban area) of 50,000 or more inhabitants, plus the adjoining counties which have similar characteristics and in which at least 65 per cent of the population are engaged in non-agricultural activities.'

But we must not delude ourselves; the very lack of precision that we have encountered in defining the town leads us to suppose that these conceptions, which differ from those of earlier times, can evolve still further. The urban phenomenon is subject to renewal; it corresponds to and expresses a certain form of civilisation.[1] The definition cannot be the same for all times and all countries. Perhaps we shall see in our own countries the completion of one of those cycles which cuts across the urban phenomenon. It is by no means sure whether, in the new or emerging countries, the cycle which appears to be beginning will not be interrupted before its completion under the influence of other civilisations.

We cannot, therefore, do better than to replace the towns in their setting and group them according to the civilisations where they belong. We shall then be able to examine the problems which are common to them all.[2]

[1] G. Roland Martin, *L'Urbanisation dans la Grèce antique,* Paris, 1956.
[2] The concepts of agglomeration and conurbation are further discussed in Part Three.

CHAPTER 3

THE CARTOGRAPHIC REPRESENTATION OF TOWNS

Urban cartography presents a certain number of problems of its own, for it is a question of representing discontinuous phenomena, and these phenomena cannot be expressed simply by their extent in actual size on the scale of the map. On large-scale maps this extent is probably quite significant, but it is not sufficient to express the importance of the town; even more so on small and medium-scale maps, where the town can only be represented by a sign (Fig. 4).

Let us consider the simplest case, where it is only required to localize the town and express its importance by the number of its inhabitants.[1] A rough estimate is usually given by determining only the order of size. Towns are represented by different signs as having for example less than 10,000 inhabitants, from 10,000 to 100,000 inhabitants, from 100,000 to 500,000 inhabitants, and over 500,000 inhabitants. If all towns are indicated by circles of the same size, the colours of these circles can be varied. It is easy to find a scale of four or five shades from pink to purple or, in black, to use stippled, ruled or solid black circles. However, the order of importance is most evident when the symbols are of the same colour, but different in size.

Circles of four or five sizes may be used; the impression is increased by varying the shape of the signs, circles for the smaller quantities, squares for the larger ones. Care should be taken (which it often is not) to represent larger numbers by larger surface areas.

However, these are only approximations. Generally speaking, a more exact representation is attempted, and to indicate this by juxtaposing the figure of the number of inhabitants is a common though not a very expressive method.

The disadvantage is that the symbol (square or circle) for very large towns covers a vast surface area, out of proportion to the extent of their territory, and overflowing on to the neighbouring areas. If it is simply a matter of a circumference, this does not matter very much, but the symbol is less expressive. If it is a solid circle it becomes impossible to represent neighbouring towns covered by this circle. In this case it is necessary to reverse the shades of colour, and indicate the neighbouring towns by little white circles piercing the large black circle (Fig. 4. 7).

[1] This refers, of course, to the agglomeration, not to the administrative town. Roger V. Brunet, *Le Croquis de géographie régionale et économique*, Paris, 1962.

THE CARTOGRAPHIC REPRESENTATION OF TOWNS 33

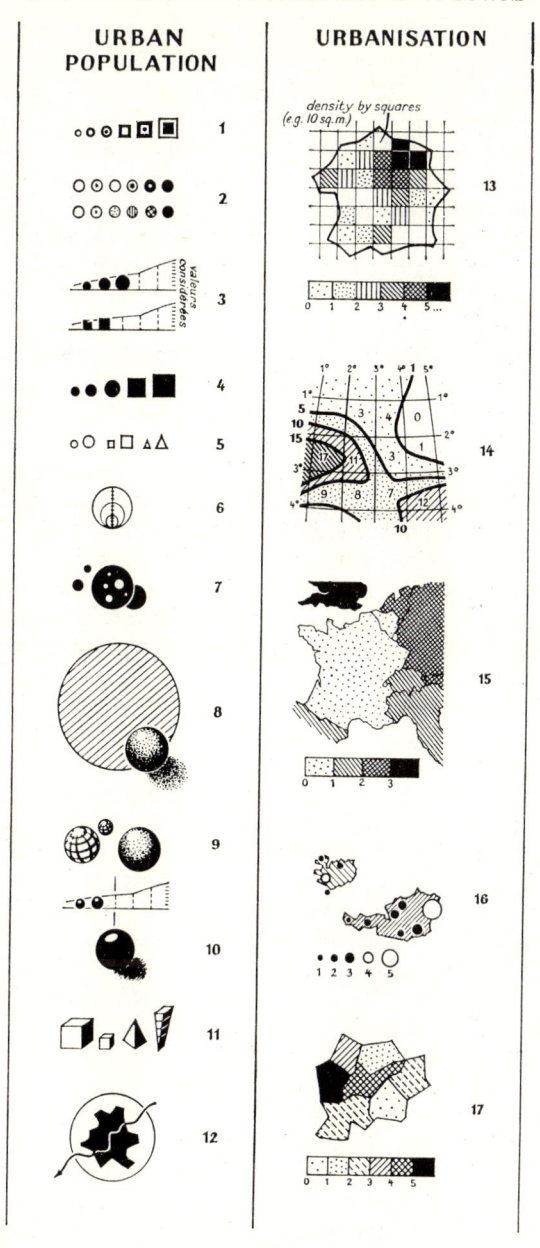

Fig. 4. Various methods of cartographic representation of towns and urbanisation

To reduce this disadvantage, one could use circles (or squares) the surface of which is in proportion not to the number of inhabitants, but to the logarithm of that number. But, this method gives a somewhat false impression.

Spherical representation, used extensively, seems to have been introduced by Sten de Geer.[1] It consists of suggesting the dimensions of a figure of which only a part appears on the drawing. Thanks to an optical illusion, the impression of a complete sphere is given, say $4\pi R^2$ (R being the radius of the apparent circle) while the drawing only covers the dimensions of the circle πR^2; in this way one can quadruple the size—or the illusion of size—without taking up more space on the map. There is an even greater advantage: the sphere gives the impression not only of a surface. If we analyse our reaction, we will notice that the image of the sphere suggests volume, that is, a size measured by $4/3\pi R^3$; so our impression will depend on the cube of the radius. With small quantities, this method does not make much difference. (With a circle with a radius of 3, the surface of the sphere is measured by the number 113·04, and the volume by 93·04.) As the length of the radius rises, so the volume of the sphere increases: with a circle with a radius of 10, the surface of the sphere would be measured by 1,256 and the volume by 4,188; thus the illusion is three times stronger. Therefore quite often only large towns are represented on maps by spheres.

Thus, the sphere has become a widely adopted means of expression as a town symbol. And the smaller the scale of the map, the more convenient this symbolic representation becomes, for it is quite impossible to give an idea of the importance of a town in the minute space which it would occupy by right.

So we come back to the question of giving the illusion of a sphere. This can be obtained (as is the current procedure in America) by drawing a network of curved lines, parallels and meridians. The impression of a ball can also be given by the gradation of light coming from above and to the left; and, an extreme example of simplification, merely by drawing a white ellipse at the top left of the black circle. Illusion, helped by usage, is sufficient. Moreover, the illusion can be increased by marking the shadow which the sphere would cast—a process used by W. William Olsson (Fig. 4. 8–10).

Representation using a cube instead of a sphere gives the same advantages; but while it is easier to draw a cube than a sphere, the cube has the disadvantage of being more cumbersome, for, in order to give the idea of volume, the figure must exceed the limits of the square, whereas the volume of the sphere remains inside the circle.

[1] 'La distribution de la population en Suède', *La Géographie*, 1922.

In all the preceding methods of representation we have had to ignore the shape of the town. On his map of Finland, Helmer Smeds keeps the representation of the town, within the scale of the map, to its true size, but surrounds the town with a ring, so that the circle is proportionate in size to the population of the town. This method of representation is suitable on a map to the scale of 1/200,000 of Finland, where large towns are not too close together and where the rings are unlikely to interfere with one another too much (Fig. 4. 12).

Some authors have suggested various methods of urban representation which do more than simply express the size of population. Jaromir Korvcák tries to show the numerical value of the town in a wider concept. Prague is the capital of Bohemia, from which most of its population is drawn, and which forms the basis of its economic power. Now, Prague contains one-sixth of the population of Bohemia; therefore the town is represented by a circle of a size corresponding to that figure and the other towns of Bohemia by circles proportionate to that of Prague.[1] It must be remembered that this is an exceptional case, with Bohemia a well-defined area in which development converges on Prague.

We may study later certain methods of representation corresponding to the functions of towns, or their zones of influence, but if we are considering towns in general, there are still other methods of cartographic representation which have been used, particularly to indicate urban density (Fig. 4.13–17).

The country can be divided into a number of equal squares, each containing a certain number of towns, and the squares are tinted according to the number obtained. This procedure is not much more expressive than direct town-by-town representation.

The use of choropleths has also been tried. One looks, for example, for the number of towns in a given square (say, for example, a square degree of surface, or 4,772 square miles). The density arrived at is placed in the centre of the rectangle after being converted into conventional units, figured from 0 to 100. Then the points of equal density are joined by curved lines. This density expressing the *number* of towns is probably artificial, as it does not take their size into account. Wanda Rewienska draws another conclusion from these curves by establishing the density of towns of under 5,000 inhabitants and towns of over 5,000 inhabitants, contrasting areas of small towns (in Poznan province) with areas of large towns (Upper Silesia, and the Warsaw region).[2]

[1] Jaromir Korvcák, 'La comparaison géographique des grandes villes', in *Lautensach Festschrift*, 1957.
[2] Congrès international de géographie de Varsovie, 1934, vol. 3.

Objections can be made to this method of representation which in certain cases is quite expressive—the same objections as are made to the use of choropleths in other branches of human geography: arbitrary division into squares, and lack of continuity in human phenomena. It can only be used for very extensive areas.

On the other hand, attempts have been made to represent not the number of towns, but the degree of urbanisation. There are maps in use on which the states of the globe figure in different shades of colour expressing their degree of urbanisation (according to standards which, unfortunately, vary from country to country, as we have seen).

We must remember, however, that there are towns and towns. It would be impossible to compare the urbanisation of Austria (with one town having over 1,500,000 inhabitants) with that of Iceland (with only one town of 72,000), so it is necessary to establish coefficients expressing the percentage of population living in towns of over 100,000, over 500,000, and over 1,000,000 inhabitants.

One should also take into account the general population density. This density could be made the basis of the map or combined with the urbanisation shades, as Emmanuel de Martonne did when representing the Rumanian population.

W. Winid[1] tried to lay down an urbanisation index which would take into account these various points. He composed it of three elements: first the frequency of towns, that is to say, interurban distances; secondly the density of the rural population whose requirements the towns have to satisfy; thirdly, the importance of towns, according to the numbers of their inhabitants. By bringing these three elements together, he obtained a trigonometric quantity; the urbanisation percentage was given as the size of an angle of between 0 and 100 degrees.

Finally, it must be realised that all maps of this type need to cover a sufficiently large area if we wish to avoid making a map which merely indicates the location of towns.

[1] 'The scope of urban geography', *C. R. du Congrès international de Géographie*, Warsaw, 1934, vol 3, and *The Methods of Presentation of Urbanisation*, 1934; cf. Margaret I. Fead, 'Notes on the development of the cartographic representation of cities', *Geog. Rev.*, 1933.

PART ONE
THE WORLD'S TOWNS

THE WORLD'S TOWNS[1]

If one asked any traveller to describe the ideal 'town', he would immediately admit his inability to do so. How, indeed, could one compare Jakarta or Dakar with Florence or Los Angeles? What have they in common, except for their general difference from the neighbouring countryside?

But that is not the only difficulty. Within one uniform type of civilisation, within the same country, there are extreme differences, as for example, between Nice and Lille, Privas and Paris, Merthyr Tydfil and Crawley, yet they are all western European towns, and in a general classification would come under the same heading.

There is no limit to their diversity.

There are towns of beautifully carved stone, the golden limestone of Dijon, Bath or Assisi, the grey limestones of Chartres, the pink sandstone of Glasgow or Strasbourg, the green scaly slate of Keswick, the slabs of black lava of Clermont Ferrand. Only the ancient quarters or sometimes even civil and religious monuments have this nobility of stone, while modern houses are built of cheap materials: from this comes the brilliant effect of the cathedral of Amiens built of chalk and dominating the low cob and brick houses. But whole cities, situated far from quarries of hard stone have succeeded in the harmonious use of manmade materials. For example, in Honfleur and Colmar, the use of plaster framed by dark wood, the crude brick which is so becoming to the great cathedrals of Toulouse and Padua and which gives the towns quite a vivid pink hue; or, in a modest way, the unfired brick, dried in the sun, which is used in the very arid countries of Egypt and the Saharan perimeter and which matches the towns with the ground from which they come, yellow like the alluvial mud of the Sudan and red like the clay of the Moroccan Atlas.

There are cold, gloomy towns, and towns of blinding heat; towns in which fine oceanic rain washes the blue slate roofs or enlivens the red of the flat tiles, and those in which the burning sun fades the small round tiles; the roofs of Angers are grey-blue and reflect the colour of the sky of Anjeu, but the roofs of Rome are as orange as the whole town at sunset. In the Icelandic greyness, the houses are crowned by brilliantly painted steel sheets which give them a festive air, whereas in the dryness of the Sahara, above the geometric line of white terraces the palm trees' tufts stand out against the blue of the sky.

[1] For the bibliography of this section see the references relating to each part of the world given in J. Beaujeu-Garnier, *The Geography of Population* (Longmans 1966), and also the footnotes to the pages below.

There are mountain towns, perched on high or shut in, but always closely confined, and those of the plains, which stretch indefinitely; towns buried in parks and gardens, and towns bristling with factory chimneys and clouded with dust; towns which shoot skywards, and towns which crawl along the ground; there are towns of yesterday and towns of today.

A town is the fruit of a whole natural and human complex. Like a flower growing in a garden, it draws its characteristics as much from space and the soil as from the climate and the efforts of man; it is born, develops, expands, becomes enormous, often changes and occasionally dies. A town is a living thing, not only through the activities of its inhabitants, but having a profound life of its own. One could speak of towns in the same terms as one does of plants or of mankind.

One is often aware of a presence around one, a friendship, a sense of intimacy with certain towns. The feeling you get from a tropical town, with its strong smells, its stifling dampness, and the impression given to a dazzled and sometimes frightened European of being merely tolerated, is so different from that which you sense in the normal and harmonious calm and the soft, grey-blue light of a little town beside the Loire. And yet they are towns... towns... the word is still the same, but how different they are in reality!

Is it possible to introduce any form of classification into this diversity and to describe the great areas of urban life in the world? This is what we have tried to do in the following pages.

CHAPTER 4

EUROPEAN TOWNS

Europe occupies a place apart where urban civilisation is concerned. An old continent, very densely populated, it is noted as much for the intensity of its urbanisation as for the richness and variety of its types of cities. Here towns are very old: founded by ancient civilisations, they have multiplied as the populations grew, and diversified with the general economic evolution. Some have lasted 2,000 years; others have disappeared; new ones are born every century, more than ever, probably, over the last hundred years. And as a result of emigration European civilisation has spread across the globe to every other part of the world.

Thus Europe can be distinguished by the complexity of its urban network and the very wide distribution of the towns which can be found over its whole area, with the exception of the extreme north.

It seems that a difference of latitude is accompanied by a marked change even in the comparatively small area of European territory. Northern towns, central European towns, those of eastern Europe and those of the Mediterranean world have their distinct characteristics.

THE TOWNS OF NORDEN

Northern towns have had to adapt themselves to the cold climate imposed by their latitude. At about latitude 60° conditions are severe enough to make life difficult. There are few reasons, therefore, for the development of towns, and urban life is largely represented by trading stations and road posts. One such reason has been the exploitation of gold mines which, in spite of the natural conditions, has made the fortune of Dawson City. There are few towns in the subpolar regions of America and Asia—Port Churchill in Canada, Berezovo, Turukhansk, Yakutsk, and Oymyakon in Asiatic Russia. They are sparsely populated, and look more than anything like big market towns.

On the other hand, in Europe, the proximity of the North Atlantic has favoured the development of an urban civilisation. Streets full of illuminated advertisements north of the Arctic Circle are an impressive sight, contrasting on the other side of the Atlantic with the icy wastes of Baffin Land or Greenland. We shall therefore consider above all the towns of the Scandinavian countries ('Norden'), because only there shall we find the examples with which to compare towns further south.

Although the North Atlantic drift makes the winter temperature less severe, it does not altogether alleviate the rigours of the climate. The mean winter temperature, in the area of the 60th parallel, is 25° F (−4°C) in Oslo and 21° F (−6°C) in Helsinki. It is 14° F (−10°C) at Oulu in Finland (latitude 66° N). The extremes of temperature are terrible. At Östersund in Sweden (latitude 63°) the temperature fell as low as −42° F (−41°C) in 1892, and −22° F (−30°C) in 1954; sometimes from the month of October onwards the thermometer falls to 14° F (−10°C) in that area.

The first consequence is that winter is a long freeze from October to April, and often longer. The covering of snow lasts all winter, and the clearing of snow becomes a heavy duty for municipal authorities; it has to be loaded on to lorries and carried away after every snowfall; in the smaller, less frequented streets there are walls of snow and only the middle of the street and the access to houses are cleared. Outside the towns clearing is done by snow-ploughs. In spite of this the roads remain frozen and slippery, and if there is the slightest slope traffic circulation becomes difficult. In spring everything thaws, but the asphalt appears swollen as a result of the freeze and roads have constantly to be resurfaced. Drains have to be deeply buried as a protection against frost and even hydroelectric installations have to be built underground, as at Porjus in Sweden.

On the individual or family level, houses have to be secure against the cold. All of them have double windows, between which pots of flowers seem to beckon the spring. The walls must provide protection from the cold. Wooden houses, many of which still exist, have double walls lined on the inside with moss or peat. The doors are always sheltered by wide porches. Next to the houses are huge, orderly woodpiles for winter heating. In modern blocks of flats, central heating is the rule. Out of 56,906 homes built in 1956 in Swedish towns, 56,855 had central heating installed. Often this heating runs on electricity, which is not expensive.

Unfortunately, sunlight is pale and seldom seen. There are days when, even in fine weather, it is totally absent. North of the Arctic Circle the night lasts for twenty-four hours in the middle of winter; workmen work by the light of floodlamps. But the light and heat of what sunshine there is must not be wasted; as the sun is always low on the horizon (one drives a car all day long with the sun in one's eyes) it must be let in. The little wooden houses used to stand wide apart the better to catch it; modern blocks also have wide open spaces left between them and are always built facing south, even if this necessitates building them at an angle to the road. Thus a developing town requires plenty of space; the built-up area grows faster than the population, and it is not unusual to see modestly populated towns covering vast areas.

It should be added that the streets are not particularly attractive, except for a short period during the summer. Life is not lived much in the open; people prefer the home comforts provided by a generally high standard of living; this is a new reason for the increased extent of built-up areas, which in turn calls for a well-equipped public transport system.

The characteristics which these northern European towns have in common are not all due to the climate. In the first place, all except the towns of Scania and Denmark are built on an ancient rock base, raised and broken up to a greater or lesser degree. The rock can be seen everywhere—in the park at Helsinki, in front of the Royal Palace in Oslo, in certain streets in Stockholm. This makes it difficult to dig underground. Houses do not usually have cellars. Dynamite is used to blast out the foundations of new blocks, and traffic has to be halted in the neighbouring street when a charge is detonated. The construction of the underground railway in Stockholm was a gigantic operation.

Above this rock base are glacial deposits. Certain of these morainic irregularities are important, such as the eskers formed at the mouths of subglacial valleys or in crevasses of dead ice. These long, narrow ribbons, sometimes forty or fifty feet high and over fifty miles long, have been valuable to urban settlement. They form natural causeways across the marshy regions and thus became the basis for traffic routes. Crossing the valleys, they provide crossing points and bridge-heads. Uppsala was established on the side of the esker which runs along the Fyris; Stockholm was built on the esker separating Lake Mälar from the sea gulf, and, commanding the passage to the sea, it offers shelter from storms at the same time.

However, this largely moraine-covered base has been affected by the slow isostatic recovery from its burial beneath the weight of the Pleistocene ice sheets; this is a phenomenon which has had a tremendous effect on the entire history of the towns; it must be realised that there have been isostatic uplifts of about fourteen inches per century in the latitude of Stockholm over the last few centuries. The towns are rising above the sea, which is retreating; ports have to move, following the changing coastline; a Scandinavian port is a seaport zone marked by three or four successive harbours; if the town has remained, it is lengthened by an outport and only receives yachts, like Turku. Rivers become incised and the limits of navigation change, to the ruin of up-river towns like Gamla Uppsala. In the seaports canals dry up, are filled in, and become streets and avenues. Gothenburg owes the beautiful avenues in the heart of the town to what were once its canals. They have played the same part as the ancient, disused fortifications which provide the basis of some boulevards in France. Some of these towns are like partially raised Venices, and only the more southerly towns of Denmark have been spared this upheaval.

Another aspect of these northern towns is due to the forests which cover a good proportion of the country, 66 per cent in Finland, 23 per cent in Norway and 24 per cent in Sweden. They are forests of resinous trees with good, straight trunks, easy to work, so the houses were all built of wood, except in Scania and Denmark. Most often they were constructed of planks forming double walls, but sometimes they simply split the trunks lengthwise, which gave a massive impression of superimposed logs, particularly in Finland. Many houses are still built of wood, even in big towns, and consequently there is a permanent fire risk. Every town has been on fire several times; until the seventeenth century this was quite a normal event, and it has happened again in more recent times. Turku was almost entirely destroyed by fire in 1827. In hotel rooms one can still sometimes see the rope which acts, in case of need, as a fire escape.

Since their foundation towns have had to adapt themselves to these physical conditions. But their foundation was late. In fact, the Scandinavian countries were slow to become civilised. Their prehistoric period lasted until the tenth century. The first mention of Oslo and Stockholm was made in the eleventh century.

In those days the sea was the only reason for the foundation of towns. All these races lived along the coast or in coastal archipelagos. The sea gave them their only chance of expressing themselves. They entered the world of commerce timidly. The peoples of the Mediterranean came to them for amber and fish. In the Middle Ages, Hanseatic trade linked the countries of the North Sea and those of the Baltic. One of the biggest of the Hanse towns was Visby on the island of Gotland in the middle of the Baltic.

These trading towns became industrialised. Fish had to be prepared before being shipped: sun-dried fish and salted fish. This is still true, but industry has become more complicated. Not only do they smoke and salt herring and cod, they also prepare all kinds of canned goods, and extract medicinal and industrial oils from these fish. The remains are processed into animal foods and manure.

So the sea is the origin of many of these towns. The wood comes from the interior, but it is worked and transformed into cellulose when, by floating, it reaches the sea. This is the reason for the almost continuous belt of towns along the North Sea and Baltic coasts: Helsinki and Turku in Finland, Luleå, Piteå, Umeå, Stockholm,

Malmö, Halsingborg and Göteborg in Sweden, Oslo, Bergen and Trondheim in Norway, Copenhagen and Odense in Denmark, not to mention Reykjavik in Iceland.

On the other hand there are hardly any towns in the interior of these countries: only Tampere in Finland and the towns of the central Swedish depression which, thanks to the lakes, are linked by waterways to the sea. But some Swedish iron-mining towns and some towns founded on recent industries should be added to this list.

These northern towns, founded above all on maritime trade, have an international character. Many of them had this character imposed on them by their geographical position. They mark out an international route. For example, Fredericia, Nyborg, Korsör, Copenhagen and Malmö on the great route which links Sweden to the continent.

It is often easier to communicate with foreign countries than with one's own hinterland. Especially before the railways were built, Oslo found it easier to communicate with Copenhagen than with Trondheim or Bergen. As the Baltic is frozen all winter, Swedish commerce is limited to the North Sea. Iron ore from Kiruna passes through Narvik, and the Swedish Norrland seeks an outlet through Trondheim in Norway.

The history of these countries has done much to develop this international character. The countries which today form separate nations were once united, in various combinations. There was a period in Danish history when, under the Union of Kalmar, Sweden, Norway and Denmark were joined under Danish rule. Sweden freed herself from this domination after two centuries of war, but Norway remained Danish until 1814, and for a long time Copenhagen continued to be the cultural centre for Norway. In the nineteenth century most Norwegian books were published in Denmark.

After this, Sweden became the great nation. She owned the south coast of the Baltic and Finland. The Baltic was a Swedish lake. It was not until the beginning of the nineteenth century that Finland was conquered by Russia, but a whole area of Finland, Västerbotten, uses the Swedish language. Swedish is still an official language in Finland and the street signs are bilingual. Norway, for her part, remained a Swedish dependency throughout the nineteenth century.

This demonstrates how urban hierarchy overlaps national frontiers. Copenhagen is the metropolis of southern Sweden; people come there from as far away as Malmö for shopping or entertainment. All these towns seem to have remained faithful to the international vocation of the Hanseatic League.

There is yet another characteristic without which we cannot understand these towns. Not only do they go back no further than the eleventh century, as we have seen, but many of them were founded in the seventeenth century (Kristiansand, founded by King Kristian, and Frederikstadt by King Frederick in Norway). Above all, most of them remained small towns until recent times. The urban movement came much later than in western Europe; it usually dates here from the twentieth century. At this time two great categories of industry were developed: the iron industry and the wood industry. Iron had been exploited for a long time in Bergslag, but the exploitation of Lapland iron dates from the twentieth century, as do its dependent towns. Even more marked was the development of the manufacture of cellulose from the vast forests which, until then, had mainly been used for purposes of heating and building. Many medium-sized towns have specialised in this industry, but its riches have spread throughout the country and every town has received from it a new lease of life. As all this took place at a time when communications were easy and when electricity was the universal source of energy, many small and quite well equipped towns sprang up. Rjukan in Norway has become the centre of the nitrogen industry.

In this way completely new towns have appeared, like Umeå, planned like a chessboard. Above all we find towns which have been enlarged recently. In certain quarters, mainly those of the centre, they have retained their eighteenth and nineteenth-century wooden houses, for there has not been serious fire damage for about two centuries, and except in Norway there was no war damage. Surrounding these older buildings new ones have grown up in the modern style, often in reinforced concrete. A thirty-year-old house is considered old even in Finland, and the rent is reduced. One Swede in six lives in a house less than ten years old, and the distribution of buildings in Sweden is as follows: 30 per cent before 1900, 49 per cent from 1900 to 1940, 21 per cent after 1940. So people move house frequently in search of greater and greater comfort.

These towns also reflect a civilisation in which communal life holds an important place. Modern houses have communal laundries and ironing-rooms, bicycle garages, and ski-rooms. In Finland every block of flats has its communal *sauna* or steam bath. On the ground floor games rooms with billiards and table tennis are provided, and day nurseries are organised for the children. All this is the result of a civilisation in which all the women work, and it is almost impossible to find servants. It also satisifies the need to keep the children occupied together through the afternoon, as lessons end very early.

But in the sphere of the town, communal life is also highly developed. Hospital care, under the best conditions, is almost free. In Sweden 97 per cent of beds for the sick are to be found in public hospitals. When a child is ill he is sent to hospital for a week or a fortnight, regardless of social class. As a result we can see the growth of vast and luxurious hospitals, and a large number of schools of nursing. At Helsinki one of the biggest estates in the town is made up of the school of nursing and its park, which belong to the hospital administration of the town.

Interest in sport is evident. Oslo is dominated by the huge Holmenkollen ski track, and in the centre of the town is the stadium with bronze plaques on the doors bearing the names of Olympic winners. In a square in Turku stands the statue of Nurmi, the runner, and the highest building in Helsinki is the tower built a few years ago on the occasion of the Olympic Games.

Thus it can be seen that these northern towns do more than just reflect a very special physical environment. They cannot be understood unless one takes into account the human environment which formed them and which they serve.[1]

However, there are some which must be dealt with separately—those of the extreme north existing under conditions of extreme difficulty. On the Swedish side these are the iron-mining towns of Gellivara and Kiruna. The latter has over 26,000 inhabitants, housed under the care of the company which exploits the iron mines; this company gives life to this abnormal town, where, in winter, work is carried out by floodlight, where the drains have to be dug very deep and where wages are very high to allow the workers to dress warmly and to eat food brought at great expense from the south.

On the Norwegian side[2] towns can be found still further north. Vadsö, Hammerfest and Vardö are north of the 70th parallel. It is true that the proximity of the warm sea makes the temperature milder, and that at Tromsö the lowest temperature recorded is only $-15°C$. The snow lasts, but is not very thick. It was mainly fishing that gave rise to these towns, and this is still their principal activity, with factories for fish-freezing and the utilisation of waste-products. The evolution of the fishing industry, which is becoming more and more concentrated in towns, increases this characteristic.

The abundance of natural resources (iron and fish) established these towns. Today they seem more necessary than ever. Life in these regions of the extreme north is hard. Men who toil along the shore or in the forests dream of the south lands where life smiles and work is less painful. To avoid the desertion of the far north all the states try to develop activities in the towns to retain the population, and to establish industries and cultural and tourist centres, as at Oulu in Finland, Luleå in Sweden and Mo i Rana in Norway.

[1] Cf. *Norway North of 65*, 1961.
[2] Maurice Allefresde, 'Villes arctiques de la Norvège', *Rev. Géog. de Lyon*, 1960.

Iceland presents a special case. This island, which is tangential to the Arctic Circle, and washed by the ocean, suffers less from the extreme temperature than other lands in that high latitude, but it has vegetated for a long time because of its lack of natural resources. Its fishing rights were reserved for the Danes. Urban life was almost non-existent. At the beginning of the twentieth century, Reykjavik had a population of only a few hundred. Then came the boom. The regained fishing rights enriched the island. Its privileged position on intercontinental lines attracted an American airfield and dollars. The capital profited most from the boom; nearly half the population rushed to Reykjavik, which was suddenly transformed into a modern city with 72,000 inhabitants; the town next in size, Akureyri, has only 7,000 inhabitants.

Thus, in the towns of northern Europe, political and physical conditions have united to intensify urbanisation.

WEST-CENTRAL EUROPE

On a map showing the distribution of towns in the world, the European area between the 42nd and 57th parallels, taking in the United Kingdom, the Common Market countries, Switzerland and Austria, seems distinctly privileged. This intense urbanisation manifests itself in the midst of the greatest and mostly evenly distributed population densities on the globe.

There has been no climatic hindrance to urban development, while a large number of natural conditions have stimulated man's enterprise, provided raw materials, water, and the necessary transport, and favoured the growth of an already ancient and yet still vigorous civilisation. The heat is not overpowering, nor the cold excessive. Rainfall is sufficient without being devastating or unevenly distributed. Man feels himself at ease and invigorated by the alternation of the seasons. He is not obliged to spend a great part of his strength on solving what are in other climates insoluble problems, such as the lack or excess of water, paralysis through frost, or burial beneath an exuberant and hostile vegetation. As well as the ocean, the large, regularly fed rivers, always ice-free, provide water and means of transport. The ground is firm and space penetrable; man can build roads and railways, dig the foundations of houses and select his building materials under good conditions. Lastly, coal, the first source of the creative energy of modern economy, is abundant.

So towns may be found everywhere, and the first characteristic of the urban life of west-central Europe is a well-balanced hierarchy. Unlike certain parts of the world where medium-sized towns are few, the pyramid here is solidly based: rural markets scattered throughout

1 Reykjavik: quarter near the port

2 Trondheim: warehouses in the port

3 Leeds: working-class housing and industry

4 Vällingby: shopping centre in the new town

the countryside; isolated centres situated at crossroads of communications centralising a wider trade, like the towns of western France, and the country towns of south-eastern England, or central Germany; gateways to mountain life, such as Grenoble and Innsbruck; important centres of industrial and commercial activity, and so on.

There are very few places far distant from a town. In England and Wales, there is no place situated further than 125 miles from a town of over 100,000 inhabitants; in the pentagon formed by joining London, Bournemouth, Swansea, Liverpool, and Newcastle, the distance falls to less than sixty miles. In this same area of England and Wales there is one town of 100,000 people for every 883 square miles; in West Germany, one for every 2,056 square miles, and in France one for every 9,227 square miles. It is true that there is a considerably larger number of medium-sized towns in France.

This abundance of towns in no way excludes the importance of great cities. On the contrary, it is possible to travel for sixty miles across uninterrupted built-up areas, in the United Kingdom where the six great conurbations of England and Wales include 40 per cent of the population, or in Scotland where Clydeside alone houses two-fifths of the Scottish population, and in France where the Parisian agglomeration contains 30 per cent of the French urban population and 17 per cent of the total French population. Out of the seventy-one towns of over 100,000 inhabitants in the United Kingdom, seventeen are part of the vast agglomeration of London.

Cycles of development. The events and trends which have resulted in this intensity of urbanisation and the complexity of its network can be divided into cycles.

The first cycle was progressive and lasted several centuries. It answered the needs of barter in the tradition-bound communities relying on rural occupations and handicrafts and having at its disposal only a very ineffective transport system. It resulted in swarms of small rural markets scattered amongst the working villages. In the Middle Ages these markets served at the most an area of six or seven miles' radius.

The second cycle opened with the development of industry, and the improvement of transport systems. The end of the eighteenth century and, even more, the beginning of the nineteenth, saw the creation of new centres for the exploitation of mineral resources such as coal, or the concentration of rural craftsmen in factories, as in the textile industry. At the same time the appearance of modern transport systems, especially the soaring growth of railways, allowed more rapid carriage of larger loads over greater distances. The result was an increasing contrast between the crowded areas favoured by the new activities, and the rural areas where urban life grew drowsy and self-centred, and sometimes disappeared.

The third cycle was initiated by the increasing powers of attraction and absorption of the large centres of economic activity. First they expanded immoderately and burst their traditional administrative limits. Then they overflowed into huge suburbs completely surrounding their former nuclei, and spread further along the main arteries of communication. In the United Kingdom, between 1919 and 1939, four million houses were built in the country surrounding the towns, and by 1951 nearly a fifth of the population living in rural areas were no longer working in agriculture. This favours the development of the daily to-and-fro movement between home and work place. West-central Europe is one of the parts of the world most affected. The confusion of the zones of influence of the industrial towns of the Ruhr, of northern France, or of Lancashire or Tyneside is almost inextricable, while around the biggest concentrations there are considerable movements of flux and reflux.

In this part of Europe a new era is beginning, perhaps a fourth cycle enticing people towards urban life. It is an attempt at voluntary reorganisation of urban life, from the point of view both of the grouping of buildings and of man's way of life. It is no longer simply a question of satisfying economic needs, but more of satisfying the tastes and needs of the people. This has resulted in a phase of urban renovation, or even creation, which characterises the second half of the twentieth century.

Only Europe has known such a complex evolution, but this idea of 'urbanisation cycles' seems so characteristic that on it rests the entire zonal description of world towns attempted in this chapter: we shall return to this point later.

The contemporary urban scene in west-central Europe reflects these vicissitudes. Ancient towns, posts of military attack and defence, rural markets and mountain crossroads languish and become depopulated unless their picturesque sites and the beauty of their historical monuments attract the tourist trade. The Massif Central, the inland areas of Provence, the foothills of the Alps and parts of Wales offer examples of this sad decline.

Yet a certain number of towns belonging to the first cycle have escaped this stagnation. Situated on mineral deposits or on important transport routes, they were transformed during the nineteenth century. With the beginning of industrialisation, the tempo of their growth changed, as did their appearance. The original plan was upset and the town-planners have left nothing of the ancient heart of these towns save perhaps two or three typical monuments, and a few old houses lining the narrow streets. Sometimes all the old buildings were demolished, leaving only the monuments standing in the middle of a park. With each stage of economic development new buildings have been added, sometimes over a large area, representing a wide range of architectural styles.

The commercial quarter and the industrial establishments are usually systematically grouped in the central part of the agglomeration: in the nineteenth century, factories, warehouses, shops and dwelling-houses stuck as closely as possible to the town centre, forming a mixture whose density was equalled only by its unhealthiness. Lille, Cardiff, Carlisle, and Douai are towns of this type.

Other towns have progressed since their foundation in the second cycle of urbanisation. Created by the agglomeration of factories and workers, their origin is reflected in a landscape devoid of beauty or grandeur, overpowering in its monotony and sometimes in its dirt and hideousness. Groups of dwellings in mining areas, craftsmen's villages turned textile towns, railway junctions—these are functional agglomerations like formless ant-hills. The only thing that mattered was to house the workers close to their work. The builders did not take the trouble to plan a spacious monumental town centre, nor did they allow land for public gardens or broad streets. There was no preformed plan, and little functional segregation. The station, the dockside, and the factories formed the only rallying places. Commercial, administrative and cultural amenities were reduced to the barest minimum. The heart of the town was a square surrounded by banks, shops, cinemas and municipal buildings, situated at big road junctions. Roubaix, Tourcoing, the textile towns of Lancashire and Yorkshire, great railway centres such as Crewe, Swindon, Vierzon, and Creil, and the countless small towns born of coal mining, present the same characteristics.

This constant economic expansion, progressing alongside the development of other techniques, the diversification of industries, and the growth of social and architectural awareness, has resulted in more and more towns being roused to industrial activity. Now it is the big agricultural market towns, and a few tradition-bound towns famed for their cultural or administrative functions or residential qualities which are in their turn attacked by the wave of industrialisation. Le Mans, Toulouse and Norwich belong to the first group. In the second, Oxford has witnessed the growth of the great car factories; Nice, for many years unrivalled queen of the Côte d'Azur, has become a great industrial centre. The tourist industry creates its own capitals: Cannes, Biarritz, Les Sables-d'Olonne, Blackpool, Brighton, Chamonix, Vichy, Zermatt—so many names which only a century ago were no more than market towns, little known or frequented only by a small elite, but nowadays submerged every year by tens or even hundreds of thousands of visitors. To this recent increase of towns must be added the 'dormitory towns' where people live and sleep, but do not work: great residential suburbs with few occupations except in the tertiary sector but extremely well linked with neighbouring industrial centres.

Alongside the results of this teeming and disordered urbanisation, mention must be made of the latest progeny of western Europe, Britain's new towns—Crawley, Peterlee and others—and certain centres built near recently discovered riches, like Mourenx near the Lacq oil and gas field. These comprise an ordered assembly of new buildings for living, surrounded by lawns, and with wide streets, served by every possible modern technique.

The dominant influences. A quick look at the map shows the dominant influences. The biggest masses of population and the most important towns are to be found on the site of the great mineral deposits, in contact with the sea, and along the big rivers.

In this part of Europe coal plays a part unequalled anywhere else. At the time when the industrial revolution was taking its first steps in the region, coal was the basis of the transformation. The nature of these deposits, the fact that the methods of exploitation of those days needed a large labour force, and the local systems of transport favouring the concentration of industries using coal on the spot, combined to bring about the foundation of countless mining towns which multiplied along the outcrop of the seams, while the most ancient towns retained the industry. Thus in the Ruhr, coal mining is at present carried out in the north while the great textile and metallurgical centres are to be found in the southern regions, which not long ago mined coal. It may be added that European coal represents 30 per cent of the world's total production.

In these multiple and widespread mining areas, as in the iron-fields of Lorraine, one cannot speak of towns so much as of agglomerations, of layers of population. The geometrical alignments of the miners' dwellings are disposed around pyramid-shaped spoil-heaps and smoking chimneys: one town follows another without any marked difference of style. These characteristics reach their height in the famous region of the Rhondda where, in a narrow and deeply cut valley the 'Rhondda Urban District' stretches for a dozen miles. This district housed 141,346 people at its maximum in 1931, and still contains almost 110,000. There is no town centre, nor even the beginning of a hierarchical organisation.

Rich in manpower, coal and industrial enterprises, Europe needs multiple contacts with outer continents as an outlet for her sons and her manufactured goods, and as a supplier of foodstuffs and raw materials for her abundant population. Broken up by arms of the sea, she was faced with the necessity of linking her various separated parts by an intensive system of water transport. The North Sea and, especially, the English Channel are among the busiest seas in the world and along their shores at the opening of deep estuaries stand four of the world's fifteen largest ports: Rotterdam, London, Antwerp and Hamburg. Of the seven great conurbations of Great Britain, five depend more or less directly upon the coalfields, and two on maritime estuaries.

Finally, rivers, remarkable for their dependability as means of communication, play an important part. Not only do they provide openings to the sea where seaports are situated—Rouen, Le Havre and Bremen, though not as famous as the four mentioned above, are examples of their importance in this context—they also have a significant role in the general movement of traffic along their courses. Paris is situated in the Seine valley and the Rhine has along its banks the greatest number of large towns in proportion to its length of any river in the world.

Characteristics. All these cities have characteristics in common. They belong to ancient countries, highly developed technically, and relatively prosperous. They are well equipped with means of transport, and railway stations, docksides and access roads hold an important place. Private and public motor traffic is intense and every sort of amenity is well developed; educational and cultural institutions and hospitals are to be found everywhere. Wide public squares and monuments left by successive centuries are the scene of many happenings. History is apparent almost everywhere, albeit unobtrusively. The line of an avenue traces that of an ancient rampart. The towers of a venerable cathedral dominate the busy centre, and some of the old quarters are carefully preserved. But all around most towns spread the generally monotonous suburbs, sometimes verdant, but stretching inexorably and almost joining together. In Belgium, the Rhineland and some parts of England one has the impression of never quite leaving the town.

Comfortably off, if not rich, these towns are also intensive working centres. Factories are numerous, sometimes stretching along the roads or railways, and sometimes scattered about the very heart of the city. Commerce is strongly represented; the population has many needs, and neither goods nor customers are in short supply. Free enterprise favours tempting displays and the predominance of numerous small businesses can be seen from the multiplicity of shops and the originality of the shop fronts. The personal relationship between shopkeeper and client has not yet been killed by the big impersonal chainstores and supermarkets. At night towns are lit up, but do not become rivers of light, and neon signs are still comparatively discreet.

The crowds are hurried, and walk quickly. The climate does not invite strolling, but processions and sporting activities are popular. The country fairs of the north give way to festivals of folklore and outbursts of exuberance. Fairs and exhibitions go hand in hand with a general animation, flags flap in the wind along the shopping streets. Collective movements of population are considerable: the journey to work in the morning, the return home in the evening, and mass departures for the country to spend the weekend in the open air. All who can manage it have comfortable apartments or houses, and protection against the cold seems to be more important than the struggle against the heat.

Of course the working-class neighbourhoods do not always answer this optimistic description, and the sale of drink is too often excessive, but the general improvement in the purchasing power of money and the development of social life continue to progress. The new blocks of flats and individual houses are remarkable for the care taken to make them more light and airy, and less like barracks. The rebuilding of the dismal parts of the East End of London and the new factories with lawns and flower beds in front of them whenever possible, contribute to the impression of an urban life which, having developed mainly at a time when the individual counted for little, tries now to please him more and more.

MEDITERRANEAN EUROPE

The Mediterranean countries present a totally different atmosphere. This is one of the most ancient areas of civilisation in the world. The cities of ancient Greece and their spread to distant colonies, and the organising power of Rome created the urban nuclei which have lasted until the present day. The sites have remained unaltered, even the ancient town plans are sometimes still distinguishable, and there is a heritage of monuments, and even of the same urban style.

Taken as a whole, Mediterranean towns, with the exception of a few recent creations of which Madrid is the most famous, are the children of history. Who has not discovered with surprise in the storeyed houses, the *emporia* and the fish market of Ostia Antiqua the foreshadowing of a little Naples or Genoa, in the *Tabernae* of Pompeii the exact equivalent of the everyday shops of Rome, and in the sumptuous houses of Herculaneum a foretaste of the most famous villas of the Renaissance? Who has not felt great emotion on visiting those sites of Roman piety where a modern church has been built, like Saint Clement, over four levels of primitive sanctuaries where, superimposed on the Roman shrine are those of eastern cults, first that of Mithras and then that of Christ, whose permanence is represented by several successive buildings? It is enough to have seen the markets of Trajan, where successful modern exhibitions are still given, surmounted by the loggia of the Knights of Rhodes with the more recent superstructure where many families live very close to the Corsos—the pulsating arteries of modern Rome, running parallel to the ritual routes of the ancient city—to understand that history has never stopped, that there was not a sudden break but a gradual evolution and adaptation, and that Italian towns derive a little from antiquity and a great deal from the Middle Ages.

The narrow streets, the massive blocks of stone, the palaces and the countless churches are part of the very fabric of these cities, venerable and yet always overflowing with life, whose names evoke both the riches of the past and the marvels which draw crowds of tourists, pilgrims not only to the religious sanctuaries but also to the places of eternal beauty: Florence, Verona, Venice rising inimitably from the water with the network of canals reflecting its palaces, Siena and its square in the form of a giant shell which is still the scene of the colourful *Palio* games, Ravenna and the inexpressible luminosity of its mosaics, Alberobello with its extraordinary *trulli*, perhaps keeping alive the memory of a prehistoric population.

If Italy owes the fact that she is a veritable museum of cities to the building genius of her inhabitants, the strength of the materials used, and her preserving climate, she is not alone. History also dominates the towns of Greece. The Acropolis is still the mutilated heart of modern Athens, which surrounds the remains of the past and whose fortune still lies on the Piraeus. The quarries of famous Pentelican marble still gash the mountainside. Corinth still watches over its isthmus across which the ancient Greeks dragged their boats and which is now cut by a canal. Modern Sparta is in the plain but still lives on the olive trees of the same valley as its forerunner. Here, however, many great names denote only ruins beside which stands a small country town. Memory and imagination hold a more important place than continuity of tradition.

Almost all towns in the other Mediterranean countries also have their origins in past centuries. Religious and political vicissitudes have covered the Iberian peninsula with monuments which are the jealously guarded treasures of every town. Granada is dominated by the majestic buildings of the Alhambra which open into the infinite delicacy of its interior patios. The little echoing streets of Cordova wind among the wrought iron gates which give access to the courtyards of the houses, to the immense and splendid mosque with its forest of columns. Avila hides in the shelter of its austere ramparts; haughty Toledo is almost encircled by the deep gorge of the Tagus; Segovia prides itself on its Roman aqueduct and the Alcazar built by the Cid's king. And so one could go on multiplying the examples.

If the past holds such an important place in these countries, it is because of their strong traditions, and also because the economic transformation is recent, not very vigorous and only partial. Until the dawn of the twentieth century modern industry had made few inroads in these countries and the movement of people towards the towns is much more recent than in north-western Europe. Over about thirty years it has caused only slight and generally peripheral changes. Rome, where the walled boundaries, dating from the late Roman Empire and rebuilt in the Middle Ages, were not fully occupied, saw in the first third of this century the development of compact suburbs of many-storeyed buildings. Madrid has witnessed the same phenomenon. Valencia and Lisbon are expanding on all sides over the neighbouring plains. Naples is scaling the heights which enclose its bay, and joining with Pozzuoli and its smoking factory chimneys. Genoa is rising over the slopes of the amphitheatre of Oro. Venice has its mainland counterpart in the industrial town of Porto Marghera. But unlike the towns of western Europe this expansion and these changes are much more recent. They are going on now, and look like additions to ancient and well-preserved centres which are important and still almost inviolate.

However, in addition to these illustrious survivors, dominated by their past, there are other cities belonging essentially to the modern era. They have for a long time been a prey to modern industry and big business and are similar to the towns of western Europe as much because of their urbanisation as because of the activities of their inhabitants. Turin, Milan and Barcelona come under this heading.

Strictly speaking, the transition is not abrupt and does not coincide with a political frontier, nor with the considerable mountain ranges on which the frontier lies. Marseilles and Nice as well as the small towns of the French Mediterranean coast belong to the Mediterranean world where outposts are to be found as far off as Arles, Avignon, Vienne, Nîmes and Carcassonne. On the other hand the cities of the Po valley, Catalonia or the Basque country are nearer, because of their human and spatial characteristics, to the cities of western Europe. To trace the limits, it would be necessary to follow all the vicissitudes of history, the stages of human conquest and modern economic expansion.

Yet another characteristic unites all the Mediterranean towns, not only the ancient cities of tradition but also those which show evidence of active urbanisation over the last decades. It is the influence of their natural environment. The towns of the Mediterranean—and not only those of its European shores—are the cities of sunlight and limited space.

The fragmentation of the relief makes large plains rare. With the exception of the Po valley and a few alluvial areas fringing the mountains, towns always have to cope with hilly landscapes: the high, deeply scored plateaux of the massive Iberian peninsula, the Andalusian valleys bristling with small local ranges, mountain peaks plunging almost directly down to the plain or the sea like the buttresses of the Alps or the Appenines. Towns have had to adapt themselves to this relief. Rome itself was born amid seven hills, but has covered many more in the course of its development.

Many towns are built in the form of an amphitheatre, dominating the end of a gulf which brings trade to their port, like Genoa, Naples, La Spezzia and Malaga, or hanging on a slope overlooking the plain like Assisi. Others occupy inland valleys—sometimes vast, sometimes hilly—and rise to their rims, like Florence, Bilbao and L'Aquila, or stretch at the foot of a dominant hill, like Granada. Athens does both. It descends from its Acropolis and rises to the surrounding mountains. Other towns, like Oporto and Toledo dominate narrow valleys.

In contrast to this domination of the environment by the mountains, we have that of the alluvial plains; these are recent and unconsolidated, built up from the erosion by rivers of terrain subject to torrential rains and in which human activity has long contributed to soil erosion; some of them, along the coasts, have grown during historic times, as at Thermopylae and in Apulia and Calabria. On these recent deposits, for a long time infested by malaria, towns struggle against the water or use it as does Venice, and occasionally suffer disastrous floods from it like Valencia. Some towns are founded there, like Pontina. Few Mediterranean towns have a dull or commonplace site.

The sun shines everywhere. The cold of the winter and the rain of the intermediary seasons do not count when compared to the heat of summer. The people of antiquity protected themselves from this heat, and today people still take precautions which sometimes seem rather one-sided, for the absence of effective domestic heating does not make a stay during the cool seasons very pleasant. On the other hand during the hot summer days one can appreciate the narrow streets, the alleys and streets with vaulted roofs, the thick walls, the buildings facing north, the high ceilings, the paved or tiled rooms, the interior courtyards made refreshing with greenery or a fountain. It is easy to understand why Roman houses enclosed the *atrium* in the same way as Andalusian dwellings enclose a patio. The terrace is still used to enjoy the cool night air in the old houses surrounding the Acropolis of Athens as well as in the big apartment blocks of Posillipo above the sumptuous Bay of Naples. Balconies are esteemed everywhere, protected from the sun by roofs, or by stone cross-bars hung with foliage. They are present in abundance on large modern buildings. People take their siesta on them, have meals and even sleep the night there. Taking advantage of the sun and the easy going atmosphere so typical of the south, the washing dries at the windows or on long lines hung across the streets. Water is scarce and is the more appreciated. Every square has its fountain playing, or its pond where the town children come to refresh themselves and splash to their hearts' content. The sumptuous fountains of the Villa d'Este and the hundred fountains of Aquila bear witness to this Mediterranean love of water and use to advantage their situation in the central mountains of Italy. The pools of the Alhambra of Granada, the gardens of Seville, and the fountains of Rome answer the same need, and the street urchins from the hovels of Trastevere can at least console themselves in the neighbouring square with the burst of hot sunlight and the splashing fountain.

In such a climate, outdoor life takes on an importance unheard of further north. Whether it is the attraction of an open-air life, or because of a characteristic common to all these peoples, who are nonetheless of very varied origins, a greater value is attached to things visible, and a lesser to that which is hidden. Housing may be stark and food basic, but dress is always impeccable and often more flashy than sensible. Men and women favour light colours and elegant style. Much time is given to appearance, and this is shown off every evening on long walks which, from six o'clock until nine, after offices, shops and factories have closed, throw crowds into the streets. A certain very busy little street in Naples is closed to traffic for the duration of the evening walk. In Barcelona the *ramblas* and in Madrid the great central avenues are covered first in one direction, then in the other, by a dense, vibrant and brilliant crowd. On the Plaza Mayor by Salamanca, as in many other Spanish towns, the boys walk in one direction and the girls in the other for hours on end without speaking to one another. Night life is equally animated, and often singing can be heard late into the night, even in big cities.

To all this activity must be added that of the tourists. These countries, with France, are the crossroads of international tourism. Every year millions of visitors go to Italy, especially, but also to Greece and Spain. Holy Week draws crowds to Rome, Seville, Granada and

Cordova, and all the great artistic towns of the Mediterranean. To this flood of pilgrims and visitors to museums and monuments, the summer adds the crowds of visitors seeking reliable sunshine, less overcrowded beaches, light wines and unusual cooking. The Iberian peninsula, Italy, the French Mediterranean coast, the Adriatic shores of Yugoslavia, and more and more the Greek peninsula, are becoming the holiday lands of Europe and even of North America. This brings with it a proliferation of hotels, tourist offices, bus services, swimming pools, stadiums and arenas.

On the other hand the industrial aspect of Mediterranean towns is far less apparent than in the countries of west-central Europe. If the great cities of Milan, Turin, Barcelona, Valencia, Oviedo, and Bilbao have nothing for which to envy the more industrial towns of the world, if great ports like Genoa, Naples and Barcelona have a busy trade and clearly distinctive installations, if the big capitals like Madrid and Rome take on the appearance of great multifunctional cities, these aspects are not characteristic of Mediterranean urbanisation in general. They are found more often in the north of these peninsulas rather than in the south which, in many respects, is already more like the African side of the Mediterranean.

EASTERN EUROPE

It would be useless, here as before, to try to make the differences coincide with political frontiers whose secular variations have been manifold. Urban civilisation is a synthesis of complex influences and cannot be parcelled out according to the whims of fluctuating administrative boundaries.

Those territories which today form western Poland were at one time an integral part of the Germanic world which, through the intermediary of the Austro-Hungarian Empire, extended also into the Balkans. Eastern Europe appears as a transition area, deeply influenced by the west but closer in its physical and human characteristics to neighbouring European Russia. It has the same vast plains allowing the same movement of people and of frontiers, the same long-lasting predominance of rural life, and the same hurried transformation, with the ascendancy of the new economic and political formulae over the last few decades.

Many of the towns and cities, moreover, were grievously damaged during the last war, none more so than Warsaw, which was almost completely destroyed, and whilst some of their centres—market places and town halls—may have been rebuilt as exact replicas of the architectural styles that had survived until 1939, they all now have one thing in common, namely the preponderance of four- or five-storey blocks of flats as the standard units of residential accommodation.

Between Bohemia and Slovakia, where the contrasts are particularly striking, urban life exchanges the characteristics of central Europe for those of the Ukraine. Thus it is possible to compare the towns of eastern Europe with those of the different parts of the U.S.S.R.

CHAPTER 5

THE U.S.S.R.

The U.S.S.R. is itself as vast as a continent. Its western boundaries penetrate deeply into the European complex, while to the east it stretches to the most distant parts of eastern Asia. The vast frozen coastline in the north lies beyond the arctic circle and to the south the country stretches as far as the 36th parallel, the same latitude as the extreme south of the Iberian peninsula. To the natural gifts of this enormous territory must be added the experience of an ancient and basically agricultural European country, hurled into the colonisation of vast Asiatic lands and later submitted to the forces of socialist experiment, aiming at rapid economic transformation and employing modern methods. That means that the U.S.S.R. encompasses all possible problems from the arctic township to the oasis, from the ancient historic city to the most ultra-modern of artificial architectural creations; it is, in fact, a kind of laboratory of modern urbanism.[1]

It is impossible to describe regular cycles of growth because the differences between the old Russia west of the Urals and the huge Asiatic stretches are immense. The former resembles eastern Europe. We are tempted to classify the latter with countries where European influence has been exercised within the framework of colonisation, it being of course understood that here we are dealing with a special case, both by reason of its size and by its patterns and dates of development.

OLD RUSSIA AND EASTERN EUROPE

The old Russia, like all the republics of eastern Europe, was, until the first third of this century, essentially rural. Up to 1930 less than a fifth of the population could be classed as urban. Not only was the proportion of town dwellers small but many towns were still nothing but large villages, market centres which sprang up in the heart of the countryside and which benefited by a certain amount of commercial activity and by the residence of prosperous landowners. The streets

[1] L.L. Troube, 'Les villes de la Volga', *Voproci Geographi*, 1959, no. 45, pp. 88–99; O. A. Konstantinov, 'Les villes de la République Socialiste Soviétique d'Ukraine', *Bull. Soc. Géog. U.R.S.S.*, 1954, no. 2, pp. 215–28, and 'Les agglomérations urbaines de l'Oural, *Vop. Geog.*, 1956, no. 38, pp. 78–103; V. V. Vorobiev, 'Les types d'agglomérations urbaines dans le sud de la Sibérie orientale', *Vop. Geog.*, 1959, no. 45, pp. 99–112; E, Lopatine, 'Leningrad', Moscou 1959.—La population de l'U.R.S.S.: symposium, *Vop. Geog.*, 1962, no. 56, passim.

were wide and on the outskirts of the town ran through farmland. The town centre consisted of single-storey houses with groups of public buildings, banks and shops, and here the main roads were better kept and better lit. There still exist innumerable examples of this type of township, containing a population of 10–40,000, improved and developed to varying degrees. They are also found in Bulgaria, Romania and Poland as well as in the Ukraine and in the central plains of Russia. A capital city such as Bucharest, in spite of the intense activity and grandeur of the city centre, really belongs to this class of town owing to the character of its extensive suburbs.

Development at the beginning of the twentieth century had, however, achieved some transformations, with the arrival of certain industries to meet the demands of rural life (construction of agricultural machinery, manufacture of furniture and ceramics). Improved communications meant that the country districts were more in touch with the market towns, and the same developments can be observed here as in the rural districts of western Europe. In the Ukraine the road communications between the scattered market towns have been replaced by a wider railway network.

Many of the semirural towns in this countryside of vast plains have developed in a characterless fashion in the middle of fields. But they are all alike in one respect, that their sites have picked out some special feature, such as the banks of a river, the slope of a hill, or a crossing point. In the eastern European republics and in southern Russia, where some mountain ranges cut across the countryside, a certain number owe their origin to the ancient fortified routes at the end of the Balkan or Caucasian ranges, the Dinaric Alps and the plateau of Bohemia. Many of these towns are grouped around ancient citadels and have extended towards the neighbouring plains. Cluj, Ploeşti and Zagreb are on the junction of roads or railways which cross the mountain heights; Krakow, in the peaceful grandeur of its palaces and rose-brown stone churches with green copper roofs, is backed by the lower spurs of the Carpathian range.

An even more important role has been played by the course of rivers; broad and majestic, they formed natural trade routes and first lines of defence. Very few of the great towns of the past have escaped their attraction. The sites chosen were defensive positions, bastions which were easily fortified and defended. Prague was built on a ridge 150 feet above the river, and here the city centre still lies, with its copper-roofed cathedral, a palace of countless windows surrounded by old houses, ancient churches and narrow alleyways. This group dominates the old trading town surrounded by residential quarters laid out in rectangular fashion and extending outwards for nearly ten miles. The same contrast exists between the fortress of Buda, with its enormous royal palace, a city of hills and built of stone hewn from the quarries of the Bakóny plateau, and the market town of Pest, with parliament buildings built along the river and surrounded by characterless districts, houses and factories round the railway freight yards.

In the U.S.S.R. there are many examples. Apart from Moscow, cut in two by the Moskva valley or the ancient and illustrious city of Kiev on the banks of the Dnieper, the Volga is an absolute chain of towns. It is on the Volga, of all the rivers of the world, that the greatest number of towns with a population of more than 100,000 are found. However, on account of its length, there is only one town of this size every 180 miles whereas on the Rhine there is one to every 83 miles. All these towns are linked with the river; the older the town, the richer and more varied are the links. The ancient city of Kozmodemyask derives its livelihood from saw-mills and fish canning factories, while its modern neighbour Vol'sk produces mainly cellulose and paper. The more the river spreads out and widens, the more it influences the riverside agglomerations. At Jaroslavl only secondary industries (flour mills and saw-mills) are linked with the Volga. At Astrakhan everything depends on river communications from canned fish to naval shipyards. It could be maintained that urbanisation of the river-banks has followed the course of the river from source to mouth. On the Upper Volga most of the towns were founded between the eleventh and the thirteenth century, and these include some of the most ancient cities of Russia. On the middle and lower Volga the creation of cities took place in the sixteenth and seventeenth centuries, after the conquest of the whole length of the river by the Russians. The steep right bank offers possibilities of fortification of which ample advantage was taken. Nearly all the old Volga towns cling to the rugged contours of this bank (which presents some difficulties for modern development), and extend by means of suburbs or satellite towns which spread across to the low left bank. Transport in the interior of these towns, the gradient of roads leading down to the river and the bridges produce many complications which, in face of the picturesque beauty of the old cities, tend to be forgotten.

In this Volga country we can appreciate the influence of the recent urbanisation movement of the U.S.S.R. There are towns created 'out of the void' and linked with the development of new branches of industry (for example, extraction of petrol at Jugolensk; production of hydro-electricity at Vol'sk); industrial and residential districts have expanded and there is a proliferation of satellite towns. At one time, before the Revolution, four-fifths of urban riverside life was concentrated on the right bank; now the two sides of the river are balanced. The eastern plain provides incomparable advantages for the siting of factories and the development of residential districts. The great modern industrial tendency is one of mechanical construction, followed by the expansion of the chemical industry. Thirty-seven per cent of the urban population of the autonomous republics through which the river flows is concentrated on its banks.

The example of the development of the Volga towns is not an isolated one. The transformation of the urban network of the Ukraine or the Urals could equally well be cited. In the Ukraine, during the time of the capitalist period, the urban network was transformed from 1860 onwards by the advent of the railways and by the development of navigation and industry. The more important towns grew rapidly and the total Ukranian population tripled between 1863 and 1910. New towns sprang up rapidly on the coalfield and around the metallurgical factories. This movement was even more accelerated after the Revolution: the number of towns with a population of more than 50,000 rose from eighteen in 1926 to thirty-seven in 1939. In the Donbass most of the towns are of Soviet origin. Laid out on geometrical patterns, these workers' towns extended indefinitely and in fact doubled the number of urban nuclei. In the old country towns the agricultural features began to fade and the small bourgeoisie had disappeared. The urban population is now composed of workers, employees and members of their families.

The same development is found in the industrial regions of the Urals; before 1917 there were only twenty-six country towns and no industrial towns. At the present time the figures are respectively 118 and 198. These new towns of the Urals generally arose out of an embryonic nucleus, often fairly old (fifty-three new towns grew up on the site of ancient feudal cities). On the other hand, many urban centres, especially those sited on coalfields, arose from nothing. As for the towns built for the industrial workers, at least half were founded either on a bare site, or alongside a rural village.

The presence of railways appears to be a determining factor as regards urban prosperity; whereas in 1917 ten Ural towns out of twenty-nine were cut off from all rail communications, in recent times eighty-one out of ninety-two have achieved it. Here rail transport seems to be a much more significant factor for urban prosperity than that of the rather scarce waterways. The basic characteristic of all these towns is industrial activity; two-fifths of them are concerned with the extractive industries of coal and metalliferous ores. In these mining districts the urban agglomerations are so close that they sometimes touch and run into each other, as occurs near Perm in the coal mining district of Kizel. On the other hand, large agglomerations like that of Sverdlovsk, which includes thirty-eight towns and eighty working-class quarters, are concentrated round the great industrial railway junctions. They form an asymmetrical star branching out along the routes in a low part of the Ural range. Imposing recent constructions stand side by side with ancient modernised centres. This transformation has spread everywhere: in the forest regions the old towns like Tsherkyn, which specialises in the construction of barges, are close beside newly developed towns like Krasnovishersk, a centre of the paper industry. Then again, in the old agricultural centres, like those of the Perm district, light industries and industries connected with food processing have been instituted.

Thus the urban network of these regions of Russia and eastern Europe, by its very complexity reminds us of western Europe. Here we see repeated the same broad base of small rural urban nuclei, either commercial or industrial, the great industrial centres, the vast agglomerations. In rapidity and complexity of expansion, Moscow is second to none. Its much more recent and hurried development, compared with the great western capitals has, however, produced a result which is appreciably different.

To the obvious consequences of this wave of recent urbanisation must be added the devastating effects of the last war. At the present time the urban population of the U.S.S.R. represents 50 per cent of the total. This means that within a few decades millions of human beings have migrated from the country to the towns. This situation produces great difficulties and the housing problem is far from being resolved. In 1926 every inhabitant of Leningrad occupied 143 square feet; in 1956 only 84. Great efforts have been made to find urban norms of accommodation which will house large numbers of people within easy distance of the town centres. The greater part of the working-class accommodation in the outlying districts of large Soviet towns is in blocks of flats which are economical of space and help to reduce the journey to work. Efforts are also being made to limit the spread of the Moscow agglomeration and to extend the network of public transport round other large agglomerations where dormitory towns are multiplying and daily commuting becomes increasingly important.[1] In the capital about the year 1926 there were only 0·7 per cent of blocks of flats higher than eight storeys. There are now more than 50 per cent of these. High buildings in a fairly ornate style are being built which are reminiscent of the early American skyscrapers and contrast with the smooth surfaces of vertical buildings in the west.

SOVIET ASIA

Beyond the Urals we penetrate a new world. Here urban life, though strong, is only sporadic. It is limited to a strip of about 1,250 miles in the west and barely 600 miles wide in the extreme east.

[1] Symposium, *Goroda Sputniki*, Moscow.

It appears in the form of vigorous nodes and groupings that spring up in the heart of a countryside that is often hostile, with deserts in the south-west and mountains elsewhere. In the north it is arrested by the rigorous climate and submerged by the forest.

The towns do not arise out of dense regional concentrations of population; rural life is often poorly developed and of recent date, occupying only a small part of the available land. The towns are concrete expressions of Russian penetration, and the proportion of the population that is urban increases the further one gets from Europe. In the far east four-fifths of the population is urban and 99 per cent of the townsfolk are Europeans. We are thus dealing with a veritable colonisation, related in the first place to the development of means of transport, and later to organised economic expansion and the systematic development of resources.

Russian occupation proceeded during the nineteenth century, to an increasing extent after the military expeditions which penetrated as far as the Manchurian borders, crossed the Caucasus and subdued Turkestan. The land hunger of the western peasants, the deportation of the revolutionaries, and the pioneering that followed the construction of the Trans-Siberian Railway, all contributed to favour a form of colonisation that was strictly controlled by the government. The Russian infiltration into central Asia during the decade 1870 to 1880, in the land of 'Seven Rivers' ended the urban impetus by a kind of colonisation which exploited raw materials. The rush of the new settlers, coinciding as it did with the building of railways, resulted in the duplication of the old cities, as in the case of the old and the new Tashkent. Before the building of the railway the towns supported themselves by the exchange of local products, by trade with China, or by gold-mining. In eastern Siberia, at the end of the eighteenth century, 46 per cent of the urban population consisted of the middle classes and traders. The construction of railways brought important changes: the beginning of the exploitation of mineral resources such as coal, the development of agriculture to feed the influx of settlers, and the founding of the first industries. In 1917 in eastern Siberia 87 per cent of the urban population was concentrated along the railway.

Certain traditional Asian towns have preserved from the past their sites and a few typical monuments, but more often the town is now in two parts, as in the old and the new Samarkand. In any case, the residential districts of the colonial period were primitive and often improvised, and have almost all disappeared to make room for new urban planning. Some activities continue to exist; big commercial centres like Krasnoyarsk, Novosibirsk, and Irkutsk have retained their important rôle in distribution which may to some extent have influenced their present industrialisation. Thus at Irkutsk, one of the large centres of the tea trade, a tea pressing factory has been built.

During the Soviet epoch the impetus of urbanisation has accompanied the discovery and development of natural resources. The extension of the towns has necessitated some sacrifices: the rebuilding of the old quarters, the adaptation of the sites and, in certain cases, a complete removal. Irkutsk, which was originally situated on an island at the junction of the Irkut and the Angara, has been transferred to the right bank of the latter and bears no traces in its outward appearance of its seventeenth century rôle as a fortress. Many new creations arose from small nuclei or even from a completely empty space. In eastern Siberia 70 per cent of the urban centres which have arisen since the Soviet Revolution, are industrial towns of which two-thirds owe their existence to the extraction and use of local minerals.

Here, as in the great western industrial regions and particularly in the Urals, a whole series of types of urban population can be seen. There are small mining centres, often with a population of less than 10,000, located on the mineral deposits and not even served by the railway. There are more important towns linked by railways, which concentrate on transformatory industries. There are forest towns, often situated on the banks of a river allowing for the cheap transport of timber; (such towns which have access only to a waterway have a population of about 10,000; those situated at a rail-river crossing may be as large as 30 to 35,000); there are local market towns at the junction of roads or rails and usually containing 20 to 25,000 people, like Kirensk or Ust'kut; finally there are the large centres with populations of more than 100,000, which combine administrative, cultural and commercial functions; the industrial activities of these are varied.

The pioneers and builders who developed these new towns remained undismayed by all the difficulties that confronted them. Climatic hazards had to be overcome in the arctic zones of the north as well as in the southern deserts and mountains. The development of the northern towns has equalled that of those in the European zones: Murmansk and Archangel are now well developed, Kirovsk was built in 1930, Vorkuta in 1943 and Igarka, Ikson and Nordvik still more recently. The last two have access to the outside world by means of air travel only, for the greater part of the year. It has been necessary to devise and create special environmental conditions so that human life may exist: the soil is artificially warmed and fertilised in order to produce cereals and early vegetables. The effort to obtain water is considerable: at Yakutsk artesian wells have had to be sunk through 650 feet of permanently frozen ground.

In the southern deserts the enemies are heat and aridity. These have been conquered by extensive irrigation works produced by colossal deployments of labour and by creating girdles of vegetation around the towns, protecting them from the dry winds. Thus Tashkent is an enormous modern town, full of trees; each family lives in a small house with its own garden and three or four fruit trees which provide shade in the hot season. Water is distributed everywhere by means of canals and is carefully allocated. The old town of narrow streets and small, single-storey flat-roofed houses, built of dried mud, has been systematically reconstructed. The university was founded in 1920; factories appeared in the middle of the desert, such as the textile *kombinat* which was one of the biggest factories in central Asia, an aeroplane factory and hydroelectric power stations.

All these modern towns are distinguished by their broad thoroughfares, large green spaces, the number and the grandeur of their public buildings sited in central positions, the absence of social segregation and the lack of architectural variety from one quarter to another. It is true that the early days are sometimes difficult and conditions and facilities may leave something to be desired, especially in such extreme climatic conditions. The Russians themselves complain about these.

All these towns resemble each other in their lack of private traffic—the broad streets are empty; in the dense concentration of shopping centres which is not always favourable to the delivery of supplies and causes long delays; in the passionate enthusiasm of the inhabitants for popular spectacles and the vast buildings which house these displays; and in the extraordinary mingling of races and nationalities in all the large towns. This last characteristic is also apparent in the towns of eastern Europe, where the mixture of peoples resulting from ancient and modern migratory movements has transformed them, far more than in the rest of Europe, into phenomenal melting pots. In Soviet Asia the great majority of the administrators and technicians are Slavs who came originally from European Russia to organise the development.

The dominant impression that one receives across the whole of the U.S.S.R. is that of a massive, rapid and still incomplete transformation of the urban network.

Chapter 6

AUSTRALASIA AND THE AMERICAS

In the massive continents of America and Australia the urban network is the result of European initiative. The results of human endeavour were added to those of the natural surroundings and led in the first instance to the construction of seaboard towns. The capitals and other towns of the indigenous population in the more densely peopled parts of Central and South America were built on the plateaux and even in the mountains, but discovery and occupation by the European conquerors provided a link with the outside world and changed the basic conditions.

The creation of ports was necessary to maintain relations with the outside world at a time when the interior was sparsely populated, and when the use of waterways was made even more indispensable by the absence of easy overland transport. Apart from that, the colonial organisation, which was destined to supply the continent of Europe with riches and raw materials for its growing manufacturing industries, was accompanied by the foundation of businesses, banks and places of exchange. In this way, the seaboard agglomerations of Latin America was born: agricultural centres, some of which were raised to the dignity of local capitals, or ports of exchange strictly controlled by the capital—Sâo Vicente, Pernambuco, Bahia (which after 1549 became the capital of the Portuguese possessions), Rio de Janeiro, Buenos Aires, Cartagena, Porto-Bello, Vera Cruz, Callao, Valparaiso. In North America, the opening of the St Lawrence routeway coincided with the early foundation of Quebec and Montreal, while Boston became the great centre of the first New England settlements. In Australia, Sydney was founded in the same year as the official occupation of New South Wales by the British, while Melbourne was founded in 1835, at a time when the first impetus of colonisation (which dates from 1831) coincided with a period of prosperity in the new continent.

There were perceptible differences in these early conditions. In the first place, the natural surroundings and the density of the indigenous population were not the same everywhere. The American Indians were much more highly evolved and racially superior to the Australian aborigines. They were also more numerous and formed more coherent and better organised groups in the mountains of central and South America than in any other part of the American continent.

Secondly, the principles which governed colonisation were not the same. The Anglo-Saxons, who were numerically superior in the north, developed Canada and the United States as areas of white settlement, with the importation, in the south, of coloured labour to assist in the conquest of the soil. The Indians were at first pushed back and later practically exterminated. The same thing happened during the predominantly Anglo-Saxon colonisation of Australia. In contrast to this, a much milder policy with regard to the indigenous population was pursued by the mainly Latin colonisation south of the Rio Grande. Coloured labour was imported here also, but it was assimilated to a much greater extent and consequently the population of these territories is much more mixed.

Finally, recent historical and economic development has shown considerable differences. The independence of the United States was established much earlier and they reached the highest rank of world states by their evolution as a federation. Australia and Canada evolved within a flexible political framework which proved on the whole effective for the development of their territories. But the republics of Latin America, although one by one they became independent during the course of the nineteenth century, still have a long way to go on the road of economic progress.

Three groups of towns can, therefore, be distinguished within these continents: the towns of Australasia and those in America north and south of the Rio Grande.

AUSTRALIA AND NEW ZEALAND

These distant islands of which one, Australia, has the dimensions of a continent, have an 'imported' white population on the fringes of the coloured world of Africa and Asia. The islands offer a simpler human environment than Asia and Africa, and in the case of Australia, an environment that is physically more hostile. The population is almost exclusively composed of white settlers originating mainly in Great Britain and Ireland (98 per cent in Australia and 90 per cent in New Zealand).

The islands of New Zealand, which are split up by central mountain ranges, offer immigrants many possibilities of settling on the edge of a seaboard resembling that of Japan, but conditions in Australia are much harder. Over two-thirds of the country there is an annual rainfall of less than twenty-three inches in a climate characterised by great heat and intense evaporation; the greater part of the continent belongs to the arid zone. The difficulties of settling in the interior are increased by an almost continuous chain of mountains along the south-east coast. Also, with the exception of southern New South

Wales, Victoria and Queensland, more than half the territory is uninhabited and two-thirds of the total population is concentrated in the south-eastern states, on about 13 per cent of the land surface.

The physical conditions of these two different areas are faithfully reflected in the distribution of the urban population.

In New Zealand the urban population has always been large, but it has continued to increase, passing 44 per cent of the total population at the beginning of the century and reaching 64 per cent at the present time. There are no enormous agglomerations; one-third of the inhabitants are contained in the four largest towns. The largest industrial centre, Auckland, has a population of less than half a million, distributed over a verdant landscape around innumerable bays. Seen from the heights of Mount Eden the scene appears like a rash of small red-tiled houses studding the green carpet, with several tall white buildings rising above them. Wellington, the capital, is less important, but more attractive, being built in terraces on an amphitheatre of hills. The office blocks in the town centre are taller; they surround the old wooden administrative buildings and the quays of a well built and sheltered harbour, while the suburbs stretch out into the green of the nearby hills. In a similar fashion all along the coasts, making use especially of the well sheltered bays and the broad plains of the east coast, many little towns are perched on the river banks and along the shores.

The pattern of life and the type of population is that of English towns which have been transplanted into a much more attractive climate than that of the British Isles. In North Island a rapidly increasing population of about 150,000 Maoris tends to congregate around the towns, especially Auckland.

In Australia the urban scene is quite different: the urban population represents 70 per cent of the total and each year the flow of new immigrants increases this preponderance as three-quarters of them settle in the towns. This higher percentage of town dwellers is accompanied by intense concentration in the big towns. The urban population structure appears to be very unbalanced and the capitals of the six states which contained 35·5 per cent of the population in 1901, contained 51 per cent in 1947 and 56 per cent in 1961. Melbourne and Sydney, which contained 26 per cent of the national total at the beginning of the century, now contain 38 per cent.

Why this concentration? It cannot be denied that the more recent immigrants, coming from more populated parts of Europe, miss the easy communications and contacts caused by the country's great distances and find them more easily in the large cities. The difficult natural conditions and the loneliness of the interior play a part, and economic factors must also be considered. The need for agricultural labour has diminished with the use of machinery. The interior of the country is difficult, and communications between one Australian district and another are often made by sea or air, while all exchanges with the outside world must of necessity employ these methods. Maritime trade therefore flourishes, and all the state capitals, which are also the largest towns (apart from the federal capital which was arbitrarily created in the interior), are great ports.

Good salaries and a high standard of living imply both the existence of a good deal of employment in tertiary activities and varied administrative and social services, as well as the necessity of living in a town in order to satisfy the demands of a well-ordered material life. Finally, the actual origin of a colonisation where immigrants settled in coastal nuclei, determined from the outset a certain form of urban life.

In the interior there are only mining centres, sometimes created and maintained artificially at enormous expense, rural market towns and stock-rearing centres separated by vast distances, and route centres which are most numerous in the eastern areas, especially in the Murray-Darling basin, where a sparse but evenly distributed agricultural population exists. Everywhere else, the only view to be seen from an aeroplane over Australian territory, is of thousands of miles of desert or steppe, occasionally relieved by mountain ranges and furrowed by cattle tracks leading to newly created water holes.

Characteristic of the large Australian towns is the immensity of their suburbs. The original site was often very beautiful, corresponding to a deep indentation of the land by the sea. Thus the coast round Sydney is like a lacework pattern of islands, nearly all of them bordered with cliffs and deep bays. The centre of Sydney itself, with an ever increasing number of grey skyscrapers, dominates the harbour. But all round the town stretch enormous suburbs with geometrically designed streets and innumerable small houses with their own gardens, reminding one irresistibly of English and American towns. The population of the city proper is not more than 180,000 though the agglomeration contains over 2,181,000. The vast monotony is broken only by industrial buildings, sports grounds, swimming pools and parks.

The Australians are fond of sport, nature and, above all, the sea. On all the coasts near the large cities, small sea and mountain centres multiply, some of them inhabited all the year around. Every weekend town dwellers stampede out to these centres for relaxation and sport.

AMERICA NORTH OF THE RIO GRANDE

North America can be compared with the U.S.S.R. both by reason of the size of its territory and also by the physical conditions which it offers for the development of towns. Here too one finds outposts on the very edge and even in the depths of the icy northern wastes—the towns of Alaska remind one of their Soviet counterparts in the subarctic zone—and tourist centres which flourish in latitudes that are almost tropical. But North America is favoured in having a more highly developed coastline. There are also resemblances caused by the unevenness of the urban network. The north-east, which has long been inhabited, densely populated and highly developed, is one of the main centres of urban life; the west, more difficult of access and penetration by reason of its relief, its climate and its remoteness, has only isolated nuclei and an urban coastal belt. In the north, the urban limit is not far beyond the frontier between the United States and Canada: 55 per cent of the Canadian population is concentrated within 125 miles of the frontier, 90 per cent within 200 miles. With the exception of some isolated mining or forestry centres like Dawson or Churchill, which are cut off during the long winter months, the majority of Canadian townsmen live within this narrow belt, in which the mean temperature of the coldest month does not fall below $-4°F$ ($-20°C$).

Within these limits, urban civilisation is vigorous. It has gradually developed in various stages and even though one finds there neither the ancient and time-scarred cities of the old world nor the exuberant growth of Soviet new towns it has, nevertheless, a great deal of variety. The fundamental feature, however, is the high degree of urban concentration achieved in a relatively short period. In barely four centuries this part of the continent has developed from nothing one of the largest urban networks in the world, both by the size of its agglomerations, their concentration and continuity—stretching almost uninterrupted for 600 miles in the 'Megalopolis' of the north-east—and by the essential rôle they play in national and international affairs.

It was here that the concept of towns built vertically, which has gradually spread throughout the entire world, was first brought to some degree of perfection. We have progressed from the days of ten or twelve storeys in Chicago in 1885 to twenty in New York in 1900, and to the Empire State Building of 102 storeys (1,250 feet) in 1931. If at the present time fewer storeys are built, the elegance of the heights springing up in the midst of parks and woodlands has created yet another urban style. It is here too that the suburbs or suburban towns like Los Angeles, stretching for dozens of miles, achieved their greatest extension. Town dwellers are scattered in ever widening circles across the countryside and wooded hills, thus producing those hybrid zones of urbanised countryside, which, according to some, will be the general pattern of the future and in which it will no longer be possible to know exactly where the town begins and ends.[1]

If, by way of contrast, one looks at the vast working-class quarters of Soviet towns, it almost seems that the Russians are attempting to disperse as little as possible from the industrial centres and here we have two conflicting conceptions of future urban development. The Russians consider it necessary to concentrate the urban population as much as possible in order to give them the impression of living in a town and of benefiting from the advantages and atmosphere of urban life. The Americans on the contrary, although working in concentrated centres, and possessing ample means of private transport, do their best to create the illusion of the green and peaceful atmosphere of the country, where they spend the night and their longer and longer leisure hours.

At the very beginning of the movement towards urbanisation, when Europeans first set foot on this half continent, there were coastal settlements. These were disembarkation points, supply bases for penetration into the interior, and markets. The first immigrants from Europe landed on the Atlantic coast and it is here that the deepest roots are found. New Amsterdam, later to become famous as New York, was founded in 1623; it was preceded by Quebec in 1608, followed by Boston in 1630, Montreal in 1642 and later by Baltimore and Philadelphia. The movement gained strength by using the natural lines of communication formed by the waterways in this vast country. It is possible to trace the stages all along the Atlantic coast and the Gulf of Mexico (Mobile, Hartford, Albany, Trenton, Richmond) right into the interior, by following the great arterial waterway of the Mississippi and its tributaries, and that of the St Lawrence and the Great Lakes: St Louis (1764) Kansas City, Detroit (1701) Cincinnati (1788). The era of canals which linked the great natural waterways marked yet another stage of penetration into the interior, especially after the construction of the Erie Canal which joined Lake Erie to the Hudson river (1823). It was the building of this canal that brought about the ascendancy of New York over its neighbours and rivals. Cleveland and Chicago came to life during the first half of the nineteenth century. Between 1900 and 1950 half the towns whose population now exceeds 100,000 were granted urban status.

The railways came in their turn, accelerating the flow of population westwards, new towns proliferated and others developed enormously. Denver and Spokane, in the interior, and all the great cities of the Pacific coast—Seattle, Tacoma, Portland, San Francisco, San

[1] Pinchemel, 'La transformation des gratte-ciels new-yorkais', *Ann. de. Géog.*, 1963, no. 390, pp. 247–9; G. A. Wissink, *American Cities in Perspective*, Assen, 1962.

Diego—rose and flourished. In the great plains, new transport communications multiplied. Chicago became the most important railway junction in the United States and soon afterwards its second city.

In this vast country the transport revolution is not yet complete: the automobile in its turn has taken pride of place, for this is the most favoured individual means of transport. At the present time there is one automobile for every four people; in California the figure is one for 2·2 people. From this new patterns have emerged: the dispersal of population spreading widely all round the great urban nuclei, the enormous growth of the suburbs, the proliferation of business undertakings. Since 1930 the number of suburban dwellers has increased three times as fast as that of the town dwellers.

Parallel with these successive revolutions in transport, the exploitation of natural resources and vast agricultural wealth now played an important part. The discovery of coal and iron, as in other continents, has created not only great cities like Pittsburgh, Duluth and Birmingham but also countless small towns stretching right into the valleys of the Appalachian plateau. Rare minerals, mined especially in the western mountains, provide a livelihood for many small towns, for example, the copper mining at Butte, or the smelting and refining of the same metal at Anaconda, Tacoma or larger centres such as Salt Lake City. Oil and natural gas, used either as raw materials or as sources of power, have assured the growth of great centres like Los Angeles, Tulsa, Oklahoma City and Dallas, the cleanest city in the world, where all power is provided by these two substances. As agriculture and stock breeding developed and spread westwards, so the growth of market towns followed. The most important marketing centres coincide with the big road and rail junctions of the great plains, but they are also found in the small valleys of the Appalachians, and in the oases irrigated by the rivers of the western mountains.

In the infinite variety of the industrial towns many different types occur: small textile centres in the valleys of New England, some of which are decaying and some changing or adding to their traditional industry paper manufacturing and light engineering; colossal port or mining organisations around which cluster the most varied industries, as in the great sea ports on the east coast, Chicago or the big cities of the west coast; or where extreme specialisation may develop, as in Detroit with its automobile industry. The industrial routes of the great valleys—those of the Ohio and Tennessee rivers, for example— are lined with towns, as are the shores of the Great Lakes, both along the southern bank, which belong to the United States, and, especially, on the Canadian side, north of Lakes Erie and Ontario (Hamilton, Toronto, Kingston). Recently created towns like Alcoa, Oakridge and Kitimat have highly specialised industries dependent upon hydroelectricity.

Finally, certain personal, family or religious reasons have stimulated the creation of urban centres such as Portsmouth, Providence, Newhaven and Salt Lake City.

Thus the origins and situations of these towns are varied. The stages of development of the different regions are equally diverse. New England, with its complex urban hierarchy, is reminiscent of Europe; some of its many little industrial centres are beginning to fall into decay, others are being transformed; the large cities are prospering from the diversification of their industries. The central east resembles nothing else in the world: it is one vast conurbation inextricably mingled with the countryside. The middle west is the area of great agglomerations, dense and almost continuous along the edge of the Great Lakes, more scattered and spaced out among the sparsely peopled rural districts of the central plains. In the south east, other characteristics are introduced by the importance of the coloured population; the people are more widely scattered and the area is economically less well developed; many small country market towns and historic centres have survived; often, like Atlanta they are kept alive by industries, especially textiles and metal manufactures. A certain 'colonial' atmosphere persists, as at New Orleans, whilst the enormous impetus of oil and heavy industry has brought about the growth of a string of coastal and subcoastal towns: Baton-Rouge, Port Arthur, Houston, Texas City, Baytown, Galveston, Corpus Christi. Florida is a special case; here urban expansion is linked with the tourist industry, and there are large hotels, private beaches, millionaires' residences, crowded places of relaxation and enormous camping sites where visitors come to seek the sunshine for six months of the year.

As soon as one crosses the hundredth meridian, one enters a territory where the population is scattered, and where the towns, even though they may be at the heart of dense concentrations, are independent centres. These may be centres of the metallurgical industry, agriculture or the tourist industry, which have stimulated an intelligent diversion of water supplies from the mountains into accessible valleys or basins, as in the case of Salt Lake City, Phoenix and Las Vegas, or busy communication centres which dominate important agricultural regions like Spokane, Sacramento or Portland; or great ports of the West Coast which are also famous industrial towns, and which, with their satellites, spread out into the narrow valleys and along the complicated bays—like Seattle, San Francisco and Los Angeles. It is within this last group that the most impressive progress has been made, and where the number of urban dwellers at the present time beats all records for rate of growth. The town population in the different regions just mentioned is significant: it exceeds 80 per cent in central east Atlantic, reaches 75 per cent in the Pacific states and 76.2 per cent in New England, while in the southern states bordering the Atlantic and in those that cover the Southern Appalachians, the figures are only 40 per cent and 39 per cent respectively.

The appearance of these towns, whose history is so short, is astonishing to the European eye. From the air they appear vast and surrounded by endless suburbs where houses of brick, wood or pre-fabricated materials, always newly painted, are dotted about in the midst of lawns and trees. No fences divide the gardens, and carefully tended roads cut up the ground into geometrical blocks. The monotony of style, so unattractive in English suburbs, is rendered attractive here by the carefully kept gardens and an abundance of flowers; almost every town has its own style. However, if one has seen one suburb one has seen them all, as they are repeated, with certain variations, from one end of the country to the other. Some towns are distinguished by the great opulence of the 'millionaires' quarter', others in the deep South by the so-called colonial style with pillar framed entrances. In the centre of the small towns is the 'main street' with its riot of neon lights, its well stocked shops and its busy traffic; as soon as one leaves the centre one is surrounded by green suburbs. In the large cities, even those which are able to expand on to the great spaces of the plains, the tradition in Canada, as well as in the United States, is to build high office buildings in the central districts; one moves imperceptibly, within the geometrical pattern of blocks, from a one-storeyed dwelling to the busy skyscraper at the centre.

All these towns have certain features in common. They are never finished; their boundaries increase indefinitely while the centre is being simultaneously demolished and rebuilt. Empty blocks and waste spaces stretch right into the centre of the biggest towns and are used as parking space for automobiles while a new building is awaited. The traffic is deceptive; it is strictly regulated and the checkerboard arrangement of the streets facilitates its canalisation and the automatic regulation at crossroads. The motorways, ever wider and more complicated, converge on the urban centres. Each densely populated area is, above all, a concentration of vehicles, and this demands grandiose and costly solutions to the traffic problem.

Thus the towns of North America bear little resemblance to those of Europe except in certain respects, such as their functions and the occupations of their inhabitants, which are not immediately obvious from outside; they are profoundly dissimilar from the point of view of origin, site and surrounding countryside. They play a considerable part in the life of the individual, both in the economic and sociological field, apparently meeting a need by enabling the individual to merge into the community, while at the same time keeping, through the suburbs, a contact with nature.

North American towns, unlike those in Australia, are not necessarily the result of colonisation but they follow, as in Europe, the pattern of a previous age with concentrations of people who, until recently, were agriculturalists. Movement towards the towns continues to increase rapidly and shows no signs of abatement.

The towns do not so much reflect a uniform civilisation as a definite regional pattern: puritan Boston, which stems from the oldest period of colonisation, is less free and dynamic than New York, a city of enormous expansion and world renown. The somewhat sleepy old quarters of New Orleans have kept something of their early Latin atmosphere, as has the crowded city of San Francisco on its peninsula. However, all these towns resemble each other to some extent by the mixture of their thronging populations; there is no big city which does not possess its Chinese, Italian or Negro quarter, its German, Dutch or Hungarian colony. This inextricable mixture plays a very definite part in the fusion and assimilation of American civilisation. In Canada, apart from the fundamental division between the French towns in the east and the English towns in the centre and west, there is room for a wide range of relative proportions of the two main groups and for the existence of numerous others.

Finally, all these towns, which are centres of attraction in prosperous countries with a high standard of living, are distinguished by intense activity. Not only are the traffic and new building projects ceaseless but the towns teem with commerce, amusements and every form of advertising. Whole districts are disfigured by the blaze of illumination and the glare of publicity. Domestic comfort is typified by air conditioning; the windows remain firmly closed to natural breezes. Nowhere else in the world does urban life appear to be so artificial.

AMERICA SOUTH OF THE RIO GRANDE

All is changed when, crossing the Rio Grande or leaving behind the satellite towns of San Diego, one reaches Mexican territory. In the designation 'Latin-American' it is the first word that should be emphasised and it would be even more correct as some specialists have suggested, to call these regions Indo-Latin.[1]

The urban population, proportionately and quantitatively much less significant, is in the course of much more rapid growth. Taken as a

[1] Ph.-M. Hauser, *L'Urbanisation en Amérique Latine*, U.N.E.S.C.O., 1962.

whole, the total population increases annually by an average of 2 to 3½ per cent, but the urban population in Venezuela has increased by 7 per cent in the course of the last few years; the former will double itself in about thirty years, the latter in ten years.

Here the town is no longer the expression of a civilisation but an almost peripheral phenomenon, one could almost say 'colonial'. Free from European guidance, with the exception of some island territories in the West Indies and continental ones in the north-east of South America, these states have not yet succeeded in shedding the influence of previous economic conditions, and this affects the siting of the urban network. A large number of the towns are ports, direct descendants of colonial bridgeheads; the list would be a long one from Pernambuco (Recife) to Buenos Aires, and from Cartagena to Valparaiso. The towns in the interior are situated in such a way as to form supporting bases for the seaports or indispensable complementary centres like São Paulo, Lima, Santiago.

On the other hand, there are towns which preserve important traces of the influence of the more vigorous of the Indian civilisations, which were not subjected to a severe racial clash with the Europeans. The denser and most advanced Indian populations were to be found in the healthier high plateaux and mountain basins, enclosed within the Cordilleras of Central America or the chains of the Andes. In the middle of these knots of native peoples, towns grew up. These are the capitals of the Andes states: Bogota in Colombia, where 85 per cent of the population live in the Andes, Cuzco, La Paz (at 11,945 feet the highest capital in the world) Mexico City and many others. Another important Indian centre comprised the reserves founded in Paraguay by the Jesuits; here were built Asunción and Concepción.

A great part of the population, which is more or less inextricably mingled with the native Indians, has practised forms of agriculture which require rural centres for the exchange of goods and in order to maintain contact with the outside world. This is the origin of large towns which resulted from the first generation of farmers; these were most numerous in the north-east and in the islands. Later came other European colonists, chiefly Germans and Italians, who developed the great spaces of the temperate zone. These migrants were also responsible for centres all over southern Brazil, on the plains around the River Plate, in the great valley of Chile and in the Argentinian oases at the foot of the Andes. Some of these towns have characteristic names like Blumenau.

The early pioneers of the interior, having followed in the wake of their herds, settled at fixed points, and these halting places can still be found in the Matto-Grosso, in the most remote regions of the Pampa and in the llanos of the Orinoco.

The quest for minerals, the search for gold, so avidly desired by the Spanish conquerors, the extraction of riches from the earth, have always preoccupied the inhabitants of these countries destined to colonial exploitation. In former times the treasure was demanded by the mother country and was despatched by galleons from the ports; today it is purchased by the great industrial powers. These natural resources have rendered practically no benefits to their countries of origin. The building of mining towns followed hard on the prospectors and one of the states of the Brazilian federation bears the symbolic name of 'Minas Gerais'. The town of Ouro Preto, with its fifty churches perched on many hills, scraped to the bone by the ruthless exploitation of gold is the most striking example of this type. Others are to be found in the world's highest mining town (Potosi, 13,450 feet in Bolivia) in the coastal towns of the Chilean desert (Arica and Iquique) where nitrates are exported, and in the oil centres which surround Lake Maracaibo.

Finally, the effort of developing a modern national economy and of conquering the interior of an empty continent has begun to leave its mark. Industrial towns of recent origin are beginning to multiply. These are generally metallurgical centres, either large and complex like Monterrey in Mexico or concentrated on one single activity like Volta Redonda in Brazil, Huachitato in Chile and Chimbote in Peru. The industrial city, hitherto unknown in Latin America, is now emerging, with its geometrical plan, its uniform houses and its great, smoky industrial plants which dominate the workers' lives. The growth of these cities has been in close conjunction either with the development of mining centres or with railway engineering and has been supported in the first instance by foreign capital. These towns bear tangible evidence of a changing attitude, or perhaps a new element, in the urban life of Latin America. They are the first landmarks of the great stages in the creation of specialised industrial towns which have left their mark so strongly on Europe and its Anglo-Saxon extensions overseas and on the U.S.S.R., and which at the present time are penetrating so deeply into Latin America as well as into Asia.

The efforts to encourage the colonisation of the remote interior are especially noteworthy in Brazil, where, outside the almost impenetrable Amazonian region, in which there are only the riverside towns and a few trading posts, there are vast plateaux covered with savanna, on which the few rare urban nuclei might be increased in number. Repeated attempts have been made to encourage the growth of Belo Horizonte, on the edge of the mining quadrilateral, and in less than a century it has become the banking capital and one of the largest towns of Brazil. Perhaps the same destiny awaits Brasilia, intended by its

5 London: the city seen from the dome of St Paul's. Bomb-damaged areas partly replaced by new buildings, partly remaining as car parks (1960)

6 Frankfurt-am-Main: reconstruction (1953)

7 Warsaw: postwar reconstruction of old market square

promoters to persuade Brazil, hitherto simply a coastal belt attached to an almost uninhabited continent, to turn inwards.

Thus new urban forms are created by specialisation as well as by siting: this development perhaps marks the entrance of Latin America into a new urban era.

In the meantime, the existing towns are increasing with giant strides and the states are unable to meet the expenses required to equip them adequately and sufficiently rapidly. Between 1939 and 1950 Mexico devoted 55 per cent of her investments to development projects in the federal district, where only 12 per cent of the population are domiciled. All the large towns of Latin-America possess miserable and unhealthy districts which have been hastily built and where the most recent immigrants are herded together. One Peruvian in seven lives in Lima and out of these 10 per cent inhabit the *Barriadas*. The situation is the same in all the great towns of the Argentine and at Buenos Aires 55 per cent of the new immigrants are without regular work. This incursion of a rural proletariat inexperienced in the ways of life in cities which are unable to offer them employment, is a characteristic feature of all underdeveloped countries. Here, on an average, the industrial population represents only 5·7 per cent of the total number of urban dwellers. According to the official figures unemployment is low (5 to 6 per cent at Santiago) but there exists a state of underemployment, an excess of wretched and parasitical tertiary employment, which is an obstacle to modern economic development.

In these circumstances it is unnecessary to emphasise that the urban countryside bears few traces of industrial building. The feature which characterises the Latin-American town is the presence of numerous churches. The Spanish and Portuguese conquerors, intensely Catholic, came here for the most part to spread the faith and they vied with each other in founding enormous numbers of churches and monasteries. On the hills which became the cradle of many Brazilian towns, graceful Portuguese churches, their white walls contrasting with dark stonework, are buried in the labyrinth of private houses and narrow streets. Such is Bahia (Salvador) on its long peninsula between the bay and the ocean; or Olinda, the forerunner of Recife, the old town of Rio and the mining hills of Minas. A town such as La Paz has the same appearance and is built on the edge of a great crag with steep streets and terraced houses climbing up to the town centre from the native quarter with its cottages made of baked mud, and Potosi with its old houses and churches built in terraces on the steep slopes. The towns of Chile, on the other hand, which have a long tradition as market towns, are more spread out, with low houses built along streets laid out in gridiron pattern.

But this historic atmosphere is breaking up more and more: one is tempted to define the present development of the large Latin-American town by two expressions: Americanisation and Africanisation. The latter applies to the innumerable shanty towns which spring up whereever land is available at a reasonable price. Here tiny 'houses' are built of dried mud, bricks, wood or corrugated iron, in places where some food may be obtained from a small cultivated patch or, in the case of the pile-dwellings, from the sea. (See Plates 19, 23.) As for Americanisation, this bursts out in the form of skyscrapers in the heart of the most venerable towns. The ancient city of Mexico with its great squares, its domed churches and its narrow streets has been rudely shattered by the appearance of skyscrapers, while enormous new residential and commercial districts are multiplying. The centre of Santiago is being rebuilt with buildings of ten to twenty storeys; at Caracas the wealth resulting from oil has transformed the city. Centres like Buenos Aires and Montevideo, great cities of the plains, which are enormous, geometrical in plan and cosmopolitan, overwhelm the countryside. But perhaps nowhere is the transformation so obvious as at São Paulo, the great industrial and financial metropolis where the skyscrapers rise before one's very eyes, while all around the red soil of the hills is laid bare ready for the future streets of the next urban wave.

CHAPTER 7

NORTH AFRICA AND NON-SOVIET ASIA

In Africa north of the Sahara and in the Middle and Far East ancient urban civilisations existed which have been radically and more or less directly changed by the impact of European influence. Here too there is at the present time a great uprooting of the rural population, attracted *en masse* to the towns, though the latter are often incapable of providing an improved standard of living.

The way in which contact with Europe originally took place was not the same everywhere; at least four regions can be distinguished: North Africa, the Asiatic-African crossroads or 'Middle East' including Egypt, the islands and peninsulas of south and south-east Asia, and the Far East.

NORTH AFRICA

Experienced in urban tradition since the days when the Phoenicians and the Romans founded famous cities of which magnificent remains still exist, the three states of North Africa, Morocco, Algeria and Tunisia possessed many urban nuclei before the time of the French occupation. These functioned as political centres and military bases and were created by groups whose power was not always firmly established. As military, trade and artisan centres and as the homes of privileged citizens with numerous servants, these towns were cramped within strong walls and were composed of square houses with communicating terraces, barely opening on to the narrow streets, but with their own courtyards and shady gardens. These nuclei are still very much alive and form a distinctive part of the urban landscape in the large cities—witness the *Kasbah* of Algiers, the *medinas* of Fez and Marrakesh—or continue to exist, unaltered in small towns, in the holy mountain cities of the interior, and in the desert, especially in Morocco which is more mountainous and was the last to be occupied, and in the Sahara which, at least until the discovery of oil, was inaccessible and unexplored.

In these ancient cities there existed a whole series of functions. To their complex economic activities of trade, handicrafts and agriculture, they provided centres of Muslim culture: Fez, Rabat-Salé, Tunis, Tlemcen are included in this first category. Secondly, there were the capital cities which were administrative centres and military bases, places of royal residence, and these, of course, differed in some respects from the first group. Among them were Fez, Rabat and Tunis, as well as Meknes, Marrakesh, and Algiers. In third place there were the ports, centres of commercial activity which looked outwards and had been long in contact with the Europeans on the opposite shores of the Mediterranean. Among these we find Tunis and Algiers, and on the Atlantic coast, Casablanca, Safi, Mogador, and Mazagan (El Jadida). Finally, in the interior and scattered about the countryside, are market towns of greater or less importance. Some of these have had a great military history of which their fortifications still bear witness, while others are simply rural towns.[1]

The European occupation upset this traditional hierarchy, especially in Algeria, which was the territory to suffer the most change. Having been the first country to be occupied at a time when respect for the past was at a discount, and having absorbed the greatest number of Europeans, it has seen the disappearance of many ancient towns which were broken up to make way for modern streets and submerged by the wave of new buildings.

Like all other colonial systems in the world, this one required military bases for its defence and markets and traffic routes for the use of the Europeans who occupied part of the territory. Thus monotonous centres arising out of military encampments, were founded on high plains such as Batna and Orléansville, as well as centres of trade and concentration of population like Sidi-Bel-Abbès or Sétif.

It was also necessary to develop some means of exchange with the outside world, to create centres of mining or industrial development, and to house a rapidly increasing population. In Morocco and Tunisia in the course of the last thirty years the urban population, a large part of which is European, has quintupled. In 1954 78 per cent of the European population in Algeria lived in the towns, though recent events have changed this picture. But the impetus remains marked and is reflected in the number of towns, in their expansion and in their appearance.

The juxtaposition of civilisations is expressed by the buildings in the different quarters of the cities. Historic palaces, modern administrative residences, mosques and cathedrals, medieval handicrafts and modern industries are found cheek by jowl. Some old handicrafts, such as carpet making, leather-working and the making of copper pots and trays, have been given a new lease of life by the tourist industry. Little shops lurk in the recesses of the *Souks* which are swarming by day and deserted by night, and there are big stores with costly façades in the busy streets of the European quarter. Everything has two faces in most North African towns.

[1] J. Despois, *L'Afrique du Nord*, Paris, 2nd edn., 1958.

In Rabat the *medina* is surrounded by spacious European quarters which in their turn are dominated by the royal palace and the old buildings of the French authorities which now belong to the national government. In Casablanca the old and the new *medinas* are crowded together in the midst of the white semi-skyscrapers of the European part of the city, the great industries and the port. In Constantine the native quarter is perched up on its cliff which is notched by the deep gorge of the Rhummel, but upstream, where the cliff is lower, there are new villas and gardens. On the outskirts of all these double towns there is one feature born of the incessant influx of the rural population which is common to them all: the shanty town, wedged into the urban mass in the most unfavourable position.

The development of almost all the big towns which have been recently modernised has been achieved by the use of river water or a series of wells. The towns are also often surrounded by market gardens and small holdings, forming a green belt. The enormous cemeteries with their little mounds of grey stones and the white domes of the Muslim tombs provide parks where the women congregate.

In the Sahara the oasis towns have also undergone a complicated evolution. Traditionally built of dried mud, sometimes the colour of the soil and sometimes painted white or pale blue like the towns of Mzab, they fight a desperate battle against the heat of the sun, and the streets are often covered so that they appear as long, dark tunnels. Here the military administration has created magnificent additions in a style derived as much from Black Africa as from Ouargla. The discovery of raw materials, particularly oil, accompanied by the gushing or pumping of artesian water supplies, has given rise to further developments: air-conditioned houses sited in green oases in the midst of palm groves. Here too the population is increasing owing to the improvement of health standards and the settlement of migrants.

But it is not only the appearance of the houses or the streets which changes, it is also the atmosphere. On the one hand there are the raucous sounds of the Arab world, the chanted recital of the Koran, the shouts of the donkey drivers and porters in the narrow streets, the odour of spices, the muffled movement of the white-cloaked people, the piles of merchandise in the teeming native markets, the more or less strict veiling of the womens' faces, the mass of small shopkeepers, the swarms of children. In contrast, the European quarter presents different aspects of the same phenomena: dense motor traffic, the sound of hooters and the smell of petrol; wide pavements and shop windows full of goods; a population dressed in black or in bright coloured garments that give shape to the figure and leave the womens' faces exposed to view. However, western customs and habits continue to spread. Segregation is often more a question of standard of living and of the degree of urban development than of race and we may wonder whether the westernisation of the towns of North Africa will not continue to gain ground.

THE 'MIDDLE EAST'

In that part of the world which Europeans still call the 'Near East' but which is becoming more widely known as the 'Middle East', Europe has exercised a much slighter influence in a setting where the towns are very ancient but also very diverse.

Among these Istanbul is the side door into Europe. This is a city which still belongs to the older continent by virtue of its comparatively cool climate, its overland communications and the wooden houses in its old quarters, which resemble those in old Slav towns. At the same time Istanbul is still part of the Muslim world, and this is apparent from the panorama of domes and minarets, disconcerting to the western eye accustomed to the triangular design of spires and roofs. The city also possesses a bazaar, colourful with copper and silver and silken goods, and heavy with the odour of spices. Cut off from Europe by enormous fortifications and from Asia by an arm of the sea the size of a great river, Istanbul represents a kind of synthesis between the western world and oriental civilisation and is the symbol of the intermediate position that one finds everywhere in the towns of the Middle East.

Bordered by an almost enclosed sea which tends to unite rather than to separate, all the Mediterranean shores have their ports. There is a multiplicity of small ones, each clinging to its little bay, as on the fretted coast of Asia Minor, growing larger and more important towards the south where the natural conditions require much greater adaptation, as at Alexandria, the great outlet for Egypt and the Delta. Some of these towns have seen the rise and fall of ancient civilisations like Byblos, where traces of successive ports going back to Neolithic times can still be found. Others have grown up with the development of transport and technology, like Port Said and Suez, guardians of a canal not yet a century old, with their geometrical layout in the middle of the desert, where water has to be brought from the nearby Nile; or Zonguldak, a smoky mining and metallurgical centre, or the fort of Aden, always on guard beneath a torrid sky.

In these countries which are mainly composed of deserts and where only 5 per cent of the soil produces regular harvests, an even distribution of population is unknown. Turkey alone, cooler and more humid, is an exception. Here, because of the mountainous relief, the towns are sited in characteristic positions: at outlets of plains or mountain passes, at the centre of fertile valleys like Bursa, where the hills slope up to the green mosques among the trees.

Towns on mountain slopes or in basins are also found in Iran, such as Tehran, Isfahan and Shiraz; there are towns in the green valleys like those of the Nile or Mesopotamia, there are oases like Damascus and ancient fortresses like Homs or Jerusalem; all these have a splendid history. They possess water which is so precious in these arid countries and are sited in the midst of palm groves and green gardens. They maintain their ancient positions and are the traditional meeting places for the nomads and the ordinary townsfolk, and through them stream great processions of caravans as well as more modern transport. An important rôle is played here by the native bazaars, with their handicrafts and commerce.

These towns have inherited from the past their sites, their old quarters which are often fortified, and the charm of their gardens and cemeteries. Their position in one of the important trading centres of the world, which was also the birthplace of the great monotheist religions, has made them into strange places of pilgrimage where, without mixing much and often with mutual hatred, representatives of incredibly diverse races, religions and languages jostle each other. As each group has its own quarters, type of houses, customs and costumes they have the appearance of mosaics. Damascus is one of the most typical; Beirut is more Christian; Jerusalem is divided. During the last few years the big towns of Egypt have seen the disappearance of a large part of their minorities.

European influence is exercised through the medium of the technicians; through the capital cities and, in certain countries at the present time, by rulers who make great efforts in this direction. This has contributed to the creation of new districts of a western pattern, the introduction of new building techniques of skyscraper type, the foundation of administrative buildings and, above all, to an urban framework and equipment (transport, drainage, water supply, road maintenance). But the great common phenomenon is Islam, which has left its mark on the towns with their silhouette of mosques and pointed minarets dominating the horizontal lines of the terraces.

The prestige of their classical monuments and their magnetism as cradles of the great religions make all these towns, but some especially like Jerusalem or Mecca, into international centres, through which pilgrims and tourists stream all the year round.

Into this classical atmosphere new influences have brought changes of varying significance. Parallel with the development of Turkey has been the foundation of a new capital, Ankara, an entirely modern city, built in the heart of the dry plateau and by the side of an old fortress; here a population of more than 500,000 is living in concrete blocks surrounded by wide streets or in tumbledown dwellings in the hills close by. Industries have been established and have resulted in the growth of new towns. Since the end of the nineteenth century urban centres have been developed among the dense rural population of the Egyptian delta; these have a uniform appearance, with their orange brick houses crowded together along the narrow asphalt streets. Mosques, with their slender minarets, are still to be seen but there are also enormous water towers placed in the centre of the city and factories and shops are multiplying on the outskirts, especially round the stations. There are agricultural market towns like Faqus, commercial centres like Tanta, but also industrial towns like Mahallah-el-Kubra, with its cotton manufactures, where small workshops and large mills are found side by side. In the vicinity of the Persian Gulf and its neighbouring countries a whole new growth of towns has been stimulated by the discovery of oil: Abadan with its enormous refineries, Kuwait, a miracle in the desert, and others.

Finally, the settlement of the Jews in Israel has transformed local conditions and great modern cities like Tel-Aviv, the capital, and Haifa have grown up in less than fifty years. These are cities of the plains, practical, unromantic and intensely industrialised: Tel-Aviv contains 45 per cent of the factories of the whole state. The development of the land has led to a multiplicity of real agricultural towns. Israel is like a piece of the western world transplanted into the Middle East.

One thing is quite certain, that towns both old and new will expand rapidly. The rural population is converging on them and new construction is proceeding apace. Unfortunately, here as elsewhere, the shanty town is a feature of this process.

SOUTH AND SOUTH-EAST ASIA

Western Pakistan forms a transition between the Middle East and India and is really more in harmony with the former. The earth coloured towns are built of mud huts in small, low blocks crowded together and laid out geometrically beside green patches of irrigated farm land in the heart of an arid environment. The population has been increased by the arrival of the Muslims who were driven out of India, and in a few years Karachi has grown from a population of 200,000 to two million. It stretches out indefinitely, flat and depressing, between the desert and the sea.[1]

[1] P. M. Hauser, *La phénomène de l'urbanisation en Asie et en Extrême-Orient*, U.N.E.S.C.O., 1959.

However, if one flies eastward one gradually reaches more fertile regions where the population is much more widely distributed.

The Indian Federation may not yet be very highly urbanised, less than a fifth of the population being domiciled in the towns, but a vigorous movement towards urban concentration is nevertheless taking place. There are now seventy-three towns with a population of more than 100,000, as against only fifty in 1941. These towns are uniformly distributed across the country, but the largest concentrations coincide with the thickly populated plain of the Ganges, the coastal districts and certain inland river basins.

The two largest agglomerations are those of Calcutta and Bombay and there are thirty other towns in the vicinity of Calcutta along the banks of the Hugli. This metropolis is a great industrial and administrative centre, but all round a centre piece of imposing palaces and marble monuments built by the British is found heart-rending Indian poverty and incredible overcrowding. A whole mass of the proletariat, who often have no shelter but the edge of the pavement, work in the factories and in the industrial cities of Howrah, Titagarh and Bhatpara where two-thirds or even four-fifths of the population consists of workers and their families. More than six million people are herded together here; the men far outnumber the women and the arrival of refugees has still further increased hazards to health. The workers live in one-roomed shacks (*bustees*), hundreds of which line each side of the narrow alley ways which are barely a yard wide. On the other side of the peninsula, Bombay seems to be in a better position. Next door to the English district with its wide and busy main streets, is a typical Indian town with blocks of flats with wooden balconies, suburban roads with small houses belonging to artisans and traders, factories, working-class districts and, of course, the sordid shanty towns.

Apart from these two great cities, the traditional towns which in bygone days were religious and political centres, have disappeared with changing circumstances while others, which have become essential to religious life, like Benares and Allahabad, continue to exist but without being very dynamic. On the other hand colonial trade development, transport and the impetus of industry have created a new urban movement. Working-class towns such as Ahmedabad are built around the big factories, and mining and industrial centres have been built up from nothing. An example of this is Jamshedpur which in 1911 was a small village of a few hundred people. It was selected by the Tata family as the main centre for their steel works and now has a population of more than 200,000.

In the Indian quarters right in the heart of the largest towns, the means of transport have remained incredibly primitive. Horse and ox carts jostle taxis and buses; the sacred cows, in spite of being more or less herded together still sometimes wander at large. Crowds throng the many ornate and majestic temples which are fragrant with sandalwood, while outside in the streets swarm hordes of peddlers.

But it is the size of their population which gives these Indian towns a unique appearance: a countless throng, crowded together, silent and tense, the men in white, the women wearing multicoloured saris; children everywhere. Whenever a factory, a dam or a road has to be built, the workers, especially the women, carry the earth on their heads in small baskets in the fashion of earlier centuries. There are not enough houses, food or work for everyone. In India more than anywhere else, the towns are too small for the population they contain.

When one moves to the towns in the peninsulas of south-east Asia, the contrast is very striking between the small inland market towns and the larger towns, which are mostly ports, often linked with the fertile and densely populated deltas, or with the river plains, where technology in the shape of the canal network has penetrated into the town. In some cases the town has moved downstream as the river has become silted up: thus Ayudhya, the old capital of Siam, now too far inland, has been abandoned in favour of Bangkok.

These towns merge with the plain and the rivers. In the heart of modern Bangkok, canals full of brilliant-coloured water lilies run beside modern streets and buildings. At present they are being roofed over, which is regrettable from the picturesque point of view but hygienically a great improvement. The general picture is one of single-storey building as at Pnom-Penh, Saigon and Singapore. Only here and there is the multicoloured fantasy of the temples pierced by a few high buildings, hotels or offices which rise above the rest. Prosperous residential districts consist of villas set in gardens, but the poor have to be content with wooden shacks often built on stakes along the reaches of the river, or with floating junks. It is indeed hardly possible to speak of towns and suburbs when, in fact, the agglomeration fades away, as around Jakarta in Java into a succession of villages that are barely connected with each other by any means of transport.

THE FAR EAST

China and Japan have never been systematically colonised by Europeans with the exception, in China, of some peripheral trading centres like Hong Kong. Their internal development, longer established in Japan and more recent in China, has produced striking results in the development of the urban network. Furthermore, the overcrowded conditions have led to the creation of towns as a spontaneous reaction to the general congestion.

China. For centuries past Chinese towns have been famous and, as one of the first Europeans to explore them, Marco Polo wrote glowing descriptions of their crafts and trading activities. These ancient cities had military, administrative and religious rôles to play, and teemed with every kind of craftsman.

The majority of the old towns arose out of local economic factors: as agricultural centres scattered across the plains, trading centres, staging points for river crossings, river ports or centres of the mining industry. In north and north-west China preoccupation with questions of defence stimulated the growth of a whole new series of strategic and military towns, which are especially numerous in Kansu. The peculiar situation of these towns has been emphasised in a study made in 1937 which estimated that out of 193 towns with a population of more than 50,000 only fifty-two were served by a railway. There were even six towns with a population of 200,000 to 500,000 and one of more than 500,000 which had no such service. This is one of those aspects of the contrast between urban Chinese life and the one familiar in the West. It has been said that these urban organisms 'do not possess an organic unity as in Europe, but seem to be a collection of bits and pieces; there are thousands of small shops and workshops, each one self-sufficient'.[1]

The Revolution was followed by the provision of transport centres and the scattering of mining and industrial towns across the country; and the same thing happened during the Japanese occupation of Manchuria. New towns were built right into the upper reaches of the rivers; the population was dressed in blue cotton and the appalling social inequality was reduced if not abolished. But the legendary legacy of the centuries remains unchanged, the number of skilled artisans is still twenty-five times greater than that of the unskilled workers and there is still a great diversity of small popular trades on every street corner.

Among the great Chinese towns two types are noteworthy: Pekin, traditionally the old capital of the interior, and Shanghai, formerly the great cosmopolitan centre, on the coast.

Pekin, sheltering behind rectangular fortifications 'gris au ras du sol et verte vue du ciel'[2] ('grey in the sunlight and green seen from the air'), is bisected from east to west by small alley-ways of beaten earth running between walls of dried mud, and from south to north by shopping streets bustling with tramcars, bicycles, cars and rickshaws,

[1] S. de Beauvoir, *La longue marche*, Paris, 1957, p. 474.
[2] C. Roy, *Clefs pour la Chine*, Paris, 1953, pp. 26–7.

all decked with wide bands of purple silk. For those who frequent the streets, it is a depressing town of stones and earth, but from the air can be seen a multitude of gardens in enclosed courtyards and the bright roof colours of the huge palaces and temples.

Shanghai was originally a little fishing village on the delta of the Yangtse river and became a town in the thirteenth century. It concentrated on trade, has been a free port since 1842, and was guaranteed by a British agreement of the same date; its first skyscraper was built in 1925. Having reached a population of seven million, at one time threatened by decentralisation, it is a town of many aspects. The central blocks of skyscrapers, built on firm foundations of reinforced concrete sunk into the alluvium, contain banks and offices. Beside the old international zone are lines of other skyscrapers and villas behind bamboo railings, all round the trading quarter, which includes the State stores and offers everything that new China has to sell. The river close by is a scene of intense activity and serves as the home of many families both in junks and in the villages perched above the water. The walls of the old Chinese town have been destroyed but they still delineate a central district where it is only possible to go on foot, where there are many canals, and where the whole of traditional China can be observed in small shops, multicoloured advertisements, tea-houses and amusements reminiscent of our own fairs. Whole districts are given over to industry.

The diversity of these towns is as great as in Europe: diversity of site and of building, and disparity of population densities. The population of Shanghai is more poverty-striken than that of Pekin; there are more children and they suffer from greater malnutrition. The amusement places are frequented almost exclusively by men. In Canton in the south, on the contrary, the women are more charming and attractive but their work is much harder. Generally speaking the people of Canton are more lively but less industrious than those of Pekin. However, one still finds an abundance of small trades, ceaseless traffic on the waterways which form the vital arterial routes of the country and the floating dwellings. All these cities are experiencing a rapid growth of their population, a growth too rapid to provide employment and housing for all, and this varied urban civilisation is spreading right into the farthest parts of the interior.

Japan is distinguished by a much denser urban concentration which represents nearly two-fifths of the total population as compared with 10–15 per cent in other countries of the Far East. In connexion with this disparity, it is important to consider just what is meant by a town.

Amongst the over-populated rice fields there exist communities of 20–30,000 people which fulfil the rôle of real agricultural market towns and can claim the title of 'town' by virtue of the size of their population. They are composed of vague and indeterminate districts lying between rice plantations; this phenomenon is found even in Tokyo.

A large number of medium sized towns with populations of 50–100,000 were formerly rural market centres and have now been transformed and surrounded by endless rows of typical small Japanese houses. In the larger towns there are industrial concentrations, and charming and graceful villas and gardens on the edge of the neighbouring hills form a ring round a European-style business district that has succumbed to the attractions of the skyscraper.

The great urban concentrations, and particularly those of Tokyo and Yokohama, are completely on a par with those in Europe or the United States: the same feverish atmosphere, the same business and commercial districts, the same mighty industrial installations gradually spreading over whole plains, the same daily commuting demanding the creation of satellite towns, dormitory towns and all the ingenuity of a new form of urbanisation.

In this land of islands and mountain peaks, urban life is almost always coastal. The majority of large towns are also great ports, which have grown up on the shores of complex and intricate bays; others are wedged into mountain valleys on the edge of enchanting lakes. Side by side with technical and economic growth, the Japanese town has been able to preserve a delicate charm. The beauty of the gardens, the profusion of flowers, the use of rare woods and harmonious colours, give the residential districts, although they may be overcrowded, a reflexion of the natural beauty of the mountains which are never very far away.

CHAPTER 8

AFRICA SOUTH OF THE SAHARA

Africa south of the Sahara occupies an extreme position in the hierarchy of urban civilisation. Indigenous towns were very rare; they were originally the administrative centres of vanished local kingdoms like Timbuctoo, Gao and Kano.[1] Before the nineteenth century some districts, like the whole Congo basin, had not reached the stage of having even the smallest urban concentration. European civilisation, colonisation and exploitation were superimposed on this base, and the first attempts to make contact with this vast continent were made from ports which served to facilitate penetration into the interior. During the second phase of inland development, some old towns were revived, others were created to become centres of the new administration. Finally, the prosperity of the territory and the discovery of important mineral deposits have resulted in the foundation of specialised towns, while the growth of trade either by established means or by newly constructed routes has brought prosperity to old trading posts.

Thus in the whole area there are few urban dwellers and few towns; only four territories have more than ten per cent of their population living in towns (according to the more or less accurate statistics that are available): in the forefront is Zambia with 18·4 per cent.

African towns may be of little importance but they play a very special rôle. They are the only places where modern industry can be found. Their commercial function in respect of agricultural produce and manufactured goods is important since it is more or less exclusive and is not fragmented by the existence of numerous small markets. The majority of them fulfil administrative, cultural and social functions, and in contrast with more developed countries, certain specialists, for example doctors, are concentrated there. Many of the towns are ports, the essential function of which is to serve a larger or smaller area of African territory according to the means of transport available. These towns accumulate a large, unproductive population made up of those who come to seek their fortunes and the dependent families of those who have jobs.

---

[1] J. Dresch, 'Villes d'Afrique occidentale', *Cahiers Outre-mer*, 1950, pp. 200–30; R. W. Steel, 'The geography of urban problems in tropical Africa', *Comm. Congrès Stockholm*, 1960; G. Lasserre, 'Le paysage urbain des Libreville noires', *Cahiers Outre-mer*, 1956, pp. 363–88; J. Denis, *Le Phénomène urbain en Afrique Centrale*, Paris, 1958; A. Seck, 'Dakar', *Man. Univ.*, Dakar, no. 9.

Finally, the town in Black Africa is the great place of contact with the white population who are almost entirely assembled there. Seventy per cent of the white population of Zambia live in the mining towns of the copper belt; in East Africa 50 per cent of the Europeans and 70 per cent of the Indians live in the eighteen largest cities. However, although created by the white settlers and containing the largest colonial nuclei, there is a majority of Africans in most of the towns. They form 90 per cent of the population in certain mining towns of the south-east and still more in towns of the west like Libreville, Bangui and Brazzaville.

All these modern African towns present a structure that is extremely fragmented; each quarter has its speciality, but a fundamental division separates the white from the African quarters. In the medium sized towns, the commercial and industrial districts, the warehouses and the workshops, are concentrated round the centres of communication, the port and the station as in Pointe Noire, Bangui, or Stanleyville. If a strictly specialised function is developed, as in the manufacturing industries at Brazzaville, Kampala and Luluabourg, this is set up in a well-defined district. In the administrative quarters are found the public services, often with a very specialised form of architecture, as at Yaoundé and Nairobi. The same is true of the military camps that are attached to almost all the big towns. This specialisation is due in part to the recent origin of some towns where growth has frequently been directed by the civil or military authorities. Kampala is an example of this type of organisation. It is built on a series of hills and on each hill there is a different specialised centre. The district of Nakasero contains the residential quarter and the public health and administrative buildings; the cathedral is built on the hill of Rubaga while the Protestant mission is on the neighbouring heights of Namirembe, and on another hill is the well-known university of Makerere. All around these hills the African houses pass insensibly from urban quarters into suburban villages.

Generally speaking, the contrast between the European town and the African districts is fundamental. The white population requires hired labour and the African population needs employment; the two groups therefore cooperate but segregation remains marked and indeed legal south of the Congo.

Above all, the towns have sought water supplies and flat, open spaces that allowed the Europeans to develop airy streets with wide pavements and shady avenues of trees. The majority of West African towns are built on islands hemmed in between the arms of a river as at St Louis, or a lagoon as at Lagos, or even surrounded by the sea as at Goree, the forerunner of Dakar, and at Conakry. The peninsulas, cooled by the proximity of the sea and by breezes from several directions are favourite building sites, as at Dakar and Abidjan; so are the stretches of sand which occupy a similar position, as at Grand Bassam and Cotonou. In the interior the proximity of the river assures the town of numerous amenities, as at Kankan, Bamako, Mopti, Niamey and many others.

The white population regards communications with the outside world as absolutely vital, and the port or the railway station is an indispensable gateway around which the warehouses and business firms play a fundamental economic rôle. Each large centre has its speciality: one collects ground nuts, another sisal, another coffee beans, bananas or minerals, while they all receive and redistribute vital imported products (oil, manufactured goods, rice). Banks and offices spring up alongside this commercial activity. The administrative quarters are always situated in a special position at some distance away, on some airy hill or river bank, in an intermediate site between the white town and the African quarters. The European residential districts are laid out in gardens; great efforts to improve hygienic conditions and comfort have been made since 1930 and these have been redoubled since 1946. They have made more or less normal family life possible for previously solitary men, who camped in the colony, often cohabiting with African women. All this development has tended to increase the number of services and specialised industries essential to such a population.

The African quarters are crowded on the edges of the white towns, surrounding them on all sides and penetrating them here and there. In many cases they are only slightly different from the villages of the neighbouring region. In the Sudan the same houses, built of rough brick with flat roofs, surrounding an enclosed courtyard, are found in both the villages and in the urban quarters, but the occupations of the inhabitants are not identical. The traders are more numerous in the towns and family and tribal life are dislocated and replaced by administrative regroupings. In general the dusty roads cross at right-angles; the huts are the same as in the country; water can only be obtained from infrequent street pumps; there are no made roads, no drains, often no electricity. However, markets, schools and chemists' shops mark the influence of the white world.

There are stages in the development of these towns and in the case of Bamako one can observe how a European town has been able to insinuate itself into the very heart of an old native settlement. In 1883, at the foot of the steep slopes of yellow sandstone which overhang the Niger, there was a village composed of four quarters: Niarela which belonged to the Niaré family, Touréla to the Tourés, now well-known in Guinea, Dravela, populated by Moorish traders, and Bozola by river fishermen. The settlement of the district by Europeans stimulated the first rush of newcomers and led to the development of districts nearer the slopes and surrounding the primitive settlements. But the Kayes railway, built in 1904, and the promotion of Bamako to the rank of capital, were determining factors in the transformation. The official buildings were built on high points while the European nucleus, dispersing the old African districts both up and downhill, increased in the centre from north of the railway station down to the river, taking in the harbour and the market. The white district now flourishes under the mango trees and the 'flamboyants' with their big red flowers, among the drab African quarters, the colour of which is that of the Niger alluvium.

The stages of development are also well seen in Conakry, an old colonial town that until quite recent years, in its broad streets full of enormous tropical trees, retained many wooden villas built in colonial style with a verandah running round the first floor of the building. Only a few new blocks on the sea front mark the beginning of a new stage. Abidjan, on the contrary, is one of the youngest towns of West Africa; it became important after the construction of the railway and the town's promotion to the rank of capital city. Here the railway station plays a decisive rôle; the town is built on a picturesque and hilly peninsula and its recent construction is marked by the absence of houses built in the colonial style and its abundance of open ground. It is surrounded by two very distinct villages which are overpopulated and where the female population have not yet escaped from their ancient bondage of plantation work.

Some cities are undergoing a complex development even in this recently urbanised continent. The triple town of St Louis is the result of early colonisation, and the first European settlement goes back to the beginning of the nineteenth century. It is built on a long narrow island in the mouth of the river Senegal. The windswept old town is meticulously arranged in rectangular formation and looks like a small French town with its main square, its balconied administrative buildings, its Restoration style church, its Post Office, terraced hotel, and big shops. Small traditional commercial enterprises line the narrow streets of single-storeyed houses built with verandahs and patios, where Africans and whites live side by side. To the west on the sandy strip of coast at the end of the Servatius bridge, the more spacious administrative district and the central market place are surrounded on the north and south by fishing villages. To the east, at the end of the Faidherbe bridge, the station is the centre of the white residential district which stretches, modern and anonymous, along the river; behind it is paralleled by villages that are becoming more and more overcrowded so that they are encroaching on to the marshes.

In the same way the development of Dakar could be described from its early Portuguese beginnings at Gorée.

The influence of the coloniser is felt right through these towns. The English architecture is colder and on a grand scale: at Bathurst the old company buildings are massive and gloomy; the European houses are not colonial in style, and the mission centres are set in the centre of English-style lawns. There are no cafés, no places where people can congregate; the town is surrounded by villages which themselves have a vaguely anglicised character; on some of the houses there are small windows with green shutters and roofs of red corrugated iron.

It is possible to characterise the main types of colonisation. For the Englishman the ideal is the cottage with a lawn and rose trees; the Belgian demands a large and comfortable villa; the Frenchman tends much more towards fantasy and imagination; the Portuguese have brought their own national style to Africa. However in the really large towns the rising price of land has stimulated the development of blocks of flats of which the upper storeys are much more easily ventilated by natural breezes: Dakar and Léopoldville have developed many residences of this type.

As for the mining towns, they have specially created camps to house their labour force. On the site of the dormitories, which are still in use in the famous compounds of the Witwatersrand, working-class quarters have been constructed, with permanent family houses and a whole complex of schools, hospitals, community buildings and sports grounds, as for example, those built by the Union Minière at Chinkolobwe.

Special mention must be made here of South Africa where the white population, the long and complex history of exploitation, the mixture of races with its Asiatic element and the existence of many half-castes are of paramount importance. The south-east, widely colonised and including some flourishing mineral fields, and the south coast, have a much more highly developed urban network than the rest of the country. Economic expansion is accompanied by a vigorous concentration of Africans round the towns and by a massive increase in the movement of workers to and fro between the reserves and the mining centres. But there is no stability in the relationship between the different races even inside the towns. Will racial and spatial segregation be one day reflected in economic segregation?

In this extremity of Africa is to be found a wide range of long-established urban structures: country markets in the eastern zones which favour agriculture and stockbreeding and which were widely populated by the Boers in the course of their great migrations to the interior, the string of mining towns, working gold or diamonds, big centres of coastal colonisation like Durban, and cities whose organisation and functions are in no way inferior to those of the great cities of western Europe—like Johannesburg and above all Cape Town. In these places the urban style and the bustle are such that one would hardly connect them with the rest of the continent if it were not for the ever present throngs of African workers.

PART TWO

URBAN FUNCTIONS

Chapter 9

THE ORIGIN OF TOWNS

The world's scattered towns, some dominating a vast area, others so close as to be almost touching one another, are of very diverse origin. We are often unable to solve their mystery. The earliest evidence we have often does no more than indicate their existence. We cannot tell how long ago those Asiatic towns described by the first travellers, or the Scandinavian towns figuring in texts of the tenth and eleventh centuries were founded. The first towns, it seems, appeared in those countries which showed technical advancement and in those regions which had found out how to organise irrigation

We know that a certain number of towns were formed more or less spontaneously: they answered a need, and building took place around a chosen point. The opportunity could have been provided by a periodic market promising a permanent site for trade. Abbeys, attracting craftsmen and traders, often established the sites of towns: the study of place-names proves this, by the multiplicity of names ending in *minster* and *moutier*. The castle, where people used to take refuge in time of war, often became a place where non-agricultural activities were concentrated, and garrisons built to guard a country remained as its urban centres, like the Roman *colonia*.

Or perhaps a village, situated near a crossroads and endowed with an enterprising population, was transformed into a little town with a brilliant future. This was the origin of the saying that such-and-such a town is a village which succeeded. The expression is incorrect, as success was achieved in spite of rather than as a result of village life, but it expresses the progressive transformation which gave birth to the town. Somewhere else perhaps, a little creek which sheltered more and more boats gave rise to a town around its harbour.

In many cases towns have been founded for adventitious reasons, arising suddenly, perhaps through the discovery of mineral deposits, as it were by the wave of a magic wand. The oil towns of California and Texas were founded on the canvas and corrugated iron put up by the first settlers. Mourenx and Hassi Messaoud became towns immediately after the discovery of gas and oil deposits. A steel town, established on the iron mines of Lorraine, grew automatically to a population of 8,000 or 10,000.

The establishment of a factory can likewise be the starting-point for the growth of a town. Sochaux grew up around the Peugeot factories and now has a population of 2,800, not counting daily commuters.

Lorient was formed around a naval dockyard opened in 1666 by the East India Company.[1]

Some townships developed around royal country residences—Versailles, for example. More recently, tourism has created similarly superimposed towns. The towns of the Côte d'Azur seem to be newcomers or intruders in the neighbouring countryside.[2]

These are some of the reasons for the growth of towns, but we must pay special attention to systematic town building. When founding a town, people were aware of creating something new, of rising above the contingencies of rural life, and from the earliest times, foundations were given a religious character. The precincts enclosed by the walls were sacred: this was called the *pomerium*, where it was forbidden to set foot. The Greeks went to consult the oracle at Delphi before founding a colony, and the Delphic priests, always well-informed, could be described as great town-planners. The Middle Ages placed new towns under the protection of saints whose names they adopted. Fustel de Coulanges observed that in ancient times a town was founded all at once and complete, in a single day. There are many examples of this. Alexander the Great consolidated his conquests by creating towns. Alexandria was designed *ex nihilo* with trails of flour crossing one another at right-angles. The Romans in their turn founded towns for veteran soldiers, to guard their frontiers. The governors of provinces placed cities in places which seemed favourable to them. Lucius Munatius Plancus created Lugdunum in this way in 43 B.C.

All political expansion is accompanied by the foundation of towns. This happened at the time of the reconquest of Spain from the Moors. In the thirteenth and fourteenth centuries the growth of Germanic colonisation to the east was accompanied by the blossoming of towns, and most of the towns of Siberia were built by the Russians to guard the crossing-points of rivers.

Towns have been founded systematically in our day too. Pierre Monbeig describes the pioneer fringe of the state of São Paulo where both town and country were planned at the same time, with the provision of special sites for market-gardening around the towns.

Monarchs, either by whim or by political design, were pleased to erect towns to be their residence or capital—as for example Madrid and St Petersburg. Modern states have copied them with Washington, Canberra and Brasilia.

Though many colonial towns were nothing more than enlarged trading posts or annexes to native settlements, there are others which

---

[1] Geneviève Lemée, 'Une ville née d'un chantier de constructions navales—Lorient'; *La vie urbaine*, 1936.
[2] Bernard Kayser, *Campagnes et villes de la Côte d'Azur*, Paris, 1958.

were created from nothing. Towns were also founded for industrial ends. The Polish state created Nowa Huta near Krakow from nothing, and many similar examples may be found in the U.S.S.R. and East Germany. In a later chapter we shall study those towns designed to relieve the congestion of the big cities, the satellites and *new towns*.

All these creations benefit from having preconceived plans and technical advantages which avoid the inconvenience of anarchic growth. At Sennestadt in northern Westphalia, before a single house was built, streets and canals were planned and laid out to provide for a population of 15,000. Governments and public authorities are not, however, the only ones to have the privilege of founding towns. When the Mormons founded Salt Lake City they wanted to have their city where they could practise their religion in complete freedom. In 1957 French citizens repatriated from Africa created a 'private town', Carnoux, twenty miles from Marseilles, which must have a population of 12,000 to 15,000.

So towns are born sometimes through the mutual consent of their inhabitants, and sometimes because of the conscious wish of their founders.

CHAPTER 10

URBAN FUNCTIONS: GENERAL PRINCIPLES

DEFINITION AND CLASSIFICATION

Men have grouped themselves in this way to carry out certain activities more efficiently. These activities make up the function of a town, its profession, the reason for its existence. This is the way it looks to outsiders: in the same way as an individual is a banker or a doctor, Detroit and Sochaux are car-manufacturing towns, Dunkirk is a port, Cambridge a university town, and Brasilia a capital.

The term function has been borrowed from physiology; it likens a town to a bodily organ; it was first used by F. Ratzel in 1891, and has been employed since then by all geographers and town-planners. This has sometimes led to very varied interpretations, and it is important to define the idea in precise terms.

First we must consider the activities of a town as an organ carrying out a function within a whole, that is, its activities as seen by an outsider. The activities carried on in the interests of the inhabitants, for their own use, so to speak, do not reveal the function of a town. This is what Gunnar Alexandersson calls 'city-serving production'. Among these activities are those of masons building dwellings for the inhabitants of the town, bakers making bread for local consumption, grocers supplying neighbouring streets, and so on. To these can be added garage-owners repairing the cars of their fellow citizens, and teachers who educate their children. All these activities are the result of urban living, and not the reasons for the existence of the town. In this connexion, the expression, 'internal function' has sometimes been used, but we consider that this term is inappropriate. It is not really a question of function. In the same way, we all potter about the house, doing odd jobs, but we do not give this as our profession. Very often, besides, these activities are the time-honoured prerogative of particular families, for example, the making of clothes or cakes.

Therefore we will consider as the functions of a town only those activities which justify its existence and development, and which bring in the resources necessary to its continued existence. Many of these resources come from the surrounding region and regional function is often spoken of. In fact it is difficult to think of a town without also considering a certain area which gives it life, and which in turn revitalises and animates. Some authors incorporate this important rôle in the definition of a town.

But this town influence, to which we will refer again, is exerted through the products it buys and sells, and through its administration. That is, through certain externally directed activities which are in fact its functions. It would be clearer therefore to speak of the regional rôle of the town, and to make a separate study of the various functions from which the elements of the regional rôle are drawn.

Gabriele Schwartz[1] distinguishes between general, centrally linked functions, functions connected with the region (*Allgemeine Funktionen*), and specific functions (*Besondere Funktionen*). Here we will keep to specific functions, reserving the regional rôle for the fifth part of this book.

We must therefore consider what these functions are, and set up a sort of catalogue of the occupations carried on by towns. According to the definition we have given we cannot completely rule out an agricultural function; this probably applies more to villages (the term would in fact only be exact in a free economy, as the supplying of its own food by a village does not constitute a function), but we have seen that there are towns greatly affected by agriculture, and even agricultural towns. The most important of these agricultural towns are those which are often denied the title of towns and described as large villages. Many Mediterranean and Asian agglomerations come into this category. More often the agricultural population works to supply the needs of the industrial and commercial population of the town. It is as if a big urban family had a kitchen garden cultivated by certain members of the family. This is a division of labour rather than an external function. These agglomerations are often merely gatherings of agriculturists and urban functions are carried on some distance away, as in Indo-China, by small administrative centres with sometimes only a few hundred inhabitants.[2] These towns were not created for agricultural production, for villages could have provided this just as well; on the contrary, it was to carry out certain forms of industry and commerce more efficiently that people gathered in towns.

Urban functions, then, are those for which towns originate.[3]

For a long time they were treated vaguely, and attempts at systematic classification have only been made during the last twenty years or so. In 1943, Chauncy D. Harris[4] established a definite distinction between: 1. mining towns; 2. industrial towns (that is, manufacturing industries); 3. commercial towns; 4. centres of communication;

[1] *Allgemeine Siedlungsgeographie*, Berlin, 1959, 1961.
[2] Pierre Gourou, *Les Paysans du delta tonkinois*, Paris, 1936.
[3] The case of the agricultural town will be considered, but is much more rare.
[4] Chauncy D. Harris, 'A functional classification of cities in the United States', *Geog. Rev.*, 1943. John Fraser Hart used roughly the same classification, 'Functions and occupational structures of cities of the American South', *Ann. of the Ass. of American Geographers*, 1955.

5. university towns; 6. capitals; 7. towns for 'recreation'; 8. towns with various functions.

For the towns of northern Sweden, W. William-Olsson proposed a plan of classification into commercial, industrial and railway towns. Among the industrial towns, mining and metallurgical towns, and forestry towns were mentioned specially. On his fine economic map of Europe,[1] the same author distinguished between service towns and industrial towns. Within the industrial category he distinguished ore-extracting towns, coal-mining and oil-drilling towns, textile towns, and towns dealing with the cellulose industries. His economic map of Sweden[2] lists centres of commerce, industrial centres (including mining and metallurgical towns, wood and cellulose towns, and textile towns), administrative centres and centres of communication. Gabriele Schwartz[3] classifies urban functions under four headings.

1. Political; 2. Cultural; 3. Economic (markets, communication centres, commercial centres, industrial centres); 4. Capital administration.

To this list we would add a military function, which probably has little value today but was the origin of a considerable number of towns in the past. It must of course be understood that these functions overlap to some extent, and we must make a special effort to determine which is dominant.

DETERMINATION OF URBAN FUNCTIONS

To determine the function or functions of a town it is necessary to compare the various activities of the inhabitants, or at any rate those activities which are directed towards the satisfaction of the external needs of the town. The problem is one of comparing the volume of very dissimilar activities.

It is possible to establish an industrial hierarchy according to the value of the output, by which industries as different as jewellery and metallurgy may be compared. To compare industry with commerce, one can try to establish the amount each contributes to the town's income. This, of course, could not be done without asking delicate questions, and in any case it would be impossible to extend this comparison to include other functions such as administration.

In practice, the only comparison possible between these various functions is one concerning the number of inhabitants employed therein. An industrial town is one in which the majority of the active population is employed in industry; the same rule applies to commercial, administrative and tourist towns.

[1] Stockholm, 1933.
[2] *Ekonomiska Karta över Sverige*, Stockholm, 1953. [3] *Op. cit.*

The census will provide the basic information. Each census form indicates the occupation of the individual. This information sometimes has to be interpreted, as some people may indicate a specialised occupation, as for example 'metal turner', while others put 'metallurgical mechanic', or simply 'mechanic', and the census taker sometimes gets into difficulties.[1]

Supposing that access to individual forms is possible, which is not always the case, research is complicated. While it is possible for small towns, it necessitates extremely lengthy enumeration when dealing with large agglomerations.

Research has been simplified in France in communities where a recapitulatory list of the inhabitants has been established. This list, which used to be compulsory, is often kept for the convenience of the municipal services, and records both the age and the profession of the individual. Further, in large agglomerations with a population of over 20,000, one can use the results obtained by the I.N.S.E.E.[2] in their investigation of a 5 per cent sample.

Let us suppose that we know the occupation of each inhabitant of the town. We begin by deducting the agricultural activities (including forestry which, in some countries, occupies an important place). The proportion of these activities should be borne in mind so as to determine which are agricultural towns, but the truly urban functions, determined by the occupations in the secondary and tertiary groups, are our main concern here. In many parts of France, incidentally, agricultural occupations often occur because there are farms within the commune but outside the agglomeration itself. This applies especially in the south of France, where the perimeter of the commune is considerably larger than that of the urban nucleus.

The next task is to classify the remainder of the population into occupational groups. The distinction between industry and commerce, for instance, would appear to be clear enough, but the census does not always establish it very easily. It is easy to find out how many people are employed by big organisations such as Electricité de France or the French National Railways. But in these organisations there are industrial workers and employees in the tertiary sector, people who work on the tracks and people who punch tickets, so it seems best to take the assessment of the whole organisation and apply it to the town under consideration. This procedure appears valid in the case of large agglomerations, but may not be so for a small centre with a specialised workshop.

---

[1] Difficulties over women's occupations arise particularly in country districts, where one may describe herself as a farmer and another as 'without profession' when in fact they are doing similar work.

[2] Institut Nationale de la Statistique et des Etudes Economiques.

This, however, is not the main problem. The principal difficulty is encountered when trying to differentiate between the proportion of people working directly for the service of the town, and those not so employed, for it is the latter who determine the town's function, and comprise what Gunnar Alexandersson calls the fundamental population, but what we prefer to call the functional population.

Detailed enquiry is necessary in every case to determine whether or not a person works for the town itself. One must find out, for example, whether a particular garage is run by a handyman who repairs his neighbours' cars, or belongs to a big firm, undertaking work for the whole region; one must distinguish between the corner grocer's shop and the store which delivers to all the neighbouring villages. Even building work can be done for outside interests. These investigations must be made in the considerable detail that is appropriate to a thesis or monograph.

For larger projects, more general methods must be used. These consist of deducting straight away a certain percentage of the population supposedly engaged in activities connected with the town. This method has been advocated by Klaasen, Van Dongen, Torman and L. M. Koyck;[1] it was used again by Gunnar Alexandersson, who wrote a critique of it.[2]

Given a certain number of towns, the percentage of the various activities for each is determined. When they are compared, one of them will have the smallest percentage; this is considered to be the minimum necessary to the internal life of the town, and everything which exceeds this percentage in other towns must be supposed to be an extra-urban activity, externally directed, that is to say, dedicated to carrying out the town's function.

Let us suppose, for example, that we are considering twenty towns with an active population of between 5,000 and 500,000. One of these towns with an active population of 40,000 has 2,000—5 per cent—of its inhabitants engaged in the building trade; that is the lowest percentage of builders for the group. We will therefore suppose that in each town 5 per cent of the builders are working locally and all the builders over that percentage are working for outside interests.

The method can be represented by a diagram as follows. Columns of equal size are arranged to represent the various towns (Fig. 5). The percentage of workers carrying out a certain activity is shaded; the lowest percentage appears in column 5: 'A' represents the level of the shaded portion of that column. If a line is drawn through 'A' at

[1] Leide, 1949; cited by G. Alexandersson.
[2] Gunnar Alexandersson, *The Industrial Structure of American Cities*, London, 1956.

right-angles to the columns all shaded portions above the line will represent part of the urban function.

Occasional aberrant cases occur, of percentages which are obviously too low because a town is not yet fully equipped, because of some particular specialisation, or simply through an error of arithmetic. Gilbert Le Guen[1] quotes the example of La Grand' Combe (Gard), a fast-growing industrial town near Alès, where only 20 per cent of the working population is in commerce or private and public service. It would be wrong to apply this percentage to other towns of the Mediterranean provinces of France and to suppose that in these towns everything over 20 per cent is part of the function. On the whole, though, the results achieved are satisfactory. Le Guen, in a

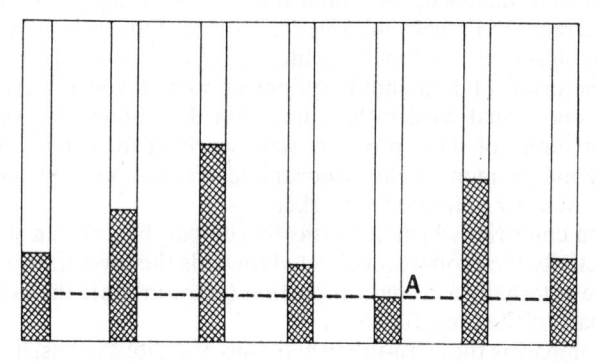

Fig. 5. Diagram representing the same urban function in different towns
Level 'A' represents the importance of that function in the town where it is *least* developed.

study of 147 French towns with a population of over 20,000, fixed at 40 per cent the section of the active population working, on average for the needs of each town. (Gunnar Alexandersson estimated a figure of 37·7 per cent for the United States.) This 40 per cent is made up as follows: building, 5 per cent; manufacturing industries, 11 per cent; commerce and banking, 11 per cent; liberal professions, 5·2 per cent; administration, 5·2 per cent; transport, 1·9 per cent.[2]

Edward L. Ullman and Michael F. Dacey adopted[3] and improved this method with reference to a certain number of towns in the United

[1] G. Le Guen, 'La structure de la population active des agglomérations françaises de plus de 20,000 habitants', *Ann. de Géog.*, 1960.

[2] *Ibid.* The same author, again using this method, arrived at a comparable figure, 42 per cent, for Breton villages of more than 2,000 inhabitants. *Norois*, 1961.

[3] 'The minimum requirements approach to the urban economic base', in *The I.G.U. Symposium*, Lund, 1960.

States with populations ranging from 2,500 to over a million. They worked out, for fourteen activities, the minimum requisite percentages for the needs of the city itself. They show that the bigger the city, the higher are the percentages; they are only 24 per cent in small towns (population 2,500 to 3,000), but reach 56.7 per cent in cities with a population of over one million. This is of course on average, the percentages varying according to the activities.

Generally speaking, a rise in the standard of living is reflected in an increase in the tertiary sector.

This method can only be of use when similar civilisations are under consideration. While it can be successfully applied to towns of the United States, it would make no sense for all North American towns if Mexico were included, nor could the same percentages be retained with reference to French and American towns. Domestic gadgets play a much bigger part in American life, and if the French figure for the manufacturing of household equipment were taken for American towns, one would receive the impression that almost all American manufacturers of these products were working for export. On the other hand, a more highly mechanised civilisation may require a smaller number of industrial workers.

Given countries which are basically comparable, we can select for each activity the proportion directed towards the town (intra-urban), and the externally directed proportion (extra-urban); the latter will form part of the town function.

By comparing the extra-urban activities of various towns, it appears that we can determine each town's dominant function. An industrial town will be one in which the majority of the working population is engaged in industry. But, this is not absolutely correct: the presence in a town of a hundred bank employees over and above the requirements of the town, indicates an important banking activity; the presence of a hundred more workers in a textile factory working for outside interests only indicates a modest interest in textiles.

As Herbert Lehmann observed,[1] if at Frankfurt-on-Main only 16·8 per cent of the active population are engaged in commerce, banking and insurance, they are still more significant in the town's economic life than the 21 per cent employed in services. This reasoning is valid with regard to externally directed activities.[2]

It could be of interest to use coefficients according to the method used by J. Klatzmann for agriculture; one hectare of vineyards could not very well be compared with one hectare of meadowland to determine the speciality of a *commune*, and J. Klatzmann adopts a

[1] 'Frankfurt a/Main', *Die Erde*, 1954.
[2] Cf. K. A. Boesler, 'Zum Problem des quantitativer Erfassning städtischer Funktionen', *The I.G.U. Symposium*, Lund, 1960.

coefficient of five for the surface area of a vineyard. Would it not be possible, in calculations of the active population, to work out variable coefficients for different occupations? This would be an interesting research topic.

The determination of extra-urban activities must allow for the classification of each town into a particular category, according to its dominant function. With some towns this presents no difficulty. There are towns where, if the intra-urban activities are subtracted, the entire population is engaged in a single activity. This is the case in certain mining towns, trading ports and administrative centres. But it is rare for no secondary function to develop alongside the principal function; the iron-mining town gives rise to metallurgical enterprises; factories are built for the treatment of raw materials brought into a port; the administrative centre becomes a commercial centre for the neighbouring region.

These subsidiary functions can develop to such an extent as to outweigh the principal function. Marseilles is now as much an industrial as a commercial city. At Uppsala and Oxford the university function has been overtaken by an industrial function.

We shall study each of these distinct functions, and observe later how they are combined.

FUNCTION AND GEOGRAPHICAL SITUATION

The town which was founded or developed to carry out a particular function is, of course, situated at the place best suited for the exercise of that function. This place characterises the geographical situation (or position) of the town. It answers certain more or less exact requirements. An estuary port is situated at the highest point reached by the tide to take advantage of the sea downstream and at the same time the lowest easy crossing point of the river. The junction of two rivers creates a favourable commercial situation through the convergence of navigable waterways or transport routes. Care must be taken to distinguish this situation from the *site*, that is, the most convenient place for daily life. The situation commands; men use it as best they can for their own ends, and the site is the result of this arrangement. In a valley providing the situation, a gravel fan rising above land liable to flood presents a favourable site. It sometimes happens, however, that the requirements of the situation do not allow much choice of site. This is the case with fortress towns which are forced to pile up on the hill which guarantees their military function.

In the following review of the various functions, we shall in every case take into consideration geographical situation in relation to function.

REPRESENTATION OF URBAN FUNCTIONS

Urban functions are the object of graphic and cartographic representation, and many attempts have been made to represent them, some more successful than others.

A town can be defined graphically, according to its essential function, by the place it occupies inside an equilateral triangle whose sides represent three variables, each representing a function.

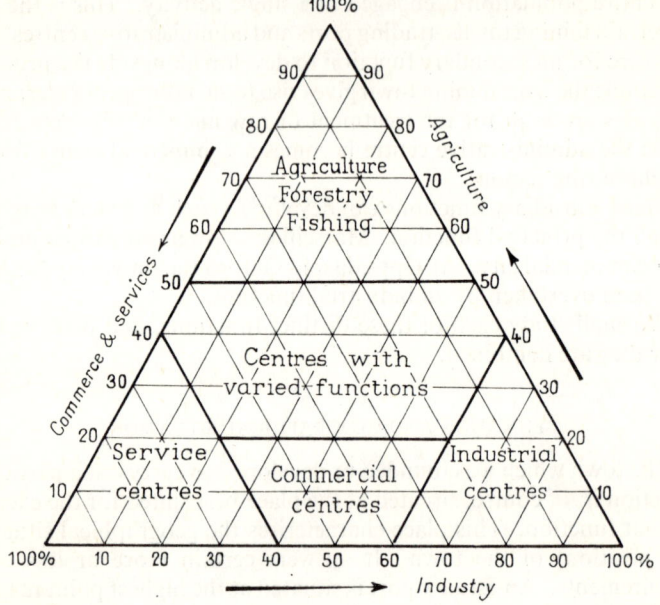

Fig. 6. Diagram showing the distribution of towns according to their functions

After Gerd Enequist, *Atlas över Sverige*, map 59–60. Note that a space has been reserved for agricultural centres.

To obtain three variables, the primary, secondary and tertiary activities are often considered. However, except in certain regions, primary activity, which tends to be the production of raw materials, is considerably less in towns; any agricultural activity is rare, and only fishing ports have much primary activity; mining towns present a problem for they are often considered to be industrial towns, yet they are mainly producers of raw materials.

More often, to obtain three categories, industry (including mining), commerce, and services are taken, the last term covering everything not included in the other two. A town where these three functions were equally represented would be found in the centre of gravity of the triangle. If not, its place can be determined by the point 'O' at the intersection of the parallels XY and ZT to the bases of the dominant functions.

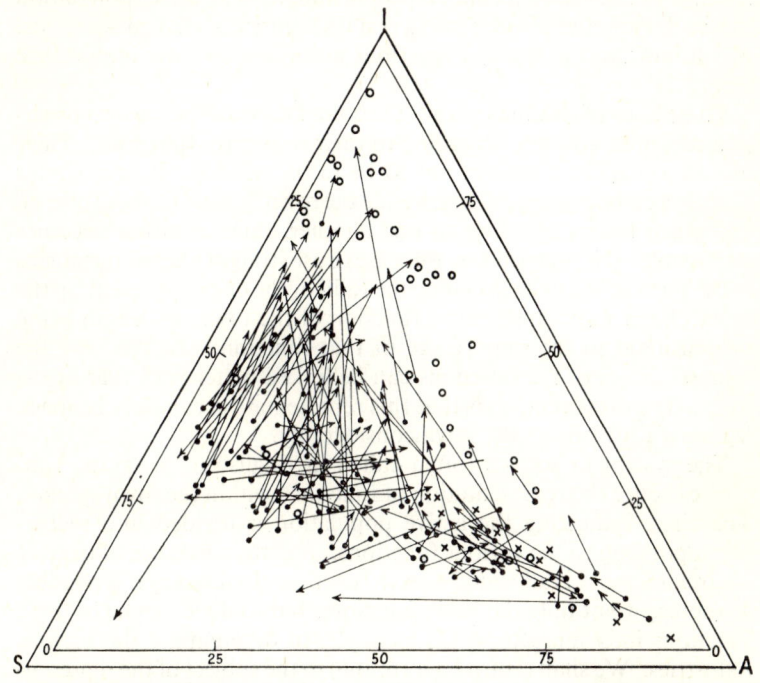

Fig. 7. *The development of town functions in Rumania from 1930 to 1956*

After I. Sandru, V. Cucu, P. Poghirc, 'Contributions géographiques à la classification des villes de la République populaire roumaine', *Annales scientifiques de l'Université Al. I. Cuza* of Iasi, II, natural sciences, vol. 8, 1961, fasc. 2, pl. IX. I = Industry; S = Services; A = Agriculture; ● = Towns in 1930; ○ = Towns in 1956; × Vanished towns.

This system has often been used, and is in current use in Scandinavia. Examples are the diagrams of Oïva Tuominen for Finnish towns, and those of Gerd Enequist for Swedish towns (see Fig. 6).

This system has the advantage of allowing one to form a mental picture of all the towns of a country by classifying them according to their functions.

I. Sandru has even given the diagram a historical significance. Each town is represented by two dots, one marking its place in the triangle in 1930 and the other in 1956; this indicates the development of functions (Fig. 7).

This process cannot of course be used for cartographic representation. In this case it is convenient to adopt for each town a figure constituting a self-contained representation of the functions of the town. For each town a column proportionate to its active population can be drawn, and divided into a certain number of shaded segments, which indicate the various functions according to their importance (Fig. 8.7).

These lines of columns permit the comparison of the various towns of a region or country according to their respective functions. Their use on a map is, however, not to be recommended. It necessitates placing a cumbersome column beside the town symbol (dot or sphere), and when towns are close to one another, representation becomes impossible. It is preferable in this case to use straight lines or rectangles in the form of sticks lengthening the radii of the circle representing the town. Each stick represents a function of the town, its length being proportional to the importance of that function.[1] In this way the representation of the functions can be linked better with that of the town. The disadvantage is that this representation can only be made legible if it overflows the town considerably.

The system of sectors (often called in English a 'pie-graph') indicates better the representation of functions within the town symbol. The circle indicating the active population is divided into sectors corresponding to the various functions. A certain number of degrees of circumference is allocated to each function. In this way it is possible to represent not only the basic functions, but to divide some of these functions into subsectors; for example, to differentiate the various industries. We shall return later (p. 166) to the subject of the representation of industries (Fig. 8.4).

As the circle divided into sectors only includes the active population, it is of interest to enclose it, on the map, inside a larger circle representing the total population of the town (Fig. 8.1).

All these methods of representation allow the various town functions to be shown, but on a map the most important thing is to give the dominant function. In this case the town is shown on the map as a circle or sphere shaded with the colour corresponding to that function (Fig. 8.2).

It is difficult to represent a town where several functions are mixed, and this has often been avoided by adopting a special shade for multifunctional towns.

On his economic map of Europe, W. William-Olsson has selected black for service towns, while red, purple, green and orange indicate towns in which 50 per cent of the active population are engaged,

[1] This system is employed in certain modern publications, e.g. *Manuel de 'enquêteur*, by L. J. Lebret, 1952.

URBAN FUNCTIONS: GENERAL PRINCIPLES   117

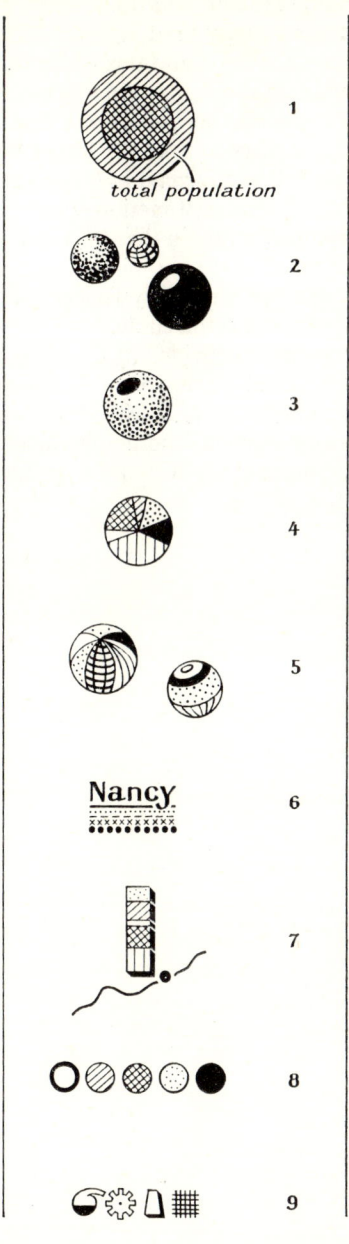

Fig. 8. *Cartographic representation of urban functions*

respectively, in metallurgy, coal and oil, textiles, and cellulose; blue indicates that no single activity predominates. The vividness of the colours makes striking contrasts and throws up clearly the industrial groupings of England and the Ruhr. An ingenious device, not to be found anywhere else, allows the indication of secondary functions. For this the little ellipse placed at the top left of the circle to make it look like a sphere is used, and instead of being left white it is shaded with the colour of that secondary function (Fig. 8.3).

On the map of urban structure in the *Atlas de France*, each town is indicated by a circle or a square in different shades according to the principal function, and the secondary function is represented by a ring of another colour around the circle.

An interesting variation is provided by the *Atlas of Britain*, Oxford, 1963 (Plates 132–45); total employment is divided into six equal sectors, each of which is shown in a different colour, with the radius of each sector corresponding to the employment in that group of trades. The visual impression, if a little gaudy, is striking and effective.

CHAPTER 11

THE MILITARY FUNCTION

The military function is often passed over because nowadays it is no longer of great significance in urban matters; however, it should be taken into account because it was the origin of many presentday towns.

It is not just a question of ramparts erected by towns for their own defence; that is only an aspect of internal organisation; the security of the inhabitants is attended to in the same way as their housing and provisioning.

Many towns were founded for a definite military purpose. The *oppidum*, the refuge, was not itself a town, but was often the basis for a town: pre-Hellenic towns like Mycenae were scarcely different from these, while the medieval fortress where peasants from the surrounding countryside came to take refuge, was often the cradle of a town.

On the other hand, fortress towns were built in colonised countries for administrative purposes, but also to guard and defend the country. The Roman colonies were of this type; they guarded the frontiers of the Empire and formed the nuclei round which towns crystallised. Besançon, Cologne, Trier and Lincoln originated in this way and are typified by their rectangular plan, a relic of the Roman *castrum*. The Arabs established similar camps to guard North Africa, and after the conquest of Algeria the French adopted this system, hence towns which were originally *bordjs*, like Fort National. We have already mentioned the Russian fortresses which guard the crossing points of Siberian rivers.

The military function also gave rise to strongholds blocking the paths of invasions. Belfort guards the Burgundian Gate and Gravelines watches over the Route des Dunes in Flanders. It is the military function, too, which maintains and develops, even if it did not create, fortified towns along the frontiers. The presence of a strongly manned garrison gives character to a town, though the fortress itself may impose certain constraints on urban development, through the presence of bare sloping ramparts, barracks and parade grounds.

These strongholds have witnessed all the vicissitudes of politics, the changing of frontiers emphasising their rôle or alternatively removing their strategic importance. It is difficult to believe that the peaceful little town of Saint-Jean-de-Losne was once famous as a stronghold when the frontier between France and the Germanic

Empire ran along the Saône. Then again, some frontiers are more active than others, and not always the same ones. Localities in the Somme valley, such as Corbie and Péronne, acted as shields in modern times against the Spanish in Flanders, as Toul and Verdun did against Germany after the war of 1870. Movements of the frontier sometimes gave the same town quite opposite objectives, as with Metz and Brixen-Bressanone.

The advance of military technique has greatly affected these towns. The fortress town perched on a hilltop gave way to the fortified town protected by a ring of small forts, like Verdun. Then the progress of artillery and the coming of aviation removed completely the value of defences massed at any one point; nuclear warfare makes these particularly vulnerable. The military function has been erased from the list of urban functions. Nowadays there remain only garrisons stationed without military objective in certain ancient fortresses. Towns which grow up around atomic factories can hardly be considered as having a military function, and rocket launching stations cannot be considered as towns.

In urban geography then, the military function occupies its place chiefly because of the traces of it which remain, and they are numerous in most European towns, or in those of the continent at any rate. This is the explanation for towns grouped around the fortified hill which was their origin, like Belfort, Nuremburg and Edinburgh. The plan of many towns is due to their military past.

Among towns with a military function are naval bases. These are of two kinds. Some were built to defend the coastal frontiers as fortresses defend frontiers on land. These are sea front ports. Their principal function is to shelter the fleet which patrols them. So that the fleet can take refuge there more easily, they are built as far out from the land as possible. Le Havre de Grace was founded in 1517 by Francis I to receive the Channel fleet, and placed at Bec de Caux in front of the difficult passages in the Seine estuary. In the same way Napoleon I thrust Cherbourg out into the Channel on the Cotentin promontory, and a jetty had to be built to protect its entrance. Cuxhaven occupies a similar position on the North Sea coast between the estuaries of the Elbe and the Weser.

However, safe shelters were also needed, strongly protected against all enemy attack. If they were difficult to enter, this was an advantage. The anchorage of Brest in its down-faulted trough, can be reached only through a narrow channel protected by the peninsulas of Saint-Mathieu and Crozon; moreover the entrance is difficult, being strewn with reefs, and the French fleet can carry out exercises there in complete security. In the South of England, Portsmouth, in the Solent behind the Isle of Wight, offers a somewhat similar refuge, and Devonport, on a drowned estuary, is another example. The Kiel canal could in the same way serve as a shelter for the German fleet.

Frontiers here do not depend on the whims of man, but politics can alter their value. The alliance with Britain has in times past lessened the value for France of her Atlantic military seaports and has caused her to concentrate her efforts on Toulon on the Mediterranean.

Sea front ports offer support to the fleet on home ground but the fleet also needs distant bases in the course of its movements. Some of these bases are like fortresses situated in border territory. For them the sea represents security, and the link with home, and fortifications are turned towards the landward side. Penetration and conquest start from there. Algiers was already in existence when the French made it their headquarters, but Casablanca was founded as the starting point of the French penetration of Morocco.

The bases found by chance in national possessions were not sufficient for fleets of warships. The linking of these possessions with the homeland necessitates safe routes provided with bases for refuge and provisioning. These routes were interspersed with privileged places much coveted by the great powers, such as the shores of certain straits or certain well placed islands. Great Britain, with an immense Empire to look after, built up a network of bases for her fleet which was, so to speak, the armour of that Empire—Gibraltar, Aden, Singapore on the key straits of the Old World; Malta to mark out the Mediterranean route, and the Cape on the old route to India. Each of these bases became a small British colony. Gibraltar is the best example, having to its advantage an anchorage of 42 acres and an easily fortified rock. Around the harbour clusters a town with a population of 21,000.

To land and sea routes we must add the airways. Mastery of the air presupposes the existence of bases. Technical progress, increasing the distance of flights, reduces the number of bases required, but for long journeys it is still convenient to have departure and stopping points. Such is Keflavik, an American base in Iceland. A whole small town has developed there, about 30 miles from Reykjavik, entirely dependent on the air base.

Naval ports and air stations are inhabited largely by sailors and airmen, but much more than in the case of inland fortresses, they have given rise to a whole range of industries without which they could not function. There is, in fact, no point in placing arsenals too close to inland frontiers, and St Etienne owed its fortune after 1870 to the fact that it was far from the frontier. On the other hand, ship-building and repairing yards and naval arsenals are close to the ports and air bases. Toulon and Brest are, for the most part, arsenals.

Naval bases have thus created towns, but their growth has often been accompanied by the development of other functions. The requirements of the fighting fleet are often the same as those of the merchant navy. The advantages which led to the choice of Le Havre as a port of war have shown themselves equally valuable to commerce, with the increase in the tonnage of shipping and the search for deep-water ports of call, so that the commercial port has taken the place of the port of war of which it was originally only a sideline. Military bases on the great maritime routes naturally became ports of call for maritime trade. Singapore is now the great emporium of south-east Asia.

So the military function, whether represented by continental fortresses or by seaports, prepared the basis for an urban life whose heritage can be seen today in the old countries of Europe as well as on the fringes of the continents of Africa and Asia.

CHAPTER 12

THE COMMERCIAL FUNCTION

The commercial function often seems fundamental, *Der Verkehr wirkt Stadtbildend*, as Friedrich Ratzel writes in his *Anthropogeographie*. It is hard to imagine a town in which no external trading of any kind is carried on. Some authors have even included this function as part of the definition of a town. Maurice Lombard has shown how in the High Middle Ages towns again became rural agglomerations when trade lapsed through lack of money.[1]

First there are towns which were founded for commercial purposes, for example the markets where country folk come to exchange their produce, and the ports where merchandise from abroad is collected. But even towns in which commerce is not the main activity, such as industrial or administrative centres, quickly become commercial. Business develops there for the convenience and well being of the inhabitants. Grocers, drapers and garage-owners set themselves up for the convenience of these inhabitants, who could not do without them. But the area in the middle of which the town has arisen, on which it has to some extent been superimposed, cannot help turning towards the town, even if it is only to take advantage of its internal commerce. Instead of going for provisions to another town some way off, whatever may have happened in the past, people use to their advantage the commodities offered by the new town, and this town which originally had businesses only for the benefit of its own inhabitants, begins to carry on a commercial function, developing it as a result of this new clientele.

PRIMITIVE FORMS OF TRADE

The local market. The local market must have been the first form of urban life and one which involved the organisation of human relationships. From the moment when people cease being entirely self-reliant and begin to establish peaceful relations among themselves, they want and need to trade in order to improve their living conditions, and this trade goes outside the narrow circle in which they live. The village economy of the Middle Ages could not be entirely closed; certain levies had to be paid in kind, and so certain products were commercialised.[2]

[1] 'L'activité urbaine pendant le haut Moyen Age', *Ann. Econ. Soc. Civilisations*, 1957.
[2] M. Bloch, *Caractères originaux de l'histoire rurale française*, Paris, 1931.

In this way, market towns were born. These often came to life only once a month or perhaps once a week—the 'weekly miracle' as René Robin calls it.[1] Many never got beyond the stage of being half-town, half-village, but many others were the seeds of towns. Markets were developed in contact with different regions which traded their goods; towns of the Vosges were built at the junction of sandstone and limestone, of forests and open country, like Cirey, Baccarat, and Rambervillers; towns of East Anglia, which R. E. Dickinson[2] lists, were built at the contact of clay, sand, gravel and chalk.

These towns were once carefully regulated. As people usually travelled to them on foot, their area of influence was seldom more than about five or six miles, and under ancient German law it was forbidden to open a new market unless it was over six miles from an already existing one.

The livestock markets were of particular importance; these were often held in market towns, usually every month, or four times a year. They acquired considerable importance on the fringe of mountainous areas where the Alpine pastures could be used for the feeding of large herds in summer only, and there was a great concourse of people and animals at the spring and autumn fairs.[3] The hiring of labour also became big business.

Markets and fairs seem to have reached their height in western Europe in the mid-nineteenth century; after this they underwent a considerable decline, the fairs expecially. Nevertheless, they still flourish in Asia, Africa and South America, and have remained quite active even in Europe. In 1930 in a town like Tartu in Estonia transactions were still being made involving 7,000 head of horses and cattle.[4]

Trading posts. Just as the local market was the origin of internal trade, so the trading post represents the basic form of distant trade. It arises from the contact of a primitive society with a more highly developed one. The indigenous people have an abundance of goods which to them are of no great value. On the other hand, certain implements are difficult to obtain with the primitive means at their disposal. Think of the difficulty of sewing with bone needles and fibre! One of the essential objectives of the Laplanders' long journeys to the coastal trading centres was the acquisition of steel needles and sewing-thread. For this they would travel scores or even hundreds of

[1] 'Parthenay', *Rev. du Bas Poitou*, 1959; cf. M. del Rosario Miralbes Bedera, *Soria*, Zaragoza, 1957.
[2] *Geography*, 1932, pp. 19–31.
[3] A. Allix, 'La Foire de Goncelin', *Travaux de l'Institut de Géog. alpine*, 1914.
[4] E. Grepp, *Fairs of Estonia*, 1934.

miles by sledge loaded with reindeer skins to offer in exchange. Trading posts were similarly set up in Canada for the delivery of furs brought back from the hunt by native Indians. Many goods were exchanged in this way, according to the country: arms, tobacco, and alcohol for ivory and furs.

These trading posts multiplied, each one becoming more important as the so-called civilised peoples made more profit and as the underdeveloped peoples got a better idea of how to improve their standard of living. Many of them, rough sketches of towns in zones with no urban life, became town centres when these regions developed, like Fort Hudson in Canada.

A more modern form of trading post can be seen in centres for certain agricultural products, especially in plantation countries. In Ghana, after the harvest, the Africans used to load a pack animal with ground-nuts and exchange them at a neighbouring trading post.

The trading post is always the commercial centre for two races of different civilisations. It is the simple form of urban life which a more highly developed civilisation installs in underdeveloped countries. While it remains a trading post, marketing is not organised. Trade is speculative, and some traders have taken advantage of this. To attract the indigenous people they have not hesitated to encourage certain vices, such as alcoholism, which was one of the worst aspects of colonialism, though antedating the colonial period.

To start with, then, the trading post's existence is precarious. It develops only gradually into a town, and does not always succeed in this. However, a trading post is a place where one would expect urban life to develop.

CONTINENTAL TRADE

Towns and Routes

Large-scale trade depends on the traffic by land and sea. Towns were founded to facilitate that traffic, being quite often trade centres rather than commercial towns. Each means of transport has its own technique, and from these techniques are born towns which provide for the traffic.

Road traffic in former times. In the days of animal-drawn vehicles and pack animals long journeys necessitated relay stations. Quite large bodies of people moved along some of these routes. Carriages followed one another in hundreds on our main roads at the beginning of the nineteenth century: regular services, express services, coaches and private carriages. At the end of the day everyone stopped at roadside inns. Where traffic was particularly intense, these inns were grouped in an embryo town. Relay stations were sometimes specially valuable, at a crossroads of two important routes, for example. At the foot of steep hills relay stations were equipped with extra relief horses; where the road was not safe, there were places at which convoys were formed. Innsbruck was the point of departure for the crossing of the Brenner Pass. At the foot of the Appenines were founded the towns of the Via Emilia which skirts the mountains.

River bridges were rare, and roads converged upon them, so the approaches to a bridge were always intensely animated, and inns multiplied in these places. London grew up round London Bridge not far from the ford of Westminster, with the town on the left bank and a bridgehead 'satellite' on the right bank.

The great caravan routes were interspersed with relay stations. Luban, in the Tarim valley, was a relay station on the silk route. In the same way Aleppo was an inland town for Mediterranean trade.[1] The fringes of the Sahara are marked out with similar centres such as Biskra, Ghardaia and Timbuctoo. Fez was at the meeting point of the caravan routes going from Algeria and the Sahara to the Atlantic. All these towns, linked inseparably with ancient forms of transport, have suffered a decline. The railways completely altered the distribution of towns in West Africa.[2] The importance of Timbuctoo declined, while that of Kano remained the same, and new towns appeared, like Baro in Nigeria. Often, however, a position once acquired has been maintained: Strasbourg was originally the town where the paved roads converged (Strassenburg).

In the days of coach travel it was convenient to break the journey before entering a large town and stay the night somewhere where one could rest. So annexes, destined to become suburbs sprang up all around Paris, like Bourg-la-Reine, on the road to Orléans.

Water-borne traffic. Boats often serve as travelling houses and do not need relay stations, but navigation is only suitable for fairly long journeys. On every navigable river there is a point, at the head of navigation, where river transport is concentrated. Lille was originally founded at the head of navigation on the river Deule. On large rivers there are sometimes insurmountable obstacles. Before the Congo reaches the coast there are rapids which cannot be crossed by boat, so that goods have to be unloaded and continue their journey on land —hence the growth of the twin towns of Léopoldville (now called Kinshasa) and Brazzaville, one on either side of the river.

Navigation cannot always be carried on under similar conditions. The boats of the quiet river Saône would be of little use on the Rhône,

[1] A. R. Hamidé, *La ville d'Alep*, Paris (thesis), 1959.
[2] R. J. H. Church, 'West African Urbanisation', *Geog. Rev.*, 1959.

so goods have to be transshipped at Lyons. In the same way, Strasbourg marks the junction of the little Alsatian river Ill and the Rhine. Waterways have their meeting points too, in the same way as roads: crossroads of natural routes, like the junction of the Seine and the Oise which gave birth to Conflans-Sainte-Honorine, or Koblenz at the confluence of the Moselle and the Rhine, or artificial junctions like the one at Saint-Jean-de-Losne, on the Saône, where the Burgundy canal and the Rhône–Rhine canal meet. These are the places where the wandering boatmen fix their roots and, particularly, where they have their schools.

Railway traffic. In the old countries, railway traffic in the early days followed the same directions as the road traffic. However, its needs were different. The distance covered by a train is far greater than that of a horse-drawn vehicle, but it needs more than just an inn for a stopping place. The early steam locomotive could do little more than 125 miles before taking on water and coal, and replacing the driver and fireman. It was quicker in those days to change engines. This could be done in intermediate towns, but these were not always situated in the most convenient places. A station midway between Paris and Dijon was required; it was also desirable that this station should be on the threshold of Burgundy, to allow the engine to tackle the rising gradients with a good head of steam. This was the origin of the station of Laroche which turned the village of Migennes into a town with a population of 5,226. Such relay towns were founded essentially as engine depots and residences for the engineers and their families; repair workshops and breakdown equipment were often added. Railway towns thus came into being, the product of a certain stage in technical progress and liable to be ruined by further technological advances. Electrification makes intermediate stopping places unnecessary. Laroche-Migennes is now no more than a junction with express trains running through it.

It sometimes happened too that natural conditions made it necessary to leave large towns off the main traffic routes. This happened when a line running along the coast had to cross estuaries upstream, by-passing the coastal towns. Boden (pop. 13,292) in the north of Sweden on the Luleälv estuary is the branch station for the port of Luleå; Les Aubrais (pop. 8,521) and Saint-Pierre-des-Corps (pop. 10,656) are really only the stations for Orléans and Tours.

Other railway towns are attached to freight marshalling yards. Where trade routes meet it is necessary to marshal wagons coming from different directions, or to break up a train of wagons according to their several destinations. Marshalling yards are essential links in the circulation of goods, and it is a pity that they are not always given sufficient importance on our maps. Such stations are placed all along the main lines for the collection of the products of subsidary lines. The junction of Gevrey-Chambertin, the largest and most modern in France with its electronic equipment, serves the lines to Paris, Lyons, the Massif Central, the Jura and the North. Marshalling yards are especially numerous near important traffic centres, for the sorting of the goods being sent to them. Grouped around Paris are Vaires, Le Bourget, Trappes, Juvisy and Villeneuve-Saint-Georges. Some of these have a specialist function, dealing especially with express freight.[1]

A fork in the line would not be enough on its own to bring about the founding of a town, but these are places which present themselves as natural relay stations and marshalling yards, and can give rise to towns which are entirely dependent. Ambérieu-en-Bugey, in the centre of the 'Ambérieu star' of lines radiating to Bourg, Lyons, Geneva and Chambéry, has a population of 7,159. A community of 5,435 people has established itself at Capdenac-gare, as opposed to Capdenac, a village of 887 inhabitants two kilometres away.

If railway development has been content to make use of the agglomerations already existing in the old, densely populated countries, and has transformed their villages into towns, it has also organised the urban network to its own benefit in the countries it has conquered. Vidal de la Blache wrote that North America was conquered by the railways. But the existence of a railway necessitates stations, which are the perpetrators and beneficiaries of this conquest. Most of the cities of the western United States and Canada were originally railway stations.

In the same way, the Trans-Siberian Railway produced towns like Ufa (pop. 546,000) and Chelyabinsk (688,000) on either side of the Urals. Harbin (today known as Pin Kiang) was originally only a station where the Trans-Siberian Railway branched to Vladivostok or Mukden, and now has a population of 1,163,000. Development here has been speeded up by the overpopulation of Asia.

Mention should also be made of frontier stations where national boundaries are crossed. Workers and engines have to be changed, and sometimes the difference in the width of the track necessitates also the changing of the trucks or carriages. Alongside these depots of plant and machinery are situated the inevitable customs houses. Examples along the French frontier are Hendaye, Jeumont, Ventimiglia and Modane.

Railway traffic has not only created towns it has also involved modification of the structure of ancient towns. The route taken by a

[1] G. Mathieu and D. Lefèvre, 'Deux gares de triage de la région parisienne', *Inf. Géog.*, 1954. See also S. H. Beaver, 'The railways of great cities', *Geography*, 1937.

railway had to respect agglomerations and skirt round them. It is not unusual for a station to be a mile or two away from the centre of an old town, and to have formed a new urban nucleus, provided with hotels and restaurants; between the two a new district is created, whose high buildings and rectangular street patterns express its youth in contrast with the twisting streets of the original town. Railways incorporated into a town can become a nuisance. They often retard the expansion of a town which hesitates to cross them, and roads under the viaducts are more like tunnels, like the 'vaults' of Lyons. On the other hand, the distance of stations like those of Les Aubrais and Saint-Pierre-des-Corps is the main reason for the urban expansion of Orléans and Tours.

Motor traffic. This has not had the same consequences. For one thing, it is more flexible, and, being modelled on the old road traffic, it makes use of the old towns: the coach prepared the way for the motor-car. Moreover, its widespread expansion being largely very recent, it has not yet had time to create real towns. In pioneer territories the road service station may already have gathered a small settlement around it, but in the old countries where towns are not very far apart, the transformation of some towns by motor traffic is the most noticeable thing. Small towns have become stopping places for the roadside meal so necessary on a journey of several hundred miles, and have drawn new life from it, like Saulieu between Paris and Lyons, and wayside stopping places for tourists and long-distance lorry-drivers who are reliable customers.

The intense motor traffic of America, which does not have the advantage of so many resting places, has given rise to motels, which are not exactly towns, but which have, in the West at any rate, as many rooms as any town hotel.

In less highly developed countries, with the intensification of road traffic, little roadside stopping places are springing up at crossroads. Midway between Zagreb and Mostar, in the heart of the country, one suddenly comes upon a little group of cafés and restaurants, complete with shops containing embroidery and wood sculpture for the tourists. A town will probably grow up there in a few years.

Commercial centres. All methods of travelling, together with their urban corollaries, are designed to encourage trade over the greatest possible distances. But this trade has itself created a certain number of commercial centres. As soon as conditions made long-distance travelling possible, and particularly when its security was assured, such trading cities appeared. They were served by caravans, like Palmyra, the caravan city famed for the incense trade. The routes from Egypt, Arabia and Mesopotamia ended in Palmyra's vast, colonnaded avenue. The caravan leaders put down roots there, as it were, and built the palaces and tombs which are today being revealed by excavation. It seems that in the third century Palmyra tried to create for herself a kind of caravan empire, whose capital she would be.[1] Exchange markets were thus established on the edges of the deserts. On both sides of the Sahara towns were founded on the salt and date trades; on the edge of the Syrian desert there was Damascus. Not all these towns survived; the edges of deserts are troubled places. Security allowed their development and insecurity ruined them. Palmyra was destroyed by the Arabs in 744 and never reappeared.

Fairs. In the old days large-scale trade was principally carried on at fairs. It was usually spasmodic, as only on exceptional occasions were enough buyers and sellers assembled to make it possible. Religious festivals sometimes provided opportunities. The great Greek games were occasions for immense fairs. Later the same conditions arose under the Christian Church: the Lendit of Saint-Denis was originally a religious festival in honour of St Denis; the festival of a patron saint, *feria*, is accompanied by a commercial reunion (in French, *la foire*). In Germany the fair (*die Messe*) derives its name from its religious origins.

These fairs[2] temporarily transformed small local markets into important trading centres. Their position was decided by the meeting points of the trade routes. The district of Champagne had them, as did Flanders.

The Leipzig fair dates from the thirteenth century. The fairs of Lyons, permission for which was granted by the Dauphin in 1420, and which were held four times a year from 1463, rivalled those of Geneva. The last great fairs, in the nineteenth century, were those of Russia; the famous fairs of Nijni Novgorod attracted 200,000 people, and to house them a complete temporary town was built of wood on the other bank of the Oka. This marginal town made the fortune of the permanent one. The inhabitants spent the whole year in preparation, as the inhabitants of tourist centres do now, awaiting the season.

This sort of trade was indispensable in times of trouble, and in uncivilised countries where the freedom of trade could not be relied upon for long periods. Merchants gathered at Novgorod armed with a safe-conduct pass that the warring factors respected. When permanent relations became safer and easier, this massive movement of

[1] M. Rostovtzeff, *Caravan Cities*, Oxford, 1932.
[2] A. Allix, 'Geography of the fairs, illustrated by Old World examples', *Geog. Rev.*, 1922.

men and merchandise became less common. Trade was, so to speak, shared among the towns, all of which hoped to profit from these distant connexions. Speciality fairs and sample fairs are now only one form of trade in certain large towns.

Important commercial towns. Important commercial towns have thus multiplied, adapting themselves to all sorts of conditions, and each one made the best of its geographical situation. This geographical situation was sometimes ordained by nature. The great rift valleys and river valleys have always been transport routes: roads, railways and waterways have made use of them. Towns were established at the points where subsidiary routes join the main routes, after crossing the high ground.

The long Rhenish corridor, enclosed by mountains for over 250 miles of its length, has always been a magnificent route. Along this route, Mulhouse and Basle are situated at the mouth of the Burgundian Gate, Strasbourg opposite the Saverne Gap, Ludwigshafen and Mannheim at the confluence of the Neckar, Mainz at the confluence of the Main, and Koblenz at the confluence of the Moselle. On the Saône-Rhône route, between Alpine and Hercynian Europe, Lyons is situated at the meeting point with the Rhône.

On the Danube route, which crosses Europe from west to east, Vienna stands opposite the mouth of the Moravian Gate.

Even if valleys do not offer long passageways, the confluences of navigable rivers often provide sites for towns, like St Louis at the confluence of the Mississippi and the Missouri, or Khartoum where the White and Blue Niles join.

In the preceding examples the crossroads are closely confined, but there are geographical situations less strictly planned by nature in the middle of basins or plains that are surrounded on all sides. The Paris basin is remarkable for the way in which everything gravitates almost inevitably towards its centre, helped by the convergence of the waterways. Framed by mountains, the Plateau of Bohemia, the Plain of Hungary and the Plain of Romania call for a centre for trade without there being any very definite place indicated. The same applies to Chung-King in the middle of the Red Basin and Baghdad in the centre of Mesopotamia.

Transport routes are not always dictated by natural conditions. Man has gradually overcome these restrictions, or rather, has controlled and combined them, thereby creating new restrictions.

The Great Lakes of America make a marvellous waterway, the combination of which with railways to the ports of New York and Philadelphia has created the transshipment function of Cleveland and Buffalo.

Some commercial places owe their development to new manmade routes. It has been shown how Hanover and Brunswick, south of Hamburg, presented similar conditions on the route skirting the mountainous region of middle Germany at the entrance to the route which goes round the Harz mountains to the west. Brunswick used to have a slight advantage, by reason of its industrial past, which at the beginning of the nineteenth century, guaranteed it a population of 35,000, while Hanover had only 26,000 inhabitants. But when the railways were built, Hanover was chosen for the crossing point of the north–south line, following the valley of the Leine, with the great Berlin–Dortmund line. The town of Hanover did not depend on it, but this unsolicited gift allowed it to develop to such an extent that it now has a population of 542,000 as compared to the 246,000 of Brunswick.[1]

Similar examples can be found in France where Le Mans has beaten Alençon, and Vesoul has become more important than Gray, because the construction of main lines has favoured them. Dijon had a population of only 21,000 in 1801, tripling this figure in 1891, thanks to the establishment of a railway junction, and it was the railways, descending from the Alpine passes, which made the fortune of Milan.

Chicago is fortunate in its situation in the centre of the United States and its position on the Great Lakes, but apart from this, the development of road transport must have given it an enormous advantage over the other small towns which, although in similar situations did not have the advantages of direct transport.[2]

On the other hand, political conditions can wipe out all the favours of nature. Belgrade, relegated to the frontier of the Turkish domain, was for a long time unable to take advantage of the confluence of the Danube and the Save.

Fixed by relief or the hydrographic network, or even manmade, all these localities favourable to trade were after all only opportunities offered in certain places by facilities for transport. It was still necessary for this transport to lend itself to trade, and to link rich and varied regions. The agricultural regions around Toulouse were too alike, and further away the massive mountainous regions to the north or the Pyrenean wall to the south presented few openings. Conversely, Lyons dominated the rich plains of the Saône, the highly civilised mountains of the Jura, and the hills of Lyonnais and Beaujolais. It was connected with Provence and the Paris Basin, and linked by means of the valleys of the Alps and the Jura with Switzerland and Italy.

[1] G. Schwarz, *Allgemeine Siedlungsgeographie*, Berlin, 1959, p. 350.
[2] J. D. Fellmann, *Truck Transportation Patterns of Chicago*, Chicago, 1950.

Finally, it must be realised that the presence or absence of all the advantages listed does not alone explain the success or failure of different trading centres. Human endeavour enables man to escape from deterministic influences, and this has become more and more true as the technicalities of trade have become more complicated. Money was the most fluid of all merchandise and paper currency made light of mountain ranges. Financial centres probably developed on the very spot where exchanges were made, but man's initiative also had much to do with it, and without Jacques Coeur, Bourges would not have become a great financial centre in the fifteenth century.

This is where the commercial function becomes confused by the intervention of other urban functions. Industries are established in big commercial towns, use the transport facilities, and take advantage of the neighbourhood of a big consumers' market and the funds accumulated by commerce. Political capitals often take charge of business, thereby attracting the banks.

Of all commercial functions, the financial function seems to be the one which most readily escapes an exact localisation. On the pioneer fringes, as Jean Labasse has stressed, air transport may allow a financial centre to grow before a commercial centre, and yet there is scarcely a function which seems, from some points of view, more stable or more faithful to the traditions imposed by the very complexity of the business. Thus Frankfurt-on-Main has remained the great centre for German finance in spite of the obstacles set up on several occasions by the Zollverein and National Socialism.[1]

In large cities it is often difficult to separate the commercial function from the multiple rôle of the town. But such cities, based upon commerce, are frequently sea ports, where transport by sea has favoured relations with the whole globe.

SEA PORTS[2]

Trade is based on traffic movement, and no traffic medium has played a more important rôle than the sea. As a result, trading centres have multiplied along sea routes.

In the past, the ship was the only convenient method of transporting heavy goods. The sea was always available, and the vessels were driven by the wind. Overland travel has, of course, progressed enormously. Roads have increased in number, and new means of traction have been discovered—steam, oil and electricity—which have

---

[1] H. Lehmann, 'Frankfurt a/Main', *Die Erde*, 1954. Cf. K. Inno, *Tartu as a financial centre*, Heidelberg, 1948.
[2] We shall not deal here with the layout and equipment of ports, but only with their repercussions on urban life.

permitted the handling of heavy transport over long distances, and which beat all other forms of transport for speed. But navigation of the seas responded to these advances with constant progress, making it possible to sail by night or far from the coast, and with increased tonnage of ships demanding modern methods of propulsion. Thus navigation not only represents the principal practical means of transporting goods between continents, aviation being reserved mainly for passengers, but it remains the most convenient method of transporting large loads, even inside the continental blocs. It should not surprise us, therefore, that the two biggest towns in the world—London and New York—are ports, and that in all countries ports occupy a privileged place.

The location of ports. Among the towns founded on maritime trade there are some which owe their very existence to methods of transport. Navigation over long distances required ports of call, for the taking on of drinking water, fresh vegetables and fuel. The stowing of provisions on board reduced useful space available for merchandise, so provisioning points had to be fairly numerous and all maritime trade routes abound with such ports of call. Dakar was a port of call on the route to South America, as was Cape Town on that to the East Indies.

These ports of call were numerous on the Suez route, which was one of the busiest, and some were specialised. Oran was equipped for the rapid reloading of coal for which the crowded port of Algiers was unsuitable. In the Pacific, Honolulu is at the meeting-point of the N.W.–S.E. and S.W.–N.E. routes. These ports of call are probably less valuable today. Vegetables can be preserved in refrigerated holds; sea water can be distilled, and oil takes up less room than coal. However, they are still necessary. Dakar refuels with oil at the rate of 500 tons an hour, and annually supplies 500,000 tons of oil products, 250,000 tons of coal and 200,000 tons of fresh water. Though these ports retain their usefulness, and will continue to do so, their function will tend more and more to become confused with the commercial exchange function, whether because of the development of their hinterland (as is the case of Dakar), or because this is the part played increasingly by big trading ports or their annexes.

In fact it is the commercial exchange function which decides the site of most ports. But the conditions for the localisation of these ports were quite different from those of continental markets. The site, which generally plays a subordinate part in the exercise of urban functions, was once of paramount importance, depending on the existence of a stretch of water suitable for harbouring ships, and on the approaches and ease of movement. The presence of a roadstead allows ships to shelter securely. Brest, with over 7,000 acres of anchorage space is privileged in this respect. Incidentally, the site is not only determined by the lie of the land. It is also affected by atmospheric conditions; fogs can make the approaches dangerous, necessitating tedious piloting, and this has hampered the use of Brest as a commercial port. A good port should likewise be protected from ocean winds. Algiers established its harbour in the shelter of the El Djeznaïr islands; and in monsoon countries, two complementary ports have sometimes been built, one on either side of the same promontory. But there are still more complex requirements: the site must offer easy access to shipping. Rocky coastlines caused by the recent submersion of immature relief may provide adequate depths, but they are often littered with dangerous reefs. Low-lying coastlines are only accessible if they offer a channel: large ships can only reach New York by means of the submerged valley of the Hudson which in former times had to be discovered by sounding. And on low-lying coastlines surf sometimes presents an obstacle to the passage of small vessels: for a long time the coast of West Africa was thus cut off. The alluvium of rivers can likewise be an obstacle; the port of the Rhône valley had to be sited away from the delta; and the problem of directing the huge alluvial load, which the Yang-Ting-Ho has eroded from the up-river loess plains, away from Tientsin, is a serious one.[1]

Nowadays the site is less important. Engineering has triumphed over many obstacles. Jetties projecting beyond the port permit easy entry and a longer stay; many ports owe little today to the original conditions, and Marseilles has transferred to entirely manmade docks all the activities which were formerly concentrated round the smelly waters of the old port. Moreover, some ports have been created from scratch in places where there were no natural facilities at all. This was the case with Cochin on the Malabar coast of India, and with Gdynia where the Poles dug their national port out of the fields after 1919.

The original site may appear to be an anachronism, but commercial and financial activities have grouped themselves around it and a town has grown up. The human capital it has bequeathed is worth more than the physical advantages.

And yet, whatever the value of the site, large ports do not develop unless their geographical situation is favourable; and it is the latter which has given rise to great urban concentrations. This geographical situation must fulfil two requirements, from the point of view of the sea and from that of the land.

On the sea side the requirements are diverse, unlike traffic on land which is confined to definite routes. But seas are not all equally

[1] M. Hitch, 'The Port of Tien Tsin', *Geog. Rev.*, 1935.

favourable. Tidal seas seemed dangerous for a long time and people seldom dared to sail them. Later they learned to conquer them, and even to take advantage of the tides. The periodic influx of water allowed the arrival of larger ships, and this advantage is especially marked in ports having several high tides or in those where the sea is slack for a long time (Le Havre, Southampton). Wet docks, closed by lock gates, increase these advantages. Some seas, on the other hand, are frozen in winter and navigation is interrupted; this is the case with the Baltic Sea, and especially the Gulfs of Bothnia and Finland. Leningrad is blocked in this way for five months.

Human conditions must be added to these physical conditions. In the first place, the necessity for security—in the Middle Ages pirates infested the Mediterranean; and secondly, commercial relationships. Everything depends upon the part played by the seas along the coasts of which the ports are situated. For a long time, the Atlantic Ocean offered no prospects. Astrakhan at the mouth of the Volga, one of the finest rivers in the world, cannot develop because the Caspian is an inland sea.

There are several types of maritime relations. Some seas link two areas of similar civilisation and favour reciprocal exchanges, as does the North Atlantic, between Europe and North America. Other routes, alternatively, connect industrial areas and suppliers of raw materials; such as the routes from Europe to southern and eastern Asia and to Africa. In this way seas have favoured colonial conquests and ports were born of these conquests on the shores of the continents; the majority of the great cities of America were founded thus.

Every sea has its privileged sites. Ships passing round peninsulas sail close inshore, and on the ends of these peninsulas are the sites of ports, like Cape Town and Singapore. In the same way, navigation is intensified in straits; these play a dual rôle, both longitudinal and transversal. Navigation routes are obliged to converge to pass through certain passages such as the Straits of Dover (and indeed the whole English Channel), the Straits of Gibraltar, the Red Sea, the Sound and the Straits of Malacca. The Suez and Panama canals are artificial straits. But these same straits are also frequented by ships joining one coast to the other, giving rise to ports like Messina and Reggio, Copenhagen and Malmö, Dover and Calais.

The advantages of site and situation can sometimes be found together. Istanbul benefits from its situation on the Bosphorus and from its site on the Golden Horn, as a result of which its urban life has floundered throughout its history.

Maritime advantages have sometimes been enough to bring about the foundation of a port with no continental connexions. There were thus towns which looked only to the sea and which were sometimes even situated on islands: Tyre played this part in the time of the Phoenicians; the empire of Athens was essentially maritime; in the Middle Ages the Hanseatic towns like Visby were the most important, and later Genoa and Venice. The town depended for its livelihood on the port and often fed it with manufactured goods for export—purple dyes from Tyre, and luxury goods from Venice. But these ports, poised as it were on the coastal rim and depending solely on the sea, have maintained their importance only where they have succeeded in taking root on the continent: the Thalassocracies were ruined by the development of continental powers.

Ports were also increasingly subject to conditions imposed by relationships with the land. Ports of call that were required by the techniques of navigation were only as important as their hinterland allowed them to be. Gibraltar, in a splendid position, did not become a big trading port, neither did Tangier. The hinterland sometimes consists mainly of a large town: Lima for Callao, Santiago for Valparaiso, and Constantine for Philippeville. But the size of a port is in proportion to the economic power which supplies it and which provides its outlet. Canals and railways have broadened this provisioning area; Genoa has become the port for the western Po valley, and Split for the whole of Yugoslavia. This provisioning area depends on political frontiers. In fact every country strives to direct its trade towards its own ports. France has favoured Dunkirk at the expense of Antwerp, while Trieste has been ruined by the moving of the Austrian frontier.

Moreover, the area influenced by a port and by a town is not necessarily the same. The grain port of Constanţa in Romania extends its influence to all the cereal-growing regions of Romania, while the influence of the town of Constanţa is much less widespread. The *hinterland* of a port and the *umland* of a town are not identical.[1]

CLASSIFICATION OF PORTS[2]

A preliminary classification of ports can be attempted by taking into account the situation and the site. There are sea front ports and ports on estuaries or inlets.

Estuary ports seem to be the most fortunate. They offer the advantage of a safe sheltered anchorage; the former valley, now submerged, usually guarantees a sufficient depth, sometimes as much as 30 to 40 feet, and thus allows the harbouring of large modern liners. Dredging helps to maintain or increase this depth. The penetration

---

[1] I. Sandru, 'Quelques considérations géographiques sur le développement des ports roumains', *Ann. scientifiques de l'université de Iasi*, 1958.
[2] F. W. Morgan, *Ports and Harbours*, London, 1952.

of the land by the estuary (a distance of more than 60 miles in the case of London or Hamburg) prolongs the advantages of maritime transport and facilitates the distribution of goods. The river itself takes over from the estuary after transshipment on to barges. And parallel with the river navigation, the valley favours the construction of roads and busy railways which may form vital links, as in the case of New York and the Hudson valley. The world's biggest ports are estuary ports: London, New York, Hamburg, Antwerp, Rotterdam.

On each of these estuaries there is a particularly favourable point, the highest point reached by the tide. This is the normal situation for estuary towns. It still has the advantage of the high tide, and is the point at which transshipment on to river boats takes place; here too, at the head of the estuary, the river can be most easily crossed and the lowest bridge is built.

However, these estuaries are not without their disadvantages; their navigation can be difficult and dangerous, in spite of the presence of buoys, and requires a pilot service; big ships cannot always enter them; they cannot act as ports of call for shipping lines that are not based there. There is a constant struggle against silting, and dredgers form part of the landscape of all these ports. The town's development is impeded downstream because of the width of the estuary and the difficulty of joining the two banks by means of bridges which would have to be high enough to allow the passage of shipping: the problem was solved only recently at Gothenburg and at Tancarville on the Seine. Lastly, the advantages of an estuary have diminished since the carriage of goods by rail became possible.

Opposed to estuary ports are sea front ports. These avoid the silting-up of estuaries, but are exposed to the storms and winds of the open sea, and have had to contrive artificial harbours with breakwaters. Every creek can be the origin of a port of this type. On the coast of Brittany there is a whole string of these little ports which go in for coasting trade, collecting the products of the surrounding country, sending them to England, and receiving coal or oil. But here again ports only develop to the extent permitted by intercourse with their hinterland. The indentations of rocky coastlines are easily transformed into harbours, but are often closed on the landward side.

The outlets of large valleys are thus particularly favourable. Marseilles is a sea front port at the mouth of the Rhône valley. Elsewhere, round an estuary, ports develop to take advantage of the stream of travellers and goods while avoiding the inconvenience of passage up the estuary, and profiting too from the proximity of the big town founded on the estuary. These are outports.

Such outports can be seen near every estuary: Cuxhaven, Bremerhaven, Tilbury. Relations between ports and outports are very varied and bear witness to the constant rivalry between them. It can happen that the outport completely supplants the port: in Denmark Esbjerg, founded in 1869, has replaced Ribe which was a large port at the end of the Nipsaa estuary, but which could not be reached by modern shipping. Saint-Nazaire, built with capital supplied by Nantes, is a serious rival to that city. The function can also divide into two. Le Havre, though not intended for that purpose, has become the outport for Rouen, leaving the heavy goods to Rouen, and receiving its own by rail. Some ports have succeeded, through hard work, in averting the rivalry of the outport. Glasgow, which seemed condemned at one time, made a vigorous recovery in the nineteenth century.[1] London's leadership is not threatened by any of the downstream outports, and Cuxhaven, apart from being a port of call for Hamburg, is hardly more than a fishing port and a place to visit at weekends. On the Gironde estuary, Pauillac, downstream from Bordeaux, is still active, but the outport of Le Verdon has not been rebuilt since the Second World War.

However, every technical advance puts the acquired position to the test. Through the increased tonnage of tankers and the development of pipelines, Le Havre has won the job of supplying fuel, at the expense of Rouen. On the other hand, steam allowed access to ports which were only reached with difficulty in the days of sailing ships, like Oslo at the end of its long, narrow channel.

Here again, a situation acquired and enriched through the centuries can often maintain a locality which no longer answers present day conditions. The port fixed the position of the town, but now the town maintains the port.

It has been said that in Europe, only Marseilles, Le Havre and Rotterdam are adapted to the transport revolution and are likely to become 'euroports'. None the less, traditional ports offer resistance: not all heavy goods are carried in huge ships. Then again, valuable cargoes develop, such as fruit and manufactured goods. So, in spite of the number of deep water harbours, traditional ports continue to increase their trade.[2]

Lastly there is a third category of seaports unknown to our ancestors and which tend to multiply. These are manmade inland ports, owing their existence to the digging of huge canals. In this way interior estuaries are contrived to serve the great industrial centres. Ghent, formerly a river port on the Schelt, has become a seaport. Manchester

---

[1] Its outport, Port Glasgow, was constructed in 1668; but nineteenth century dredging has made the lower Clyde 'as much an artificial waterway as the Suez Canal' (*Scottish Geog. Mag.*, 1921).

[2] M. Le Lannou, 'Les ports et la révolution des transports de mer', *Rev. de Géog. de Lyon*, 1959.

is joined by a ship canal some 30 miles long to Liverpool, which has become to some extent its outport. The finest achievement of this sort is the lengthening of the St Lawrence estuary as far as the Great Lakes; these have now become a sort of Mediterranean, but 1,200 miles inside the continent and with nothing to fear from sea tides, though unfortunately liable to a four-months' winter freeze.

These improvements were constructed—this was the very condition that made them worth while—to serve already existing centres and to contribute to their development. Nonetheless, new towns establish themselves along these new seaways: amphibious ports of which it is hard to say whether they are sea ports or not, but which have all the maritime advantages. This is the case of the small industrial towns that have sprung up along the canal between Ghent and the sea; while from the top of the viaduct whose arches span the Kiel canal, many small groups of quite new houses can be seen.

Ports can also be classified according to their specialisation, which is usually dictated by the hinterland. A certain port may serve a coalfield, like Cardiff or Hampton Roads, and another an oilfield, like the ports of Sabina in Texas, and Kuwait or Abadan in Asia; in some cases it is a question of exporting the products of coffee plantations, like Santos, or of soya like Dairen, or of wood from the forests, like Libreville. This specialisation is natural in countries which produce raw materials.

It is convenient for ports to be equipped for specialisation in the loading and unloading of certain goods. The machines required are not always the same: steam-driven grabs and conveyor belts are used for cumbersome materials like ores, suction tubes for grain, and pipelines for oil. Some goods require delicate handling: bananas are unloaded by means of insulated conveyors and then placed in special warehouses where they finish ripening. Thus at the consumer end also there is sometimes specialisation: Dieppe has become a banana port, and so has Avonmouth.

This specialisation is not without disadvantages. The specialist port runs the risk of having only a seasonal trade when dealing with harvested produce. There is no cargo for the return trip. There is therefore a tendency towards diversification permitting better use to be made of the port installations. Several seasonal products can be combined to complete the load and to take advantage of the passage of regular shipping lines. In this way the ground-nut trade left Kaolack for Dakar. It is also easier to find the necessary financial facilities in the neighbourhood of large ports.

To do this, it is not a matter of abandoning the advantages of specialisation, which is done in the interior of a port and sometimes at an annexe to it. Every large port is subdivided into a number of docks, each one of which has its particular occupation; in this way individual docks and warehouses can specialise. All along the Thames are docks which together make up the Port of London. Oil docks are usually pushed to the exterior as annexes. And these annexes can be the starting point for a separate agglomeration; around the lagoon of Berre the refineries and petrochemical industries, based on the oil annexe of Marseilles, have developed several centres: Port de Bouc, Martigues and Berre.

Last but not least among these specialisations, is the passenger port. These ports have pressing requirements, chief of which is that of speed. They need rapid links with the interior, provided by railways and even airlines. They have to jut out to sea as far as possible so as to get the maximum reduction in the length of the crossing which is made at a slower speed, and the changing from train to boat must be as rapid and comfortable as possible. Passenger ports can usually be placed next to freight ports, as at Le Havre, but in many cases they command most of the port's activities, as at Calais and Southampton.

Fishing ports should also be mentioned. Strictly speaking, these are not commercial ports. They carry out a function which derives from the primary sector; it is a case of a harvest ready for market. But these fishing ports have developed into towns and have given rise to industry and commerce: industries such as the making of fishing equipment, the making of ice for the preservation of fish on board, and the conversion of the catch into fillets, preserves, fish-meal and oil. Trade is in fresh fish which must be delivered as quickly as possible, and in by-products. At the great fishing port of Boulogne, 2,381 people are employed in the fishing industry and 3,800 in the salting and preserving industries.[1]

So fishing ports are increasingly becoming centres where trawlers deposit tons of fish for processing and delivery, specialising according to the seas they fish and their situation in relation to consumer centres. Tunny, sardine and lobster have their respective ports; the largest are those dealing in cod and herring. But ports may also specialise in the way they use their catch. Icelandic and Norwegian ports cannot easily dispose of fresh fish, lacking rapid links with the large consumer centres, so they process it on the spot. This is also the case with some badly-served French ports like Fécamp. Others have specialised in the delivery of fresh fish, like Boulogne and Lowestoft.

A fishing port is cumbersome; it must allow for rapid unloading, and the industrial installations and railway stations for dispatching the fish monopolise the space round the docks. Thus it is not always easy to combine a fishing port with a commercial port, unless, as at Boulogne, special docks are reserved for fishing vessels. Usually a

[1] L. V. Vasseur, 'Boulogne-sur-mer: aspects économiques,' *Inf. Géog.*, 1960.

fishing port does not permit other activities. Lorient, former naval port, which, at the instigation of the railway companies, took to fishing, is an example of this.

Often ports can be classified according to their relationships with the hinterland.[1] This may be defined as the area which receives the goods arriving at the port and which sends its products there for delivery overseas. There is an academic distinction between regional, national and international ports, but it is far from satisfactory. There are ports whose hinterland is very restricted. The port of Nice does little more than serve the agglomeration of Nice itself. But as soon as one tries to go into detail, it is hard to make a distinction. Transport facilities extend the limits of the region. Gothenburg, regional port for west Sweden, becomes a national port in winter when the port of Stockholm is blocked by ice. Le Havre and Rouen are the regional ports for the Paris Basin, but large quantities of goods arriving at Le Havre, like cotton and coffee, are distributed throughout France. A national port often extends its influence beyond national boundaries, and the port of Marseilles also serves Switzerland. This aspect is even more important now that the Common Market and various free trade areas are causing national frontiers to lose some of their significance.

THE TOWNS OF PORTS

Ports therefore appear to be places where urban crystallisation takes place. Around the port a town grows up, making its living from the port's activities and gravitating round it. A port today is a complex organism comprising basins, docks and adjacent buildings. Commercial life begins on the quays and extends to offices, delivery warehouses and banks.

If we wished to make an inventory, we should find first of all the services of the port itself, offices, warehouses and repair shops. In spite of mechanisation, the dockers represent a large body (4,000 in Le Havre) some members of which are permanent, but many of whom are hired as casual labour. To these services should be added the piloting and towing services necessary to every port, the shipping agents (representing nearly 3,000 people in Le Havre),[2] and the insurance offices belonging to numerous companies; the official services (customs with their officials, commercial courts and consulates); the banks and the agents of large firms. Hotels, restaurants and cafés crowd round the docks, as well as shops supplying provisions for the ships. The very existence of the port may provide employment

[1] A. J. Sargent, *Seaports and hinterlands*, London, 2nd edn., 1930.
[2] S. Raynaud, *Le port du Havre*, Paris (thesis), 1958.

for tens of thousands of people. In this respect, all ports are somewhat alike.

André Vigarié stresses the international character of the sea: 'There is a part of the human environment whose economic equipment and work is common to all great oceanic cities: the same large-scale shipping agencies, the same methods of transit and consignment, the same work for the stevedore labour force, the same activity of dockers protected by at least partially similar institutions . . .'.[1]

Moreover, ports attract industry. Sometimes ships are built there. And there are always huge quantities of raw materials. Sources of energy in the form of coal and oil are cheap, and this favours the processing industries. The commercial complex has its counterpart in an industrial complex.

Answering the needs of so many and such different activities, port towns could not all be built to one plan. Traces of the original port around which the town sprang up can often be seen. Bordeaux extends in a semicircle around the loop of the Gironde; the Cannebière near the old port is still the centre of life in Marseilles, and the Boieldieu bridge still marks the point at which sea and river navigation meet, which was the origin of Rouen. But a town must adapt itself to the port, or rather to its bulky and constantly changing structures (Fig. 9).

It sometimes happens that port and town overlap, each hindering the development of the other. This is the case in Shanghai where the town encloses the port, blocking its development. It could have happened at Marseilles if the port had not abandoned its former site to extend westward. Conversely, it is not desirable that the town should be cut off from the sea by the installations of the port: this has happened at Naples where the town forms an amphitheatre around the port, and at Liverpool–Birkenhead where the twin towns have given up to the port both banks of the Mersey estuary. The best disposition is one which allows both port and town to develop freely, each in its own way. At Antwerp the town extends upstream of the estuary while the port develops downstream; at Hamburg the docks are constructed on the left bank while the town is situated around the old harbour on the right bank.

At Szczecin the whole reconstruction plan has had the effect of joining the town to the port. On the one hand, in fact, the port (and its outport Swinioujscie formerly Swinemünde) is destined to become a very large port at the mouth of the Oder, if not the principal port of Poland; on the other hand, the present frontier between Poland and Germany runs along the river, restricting relations with the West: so

[1] 'Le fonctionnement d'un port néo-zélandais: Wellington', *Cahiers de sociologie économique*, no. 6, 1962.

144 URBAN FUNCTIONS

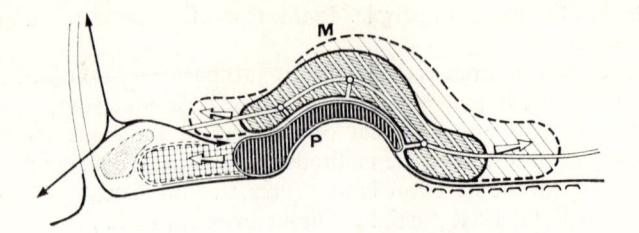

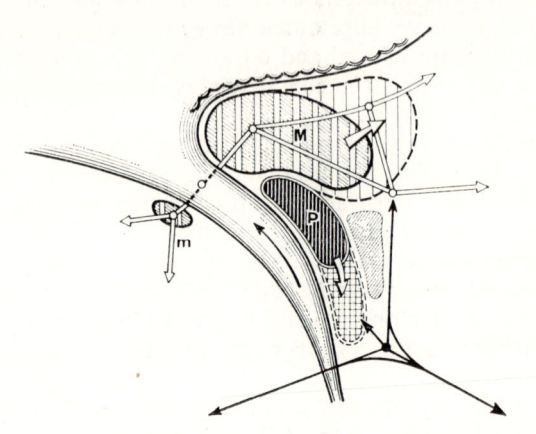

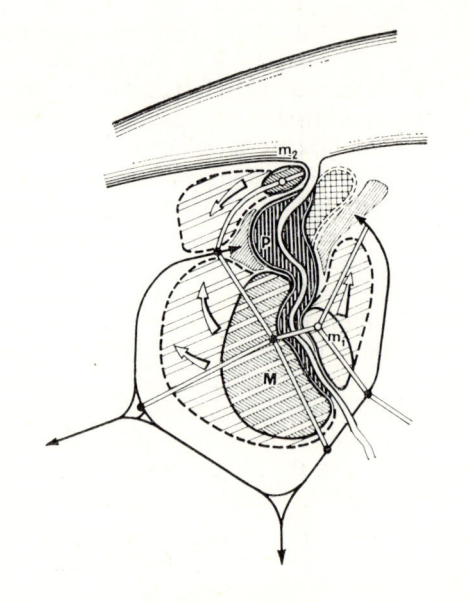

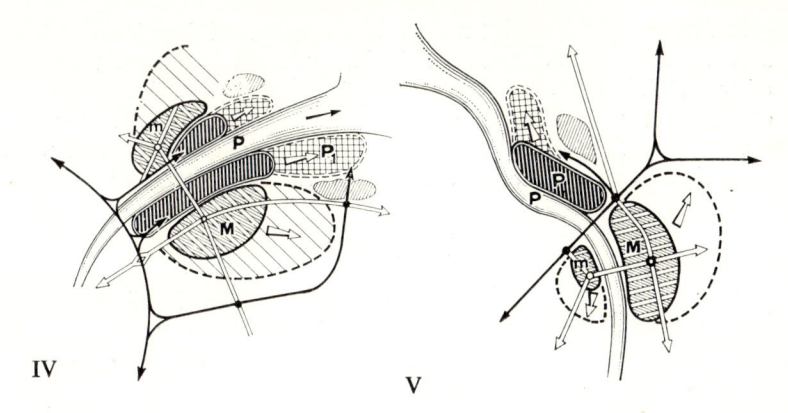

Fig. 9. Various types of development of port towns
According to Piotr Zaremba, *Les principes du développement des villes portuaires*, Polish Academy of Sciences, Paris Science Centre, no. 32.

I. Plan of a port town situated on a bay. M=town; P=port. The port pushes the town away from the bay. The town tends to develop along the shore.

II. Plan of a port town at the mouth of a river. M=town; P=port; m=suburb on opposite bank. The extension of the town towards the sea does not impede the extension of the port along the river.

III. Plan of a port town situated on both banks of a river near its mouth. M=town; P=port; m_1=suburb on opposite bank; m_2=outport district. This is a port town typical of the Far East where the docks and the town intermingle.

IV. Plan of a town and port situated on both sides of a river. M=town; P=port; m=suburb on other bank; P_1=industry. The town has no direct access to the river.

V. Plan of a port town situated on a river, the port being downstream of the town. M=town; P=port; m=suburb on other bank. The port develops in a downstream direction, and the town in an upstream direction.

See also, on this topic, J. H. Bird, *The major seaports of the United Kingdom*, London, 1963.

the town, which up to 1945 was spreading westward, is now taking a northerly direction along the banks of the Oder where an anticipated population of 500,000 will be housed.[1]

[1] Piotr Zaremba, *The Spatial Development of Szczecin in 1945-1961*. Poznan, 1962. The author has also studied the relations between ports and towns in 'Les principes du développement des villes portuaires', *Acad. polonaise des sciences*, Paris, 1962.

Sea ports, with their intensely active commercial function which acts as an attraction for other activities, are among the largest seats of urban life in the world. The founding of a port like Gdynia has in only a quarter of a century drawn to a formerly deserted place a town with a population of over 100,000. And Mark Jefferson was able to write that London had made the whole world its own region.[1]

AIRPORTS

Airports have not yet had time to exert on urban geography the influence they will certainly attain. They belong to two categories. Some are the now essential annexe of large cities. They have had to be placed some distance away from the centre on suitable available land, of which they now require larger and larger stretches. Orly airport was only able to expand at the price of costly expropriations. Most airports are between 6 and 12 miles from the town centre. Only a few still underdeveloped towns have their airports on the doorstep: Ouagadougou and Fort Lamy, less than 1 mile; Bobo Dioulasso and Bamako, 1 mile; Jakarta, 2 miles.

But this distance is 15 miles for London (Heathrow), eighteen for New York, 20 for Montreal and as much as 22 for Rome. It is true that these are the airports for medium and long journeys, and some towns have small 'pocket' airports for short distances. Nonetheless this is still one of the major disadvantages of air transport, particularly as added to the distance is the time taken up by the formalities of arrival and departure. To shorten this, the substitution of bus services by helicopters has been tried, so that travellers can reach the town as quickly as possible; some large hotels, like one in Warsaw, have provided a special terrace where helicopters could land. Around an airport one seldom finds more than the services of the airport itself.

However, travel is playing an increasingly important part in modern life. Travellers have to wait for connexions. Air transport hazards, which can hold up passengers for perhaps a day, must also be taken into consideration. These travellers have to be provided with meals and even sometimes with accommodation. It must be remembered that at one time in the neighbourhood of some railway stations there were only the buffet and a hotel for travellers, and that gradually a station district has been formed in every town. So towns are bound to extend towards their airports.

A secondary category of airports consists of the necessary ports of call, the number of which, incidentally, is constantly being reduced by the progress of aviation. The increase in the distance an aircraft

[1] *Geog. Rev.*, 1917.

can travel before refuelling threatens to ruin some calling-places, just as electrification demoted certain railway stations. Everything depends upon how much fuel can be carried. In highly urbanised countries calls are made at the airports of large towns. But long oceanic and polar journeys must not be forgotten. Gander, Newfoundland, was once a port of call on the route from Europe to North America. And at present the transpolar route from Paris to Tokyo calls at Anchorage on the extreme tip of North America. One wonders whether these ports of call, should they become stabilised, might end up as commercial or industrial towns or holiday resorts.

So it may be seen that traffic, in all its forms, is an effective stimulus for the development of towns.

THE EMPORIUM

For both continental and maritime trade, great commercial centres have grown up, great emporia where the goods of the whole world are collected and traded. They are often—but not always—ports, for the sea has always been the most convenient means for transporting goods in large quantities. In the past the Phoenician towns, Athens, the Hanseatic ports, Venice and Genoa played this part. These emporia have multiplied in modern times, and have become more complicated. An emporium must have the advantage of many and easy connexions, and must combine the various methods of transport. It has at its disposal warehouses of all sorts for special types of goods, each requiring its appropriate handling technique; in these warehouses the goods finish their processing treatment, cocoa is dried, bananas are ripened, rice is husked. These goods are warranted, and are the object of stock exchange transactions. The emporium becomes a business office where agents find out about existing stocks, take charge of transport and delivery without the goods ever passing through their hands, and banking activity becomes indispensable to the emporium.

In this way, important centres are created, the very organisation of which guarantees their permanence. Antwerp and Amsterdam receive less goods than Rotterdam, but retain their commercial and financial power.

Chapter 13

THE INDUSTRIAL FUNCTION

The manufacture of tools and of the objects necessary to daily life meets a more immediate need than trade, but it does not tend in the same way, at first at any rate, to produce urban concentration. Commerce requires a meeting point with other people, whereas manufacturing can be carried on at home, or at a more advanced stage, in the village itself. In the forest civilisation of northern countries each person carved wooden objects for himself, from spoons to complicated locks. Fabrics were made by the family: the spinning wheel was still turning in the French countryside until the last century, and though domestic weaving ceased, the spun yarn was taken to the village weaver's, as described in the novels of George Sand. In many countries of Asia and Africa materials are still woven at home. These domestic handicrafts are abandoned as a result of contact with a more highly developed civilisation. Gandhi's spinning-wheel was the symbol of resistance to the advance of British civilisation. Industry can remain rural for a long time, and can even develop and be perfected in this form, as in the case of the lock manufacture of Vimeu in France.

However, the evolution of industry favours the urban pattern, because it favours concentration. The raw material is not always available on the spot; craftsmen find it convenient to group together to obtain it in the same way that they find it convenient to group together to seek a market when production has increased beyond the needs of the village. In this way, in the Jura mountains, even in its handicraft stage, industry has concentrated around certain points, such as Saint-Claude, Morez, Oyonnax and Moirans.

It is incorrect to suppose that craftsmen as a class are opposed to urban concentration. In Germany, where this group accounts for 20 per cent of the workers, some craftsmen owe the progress of their crafts to the development of urban life: machine and motor-car repairers, electricians, opticians, printers, and hairdressers, for example. The clothing industry of Paris is to a large extent a handicraft industry; in 1938 there were 20,000 people working at home for this industry.[1]

Machines, too, have been an important cause of concentration; it is an advantage to be able to group them so as to employ a smaller number of workers to supervise them. Hence the installation of

[1] J. Klatzmann, *Le travail à domicile dans l'habillement*, Paris, 1957.

workshops bringing life to the one-time villages which have now become small towns: for example, the industrial villages of the Vosges valleys and of the Beaujolais region.

But while industries give rise to towns, towns also create industries. It is in the interest of towns to work up the products that they sell, so as to increase their value, and to extend the range of these products. Flemish industry was founded on trade. Later, in every town the manufacture of commodities for the service of its inhabitants tends to extend its market beyond the limits of the town and the industrial function develops.

So industrial towns appear very different from commercial towns; they did not grow up slowly on the increasing needs of the regions surrounding them; they were often to some extent superimposed on these regions. They can only develop if an external labour force forming a proletarian class can be brought to them. They are also more unstable, always threatened and always changing.[1]

Among towns with an industrial function we may distinguish two categories: mining towns and manufacturing towns.

MINING TOWNS

Mining towns are those in which the industrial function is most clearly discernible, as it is less closely linked with trade than in the case of manufacturing industries. Moreover the mine was in many cases the origin of the town. It sometimes happens that by a happy chance wealth is discovered beneath a commercial town: Essen, a sleepy little regional centre in the eighteenth century, found itself in the middle of a coal-mining area in the nineteenth, and the same thing happened to Douai; but usually the town is founded on the mine, even if it has since varied its functions.

Urban concentration is essential from the outset. To obtain sufficient quantities of the mineral, it is no longer enough to scratch the surface of the ground, nor to rake the bottom of lakes, as is done for iron ore in Finland. Coal and ores must be sought underground, shafts must be sunk, extraction organised and a large staff of workers collected, and a small town is established at once.

Precious metal mines attract people most rapidly because of the profits to be gained from them. Gold is typical for it lends itself to all sorts of exchange. Hence the rush to all the places where gold is reported to have been found: California, Alaska, Australia, South Africa. There were disorganised gatherings caused simply by the

[1] Rainer Mackensen, 'Industrielle Grossstadt' in *Daseinsformen der Grossstadt*, Tübingen, n.d.

coming together of people who rushed to the same place to try to make their fortune. These adventurers' towns knew triumphant successes and resounding failures. California, after being an eldorado, became a cemetery of towns.

It was during a second period that towns organised planned extraction. The anticipated profits explain the extraordinary conditions under which they were established. In Australia towns were founded in the desert and water had to be brought to them from 120 miles away by pipeline. At Johannesburg in 1884 there was nothing but a mud hut built by a Boer; in 1904, twenty years later, a modern town with a population of 159,000 had sprung up. Today Johannesburg's population is over a million.

Diamond-mining towns were founded in a similar way. After the hunt for alluvial diamonds, which were soon exhausted, their extraction had to be organised systematically. Tons of rock have to be removed, chemically treated and ground down. And, to avoid theft, the workers are housed in compounds, as at Kimberley where there is a special carefully enclosed district.

Other minerals are not so precious. They are produced in greater quantities, but there are still some among the riches of the subsoil, which have brought about similar rushes. This is the case with oil which is highly priced and easy to collect and transport. There was, as for gold and diamonds, a period when towns shot up, gathering together men armed with an old steam engine and steel rods, feverishly digging holes in the ground. In the early days, the oil towns were nothing more than huge camps, like Petroleum City and Kilgore.[1] They grew at lightning speed: Borgen in Texas, founded in 1921, had a population of 25,000 after one year.

When the surface oil was exhausted it became necessary to resort to deep boring which could only be undertaken, after costly research, by companies having the large capital required. As 98 per cent of these borings are nonproductive, temporary installations are considered sufficient at first; a town grows up when it appears that success is going to last. The oil beds are soon used up, however, because of gushing, and drilling is continued elsewhere. The subsoil of Texas is full of small oil-containing domes which are exploited one after another. One must not expect to see a town arise round every drill hole, but it is necessary to have a centre well equipped with tools and offices for organisation, research and business. Often this headquarters town in a mineral field is situated near the first installation. There are some which still have pumping stations, but Odessa in Texas is 9 miles from the nearest pumping stations—it is a mining town with a population of 73,000 without any immediate mining

[1] W. T. Chambers, 'Kilgore, Texas: an oil boom town', *Econ. Geog.*, 1933.

activities. On the other hand, Midland, an administrative and cultural centre, is quite near.[1]

Where a rich deposit is found great cities grow up. Baku in the Caucasus has a population of over 600,000; Grozny, 240,000; Tulsa, Oklahoma, was only a small market town with 400 inhabitants in 1894, but had 261,000 in 1960, and is the headquarters of the oil companies; banks have been established and airlines stop there. These oil towns which grow so fast are wealthy, and their wealth is often a great contrast to the poverty of the surrounding areas. Hassi Messaoud is in the process of becoming a town right in the middle of the Sahara; ice is brought to it by air. The only regret these towns have is the fluidity of oil. It is seldom refined on the spot, and so the advantage of industrial transformation is lost, to the regret of the Middle East whose floods of oil have not brought to it all the riches it might have hoped for. Nonetheless there are now several refineries near the oil fields at Abadan and at Curaçao in the West Indies. Oil is also used as a source of energy, transforming the oil towns of Texas into industrial towns. It must be added that because of the fluidity of oil, sea ports profit most from it, and that not only refining but also petrochemical industries are now some of the principal activities of large ports, for example Rotterdam.

Gas is exploited in a similar way to oil, but in this case there can be no preliminary exploitation stage, as gas must be controlled as it escapes, and even with powerful machinery, this is not easily done. Gas exploitation gives rise to the same sort of installations as oil, and transport facilities similarly allow it to be taken to the large consumer centres. Gas from Lacq is sent all over France. Advantage is also taken of it locally; desulphurisation is carried out on the spot, and chemical factories are set up nearby, which has brought into existence the town of Mourenx, with a planned population of 10,000.

Oil fields and veins of precious metal become exhausted, and the towns then seem condemned to die. However, a town tends to perpetuate itself, the presence of a labour force attracts other industries, the town is surrounded by a regional halo. It maintains its existence by changing its functions. The situation is very different in the case of mines supplying large quantities of cheap and cumbersome coal or ore. There is no question of individual exploitation. The extraction of coal in some parts of England in former times was in reality nothing more than the equivalent of the communal right of cutting wood in the forests of France. Extraction cannot be made profitable on an individual scale. In the pre-industrial era it was

[1] D. Weber, *A comparison of two oil city business centers (Odessa—Midland, Texas)*, Chicago, 1958.

barely possible to break even. It is moreover an advantage to use heavy ores on the spot and mining towns develop transformation industries.

The most typical mining towns are those where this industrial transformation is impossible—where climatic conditions necessitate the reduction to a minimum of staff and installations. This is the case in the iron mines in arctic or sub-arctic regions; like Kiruna in Lapland and Schefferville in Labrador. At Kiruna in 1900 there were only a few huts in an icy wasteland beyond the arctic circle; but the nearby mountain was a solid block of iron, and a town was established on account of this mountain; today it has a population of 20,000 and belongs entirely to the mining company. They confine themselves to extracting and dispatching the ore and even to do this it is necessary to work through the polar winter night by floodlight; no other industry would be profitable because of the high cost of living, the large salaries and the precautions which must be taken against frost.[1] More recently the exploitation of iron in Labrador has been started[2] under even more severe climatic conditions. The mean January temperature there is $-10°$ F ($-23°$C), and extraction is halted in winter. The town of Schefferville was built entirely by the mining company for this exploitation. Building began in 1953 with materials brought by air. The miners' houses and offices were placed right at the foot of the hill being mined, and the town grew from that point, enclosed by the lakes, as far as the aerodrome. It already has 3,000 inhabitants in summer. Only the high grade of the ore accounts for the success of these pioneering towns under conditions as inhuman as those of the gold towns.

Copper caused the foundation of similar towns in Katanga.

Coal is less precious, but is the world's most important mineral in terms of tonnage, and coal-mining towns are the most widespread of mining towns. They require a large labour force both above and below ground in spite of the effects of mechanisation. The town revolves around the mine shaft, dominated by the pit-head and the spoil heaps where waste accumulates, with the sorting and washing installations nearby. The working population must be housed, and in conditions which will keep the miners as content as possible with their jobs, to compensate for the hard work they have to do. So where possible they are provided with individual houses not too far away from the mine. These towns have an eight-hourly rhythm of shift workers arriving at and leaving the mines.

[1] G. Chabot, 'La Laponie de Jukkasjärvi et Kiruna, colonie suédoise', *Ann. de Géog.*, 1942.
[2] G. Humphreys, 'Schefferville, Quebec: a new pioneering town', *Geog. Rev.*, 1958.

The town, however, does not always remain solely concerned with coal-mining. Coal is a source of energy and a raw material. It facilitates a wide range of industrial possibilities. Coke ovens, blast furnaces, steelworks and power stations all use it on the spot. Thus it comes about that large industrial concentrations often grow up within the coalfields, like, for example, the Black Country of Britain, Northern France, the Borinage in Belgium, the Rhine-Westphalian Basin, Upper Silesia, Donbass and Kuzbass and Pennsylvania. The basins in which the coal accumulated were extensive: many pits can be sunk at a considerable distance from one another, so every coalfield is formed of a number of close but distinct towns, constituting what is known as a conurbation or town-cluster.

These colliery towns have their problems—human problems which are primarily concerned with recruiting. The job is hard and dangerous; and though the time is past when miners' sons went down the mine almost automatically, the labour force of a mine is almost exclusively male; women cannot always find work on the spot. Yorkshire has its textile factories, but in the mining towns of northern France the female labour force often has to travel long distances each day to work in the Lille region. Moreover mines become exhausted, though progress in methods of extraction sometimes allows the exploitation of lower strata nearby. Thus, in the Ruhr, the population has moved from the south to the northern coalfields where deeper strata are now being mined. But this does not always happen, and then the drama of closure because the mine is no longer profitable occurs. It is not easy to transplant workers. They are often semi-rural and cling to their small mine; the wife's job is on the spot. And then it is a great blow to the town, whose trade depends on the miners and their families. Attempts at transplantation, like those made with the miners of the Cévennes in Lorraine, have not been very successful. So increasingly the effort is made to substitute other activities for the mining. Fortunately, a town formed around a mine is often the seat of other industries and with these industries the mining town defends itself. Scranton, in the Appalachians, used to be an anthracite town, but it was also a railway junction which attracted metallurgical and textile industries, and although anthracite is no longer mined there, the town is still developing and today has a population of 126,000. Only a few token shovels-full of coal are extracted from Le Creusot nowadays, but the metallurgical industry established there by the presence of coal supports a population of about 30,000. At Graissesac in the département of Hérault, when mining ceased relief was brought by a foundry, a factory for prefabricated units and a shoe factory. Coal and iron mining towns both attract transformation industries, and the closing of the mines is only tragic in towns which have no other function, and where the readaptation of the miners must be accompanied by the reconversion of the whole region.

It has sometimes been asked whether these mining towns are really towns at all: the population is too much engaged in a single activity; when we think of a town, we expect a certain proportion of commercial activity. Pinchemel has noted the high proportion of the population employed in the secondary sector in northern mining towns—from 45 to 55 per cent of the active industrial population, remarking that 'the quality of urban life changes when the percentage of active industrial population is higher than 66 per cent'.[1] And yet it cannot be denied that an assembly of tens of thousands of men, gathered to carry out an urban function, and who have organised their lives and their dwelling places, constitutes a town.

Often, moreover, in densely populated countries at any rate, the mining town represents part of a larger urban network. Almost always a town is to be found nearby which profits from the mine, a small regional centre which derives activity from it. This centre provides supplies and entertainment that the mining town cannot; sometimes, too, it serves as a dormitory. The small town of Sandomierz in Poland was resting on its administrative traditions[2] when the discovery of some of the richest sulphur mines in the world a few kilometres distant brought it new life. Mining towns remain all the more independent, and grow all the more suddenly if they take root in a relatively underdeveloped region.

In order to reduce the expense of transporting minerals, they are often processed to a quarter or less extent at the pit-head. It may be simply a case of improving the metal content, as with the iron ore of Norway which is concentrated magnetically; or perhaps, as in Katanga, of the smelting and electrolytic refining of copper. But this transformation is often pushed even further, converting the ore into pig-iron or steel and going as far as making the finished product. Transformation industries are superimposed on the original mining activity.

MANUFACTURING INDUSTRIES

The origin of the function. The industrial function is easy to define in a simple case of mining industry. It is much more difficult to define manufacturing towns.

[1] P. Pinchemel, *Le peuplement et les mouvements migratoires de la région Nord—Pas de Calais.*
[2] K. Wilgatowa, 'Analysis of the geographic position of Sandomierz', *Ann. de l'Université Marie Curie-Sklodowska,* Lublin, 1950.

Such industries can be found in every town, answering the needs of the population, just as in the past every village had a blacksmith and a wheelwright; the only difference is that today's needs are more numerous, corresponding to a higher standard of living; it is still desirable to try to satisfy them all on the spot. But an industrial function may be said to exist as soon as the town manufactures more than its own population can consume.

This can happen simply because of the expansion of the town's services; fabrics are manufactured for the inhabitants of a town, but the shops are frequented by people from the neighbouring countryside. Small village handicraft industries are disappearing. The town flour-mills have ruined the old-fashioned water mills. The town is manufacturing for the outside world.

Commerce attracts industry. Manufacturing is carried on to supply the needs of commerce. The Flemish weaving was intended for export. Products are transformed before being dispatched again. Even in ancient times the Phoenicians dyed their cloths with purple to increase their value. Ports which receive crude oil are almost all provided with refineries. Imported raw materials are used on the spot: Copenhagen transforms oil-bearing products from the Far East into margarine. Ports easily reached by raw materials are destined, as we have seen, to become seats of industry.

Finally, there is another example of a link with commerce. Traders with capital seek to increase it; rather than place it in distant enterprises, they set up on the spot industries which they can control. This explains the development of industries at a place like Angers,[1] or in the Seine valley. Many small local market towns have witnessed the installation of a few factories as a result of local initiative.[2] This procedure is becoming more general to save certain centres from stagnation.

The extension of industry in these towns, the expansion of trade by industry, can remain only moderate and bring about a somewhat restricted and to some extent subordinate industrial function. But it can also happen that this industrial function becomes the vital one, supplanting the original function. And today many towns are considered to be industrial which were originally commercial towns.

These consequences of commerce are only advantageous to certain countries endowed with strong means of economic and even political action. This is not the case in underdeveloped countries; and underdevelopment is accounted for to a considerable extent by the fact that other countries have secured for themselves the profits of manufacturing industries. India once had a thriving textile industry, but

[1] J. Méary, 'Angers', *Ann. de Géog.*, 1943.
[2] R. Specklin, '*Altkirch, type de petite ville*', Paris, 1951.

it was systematically ruined by the British industry which, by a reversal in the Indian economy, took its raw cotton and sent back cotton goods. This is why today industrial centres are so few in Africa and Asia, although there are many large towns, and mining towns are often the only representatives of the industrial function. The development of the new countries revolves around the development of their industries. This represents their first effort after the achievement of political independence and is the direction of movement of the underdeveloped world against what has been called neo-colonialism—the monopolising of industrial production after the granting of independence.

So in many cases industry appears to be the consequence of commerce. But it also develops in an autonomous way. In the past abbeys were often small centres of industry, and these abbeys were the starting points of towns. Industrial towns are not solely the corollary of mines or of trading places. It is in fact possible to see in them the expression of two different forms of civilisation. American civilisation would be the origin of commercial towns, while in a Soviet civilisation, industrial towns would take priority. Finally it should be recognised that, in any case, industry cannot succeed unless it finds favourable conditions. We should, then, find out how industrial towns become localised.

Location of industrial towns.[1] Only mining towns are subject to factors which literally determine their location. Examples of many industrial successes in other towns due to chance may be quoted, and these support the argument against geographical determinism. The history of the rubber industry at Clermont-Ferrand is often quoted in this context. The owner of a mill which had burnt down decided, by way of a change as we would say today, on the inspiration of his wife, niece of the Scotsman, MacIntosh, to start up in the rubber industry.

The reason for the installation of the automobile industry at Le Mans was that an early enthusiast of motor-cars, Léon Bollet, lived there. Adam Opel, a sewing-machine manufacturer at Rüsselheim, bought a bicycle in 1886 to amuse his sons, who became very keen on bicycle racing and then on motor-cars in their early days, and this was the origin of the Opel motor-car firm. William Morris had a bicycle repair shop in Oxford, and from this small beginning grew the huge Nuffield organisation making Morris cars at Cowley. The Lyons photographic industry might appear to be the consequence of the chemical industry which is itself due to the necessity of dyeing silk,

---

[1] A. Weber, *Über den Standort der Industrien*, Tübingen, 1909. English translation by C. J. Friedrich, Chicago, 1928; G, Chabot, 'Géographie urbaine: la naissance et la localisation des villes industrielles', *Inf. Géog.*, 1947.

but it has been shown also that it was largely due to those two inventive geniuses, the brothers Lumière.

This shows with what caution the problems of location should be approached. However, it must be noted that happy chance is surrounded by favourable conditions, and countless obscure failures are due to their absence. So we must try to determine which circumstances favour the industrial function of towns.

The part played by raw materials should be examined first. Sometimes these raw materials exert no influence over localisation. Air, the raw material for the manufacture of nitrogen, can be found everywhere, so the localisation depends upon the ease with which energy may be found (Norway, Toulouse). Or the raw material may be very light, and expensive enough to justify long-distance transport so that it can be carried anywhere, thus leaving the labour force and consumers' markets as the preponderant factors.

Most often, however, the proximity of the raw material makes itself felt.[1] It is traditional to quote wool-working towns as owing their origin to the sheep of the region (Rheims, Sedan) or the glove industry using, as in Millau, the skins of the sheep from the Causses. Salt supplies the chemical industry of Northwich, iron mines have given rise to a metallurgical industry which has filled the valleys of Lorraine with towns. A scattered metallurgical industry, with factories and housing, was set up just before the Second World War in twenty-eight communes around the iron mines near Brunswick. In 1942 these communes were united to form the town of Salzgitter which in 1962 had a population of 113,000 and which is gradually acquiring urban facilities. Coal mines, too, are the origin of large industrial complexes.

Proximity, however, is not essential. To be able to receive raw materials easily is enough. In the Swedish Norrland wood thrown into the rivers in spring floats down to the mouth of the river, and here, rather than in the forest, the paper-pulp mills are installed. All raw materials arrive cheaply at the ports where they are transformed, as we have seen (p. 143). The part played by water as the cheapest means of transport should be stressed. It is not just by chance that in Norway six out of the seven industrial zones are on the coast.[2]

Land transport is more expensive, but the higher the value of the goods, the less this factor counts.[3] Silk from Italy used to arrive at Lyons by land.

[1] J. E. Brush, 'The iron and steel industry in India', *Geog. Rev.*, 1952.
[2] J. P. Pickard, 'The manufacturing regions of Norway', *Norsk Geog. Tidsskrift*, 1961.
[3] Changing techniques of transport over the years may appreciably modify this. Thus the potteries of Stoke-on-Trent were for a period in the late eighteenth century supplied with Cornish china clay by a route involving coastwise and

Some materials travel vast distances before being used; nickel from Canada comes to Norway and to Swansea to be refined, and sheepskins from Argentina arrive at Mazamet to have the wool removed, not to mention the raw materials from the Far East. Even a heavy and bulky commodity like iron ore moves great distances across the oceans, as from Brazil and West Africa to Britain, and from Australia and Malaya to Japan.

A division of labour may often be found. The raw material undergoes preliminary elaboration on the spot. The localisation of the town where transformation is completed then plays a much less important part. Swedish iron arrives in France in the form of pig-iron; steel from Lorraine supplies the mechanical industries throughout France. The less the importance of the proximity of the raw material, the greater is the part played by the other factors in localisation. The energy used makes some demands. The textile industry fixed its towns by waterfalls, as on the flanks of the Pennines, and at Tampere in Finland. At the beginning of the industrial era coal was the decisive factor in most localisations. It either brought about the founding of new towns, as in the Ruhr, or favoured by its presence the old industrial towns, as in Yorkshire and Flanders. And here again, ports where it could be delivered easily served in place of mines. Caen is today the first coal-importing port of France. The motive force supplied by oil drives the machines of Texas, and to use that supplied by natural gas, an aluminium factory is being erected at Lacq. This seems less applicable to electricity which can be brought everywhere easily; in well-equipped countries this plays hardly any part in the localisation of industries, but in the past, factories were placed near hydroelectric stations. Thus the small towns of Maurienne and the Romanche valley discovered their vocation for electrochemistry and electrometallurgy. Rjukan in Norway depends solely on electricity, for the nitrogen industry obtains its raw material from the air.

Raw materials and energy travel with increasing ease. Man seems thus to escape the contingencies of nature, to be master, to place industries wherever he chooses and to make whatever town he wishes profit from them. But there are physical elements which cannot be economically transported and which are vital to some industries. This is the case with water, which is changed into steam, washes out impurities, or acts as a cooling agent. Chemical industries demand

riverine shipping and overland pack-horse transport; then for a century and a half an all-water route was used, coastwise to the Mersey and thence by canal. After the second world war much of the clay travelled by lorry the whole 250 miles from St Austell; whilst in 1966 'liner-trains' were introduced on the railway, largely supplanting the road vehicles—and the canal route is dead (Editor's note).

large quantities of water (which they give back afterwards, incidentally): one ton of rayon needs 46,000 cubic feet of water; it may be observed that many German chemical industries are placed beside the Rhine, at Ludwigshafen in particular, and that French chemical industries have developed extensively along the Rhône downstream of Lyons. The fact that Chinon in France has been chosen as the site for a large nuclear centre is partly due to its proximity to the Loire. This is equally true of metallurgical industries. One ton of steel requires 65,000 gallons of water; a blast furnace uses every day as much water as a town with 30,000 inhabitants, and this water must be cold, clear, and soft; all but 2 per cent is returned to use and American metallurgical companies spend 10 million dollars every year on purifying water which is put back into circulation. This is one of the principal reasons for the siting of the metallurgical industry on the Great Lakes of America, and the Report of the President's Materials Policy Commission (Washington 1952) stated that by 1975 a good water supply would become the essential factor in the localisation of industries. Many industrial centres are beginning to dread water shortages; Württemberg has remedied this by use of a 90 mile pipeline bringing water from Lake Constance to Stuttgart across the Swabian Jura. It may well be that those parts of the world with the best water resources will offer the most favourable conditions for industry in the future.[1]

Human conditions often outweigh physical circumstances. The ease with which a cheap labour force can be recruited explains the founding and development of many industries. One reason for the rapid growth of industry at Clermont-Ferrand was that it had at its disposal the population who were abandoning the mountains. There are in fact industrial towns which owe their development to the arrival of workers. The silk workers of Lyons, driven out by the revocation of the Edict of Nantes, brought to Switzerland and Germany the manufacture of silk, and other refugees took their textile skills from Flanders to the towns of East Anglia. The textile industry of Lodz was founded near the Polish-German border, using workers displaced by the German textile industry. All along the French side of the Franco-Belgian frontier factories have been built to take advantage of the Belgian frontier labour force. The presence of industries using a male labour force attracts industries capable of absorbing the available female labour force, so that the industrial structure becomes more mixed and more balanced.[2]

[1] C. Langdon White, 'Water: a neglected factor in the geographical literature of iron and steel', *Geog. Rev.*, 1957.
[2] H. Nelson, 'Industrialisering och Stadsbyggds-bildning', *Svensk Geog. Arsbok*, 1947.

Social considerations have sometimes supplemented economic conditions in an endeavour to bring industries nearer to a possible labour supply. This avoids the uprooting of workers, and helps to bring life back to deserted regions. With this intention the textile industry was brought to the Vosges in the eighteenth century. At present there is a movement for the attraction of industry into the towns of western France; one of the most significant examples is Cholet. The combination of the textile and footwear industries allows the simultaneous employment of male and female labour. In Great Britain government encouragement has been given to new industries in the 'Development Areas' which previously had little but coal-mining.

It is not simply a question of workers and labour force. The problem of the managerial classes and scientific personnel plays an increasingly important part in the case of industries where the research laboratories take up more and more room. These people have been accustomed to life in large towns, and the comfort and facilities to be found there. They often dislike the idea of moving away, but they appreciate the advantages of a pleasant town with sports facilities. Annecy in the French Alps was able to equip its new industries because the managers and scientists were attracted by a pleasant town that was also a tourist and winter sports centre.[1]

The social question becomes even more urgent in poor areas, till now completely neglected. In all northern countries attempts are made to counteract the population's natural tendency to drift towards the less forbidding and more lively southern regions. Industries which are not always justified by economic conditions are the results of these attempts, as at Mo i Rana in Norway, Luleå in Sweden and Oulu in Finland.

Proximity to consumer centres is equally desirable. The manufactured product is sometimes as expensive to transport as the materials of which it is made,[2] so it is advantageous to be as near as possible to the customers. The French automobile industries are mostly to be found in the neighbourhood of Paris; Breton towns like Dinan manufacture agricultural machinery; the clothing industry is to be found in large agglomerations. This is even truer in certain industries like fashion and art. These need very special conditions. The proximity of a wealthy clientèle is the vital factor in the formation of a labour force. Nancy has inherited artistic industries from its past as capital, and haute couture triumphs in Paris and in Los Angeles.

[1] R. Blanchard, *Annecy, essai de géographie urbaine*, Annecy, 1957.
[2] Chauncy D. Harris, 'The market as a factor in the localisation of industry in the United States', *Ann. Ass. American Geog.*, 1954.

Nearness to capital cities is still an advantage. Large companies invest more willingly in towns which can be reached easily. The lower Seine valley has therefore been industrialised under the influence of Parisian enterprises. Greater London is the largest single industrial district in Great Britain, despite the importance of the coalfields.

Every country tries to provide for its own needs, to avoid loss of capital, and therefore endeavours to establish on its own territory manufacturing industries which do not need to import their raw materials, and just as in the past every village used to weave its own cloth, today every country tries to organise its own textile industries. This is in fact the first concern of recently formed nations, as was seen in the Baltic countries after the First World War.

Finally, certain political conditions should be included among these human conditions. After the war of 1870, armament industries were established at Saint Etienne, far away from the frontier, to make them less vulnerable.

So the localisation of industries is linked with many causes which it is often difficult to unravel. Consider the brick and tile industries.[1] The necessary clay is widely distributed, and may be of unequal quality. The technique is not complicated and can easily be undertaken by small firms; use of the products is widespread and their bulky character encourages the establishment of the industry near consumer centres. All the conditions of soil, labour force, habitat, population density, ways of life and transport come into it, and the whole geography of a region expresses itself through this industry.[2]

Attempts have been made to devise equations for these conditions. The localisation is the function of a certain number of conditions A, B, C, etc., which represent the proximity of the raw material, the labour force, etc., and a parameter m, n, o, etc. attributed to each condition to represent the intensity of the part played by that condition.

This would give, for example:

$$L = F(mA + nB + oC\ldots)$$

It is possible to produce isarithms expressing the variations of certain of these factors, for example, the cost of transporting raw materials or manufactured products.[3] Maps have been constructed

[1] A. Kuklinski, 'Problems in the location of brickmaking industry in Poland', *Polish Geog. Rev.*, 1960.
[2] This is no longer true in England, where 55 per cent of all the bricks made come from about thirty mass-production works on the Oxford clay formation near Peterborough, Bedford and Bletchley; see S. H. Beaver, 'Technology and geography', *Adv. of Sci.*, 1962.
[3] G. Törnquist, 'Transport costs as a location factor for manufacturing industry', *Svensk Geog. Arsbok*, 1962.

to express the reasons for the location of such and such an industry in relation to its raw materials. The Swedish paper industry has been represented in this way.[1] It would be possible to combine several of these factors, and to express this combination mathematically. Research of this kind can be used for organised planning.

It should be noted, however, that with the growing complexity of industries[2] exact interpretation of many localisations becomes difficult, and that in the future the order of importance of factors may well be completely reversed.

Annecy is a French town which has recently equipped itself industrially. What industries do we find there?[3] Ball bearings, aluminium rolling, electronic equipment, pillion seats, Gillette blades, measuring equipment for radio and television sets, men's shirt manufacturers, women's sports clothes manufacturers, luxury paper goods, paper for statistical machinery, synthetic stones, and the making of cheese. This last industry alone is due to the resources of the region; as for the others, they may have originally been favoured by the presence of cheap electric current which presented great opportunities; but not even this is true any longer, and how can the attraction of the physical environment be expressed mathematically?

The Burgundian Gate, a wide avenue of traffic, would naturally attract industry, but, as Michel Chevalier[4] has pointed out, social factors have also played an important part, such as the rôle of the entrepreneurs of Montbéliard and the influence of Alsace and Switzerland.

Lastly, it is very difficult, when deciding where to establish industries, to foresee the economic modifications made by changing techniques, and local conditions created by the development of the industry itself must be taken into account.[5] We can only investigate the currently favourable conditions and rely for the future on those who will exploit and adapt them. This very evolution will depend to a great extent upon the urban environment, for industry, even if sometimes it seems to dominate a town, is only one aspect, and its activity is linked to that of the whole town.

Evolution of the industrial function. What becomes of these towns in which the conjunction of favourable conditions has caused the development of industry?

[1] O. Lindberg, *Studier över den Svenska pappers industriens lokalisering*, 1951.
[2] J. Chardonnet, *Les grands types de complexes industriels*, Paris, 1953.
[3] R. Blanchard, *op. cit.*
[4] *Tableau industriel de la Franche-Comté*, Besançon, 1961.
[5] A. Lösch, *Die räumliche Ordnung der Wirtschaft*, Jena, 1944; S. Dahl, *Geography of Industrial Production*, Göteborg, 1962.

If conditions are particularly fortunate and if circumstances lend themselves to it, the industry thus established develops, sometimes to the point of monopolising the whole town. This may result from the gigantic expansion of a single industry. Eindhoven in the Dutch Brabant, is the home of Philips electric lamps, a firm employing three-quarters of the workers of a town with a total population of 166,000.[1] Ludwigshafen owes its growth in one century from a population of 2,000 to one of 150,000 to the Badische Anilin und Soda Fabrik; this firm employs two-thirds of the active population and covers an area of 1,500 acres.[2] Rochester, U.S.A., (pop. 318,000) is the Kodak town, and Atlanta (pop. 487,000) is the Coca Cola town.

Sometimes a great single industry develops through the growth of several rival establishments. Detroit, once a small fur-trading station at the junction of the American Great Lakes, has become an automobile city, with a population of over 1½ million. The key industry can even go beyond the limits of the town. Yawata, the Japanese steel town, a few years ago brought into existence nearby Tobata, which now has a population of 100,000, and specialises in the most advanced methods.

The convenience of buying and selling in bulk has caused the small towns of the Jura to specialise in wood-turning (Saint-Claude), spectacle-making (Morez), plastics (Oyonnaux). Fougères in Brittany specialises in footwear.

Once an industry is established in a town it tends to remain there and grow almost automatically, even if the circumstances which caused its existence disappear.[3] The Polish textile industry is still concentrated around Lodz, the textile town founded because of its proximity to the German frontier, in spite of the changes in the frontier, the lack of water and coal, and the absence of an adequate transport network. But Lodz has a large number of well-established specialist firms, and an industrial and commercial tradition; it is in fact a piece of machinery whose well-oiled wheels would be difficult to find elsewhere.[4]

Saint-Claude in the Jura worked the boxwood of the surrounding limestone hills and ended by specialising in the manufacture of briar pipes, the raw material of which grows in the siliceous soils of the Mediterranean regions. The inhabitants preferred to overcome their

[1] P. George, 'Les établissements Philips aux Pays-Bas', *Bull. de l'Ass. des Geog. Français*, 1961.
[2] Traband, *Rev. Géog. de l'Est*, 1962.
[3] A. Rodgers, 'Industrial inertia in the steel industry', *Geog. Rev.*, 1952.
[4] L. Straszewicz, 'Analyse des bases du développement des régions économiques tirée de l'exemple de la région industrielle de Lodz', *Regional Economic Geography*, Warsaw, 1961.

raw material difficulties rather than abandon the social and economic advantages inherited from the past.[1]

A specialised industry needs machines or products which tend to be manufactured in the same town. In Zurich and Troyes, as at Oldham and Leicester, the textile industry has stimulated the manufacture of the machines it needs. At Lyons, Macclesfield and Coventry the manufacture of silk has attracted the manufacture of artificial fibres.

Generally speaking, industry attracts industry. The factors that give rise to an industry do not apply only to that industry, whether they are sources of energy, transport facilities or the presence of a large labour force. Industries tend therefore to be symbiotic; they rub shoulders with one another. The Yorkshire coalfield had the advantage of textile industries employing a female labour force.

An industrial town is bound to expand, and expansion brings about diversification, for the town develops a population for which work must be found, and not always the same work as for the previous generation. The necessary qualifications are not the same.

With expansion, economic slumps are particularly to be feared, and this is another reason for diversification, for the disadvantages of having a single industry must be avoided. Every town tries to achieve a balance by setting up varied industries. If necessary a failing industry must be replaced by another which can take advantage of the premises, the available labour force and the industrial climate. Fourmies recovered from the depression caused by the textile crisis by installing metallurgical industries. Between 1951 and 1960, some 500 cotton mills closed in South Lancashire, of which 380 were promptly occupied by other industries.[2] Limoges was a porcelain-manufacturing town, thanks to the local china-clay beds, and exported its products all over the world. Foreign competition caused the loss of many customers, but during the 1914–18 war a footwear industry opened up there and became the vital industry.

Adaptation of a town to its function. Obviously the industrial function is echoed in all aspects of town life, but in different ways according to the various categories of industry.

Home-based industries have very little effect on the appearance of a town. The industrial and residential areas are one and the same, yet these home industries are often to be found concentrated in certain

[1] Thérèse Colin, 'Les industries de Saint-Claude', *Et. rhodaniennes*, 1937.
[2] H. B. Rodgers, 'The changing geography of the Lancashire cotton industry', *Econ. Geog.*, 1962. The same thing is true of the New England textile industry; see W. H. Wallace, 'Merrimack Valley manufacturing: past and present', *Econ. Geog.*, 1961, pp. 283–308.

districts. This is the case of the Paris garment industry. The workers, many of whom are immigrants, live mostly in the Sentier district. Work is distributed to them by small employers, and it is almost impossible to keep a check on it.[1] In the past the silk industry of Lyons was housed in the Croix Rousse district, and its high buildings lit by many windows where the silk was woven contrasted with the low, damp houses in the surrounding area where the cotton was worked. The work of haute couture and fashion is often done in workshops which likewise tend to be close together, as in the Rue de la Paix in Paris, or the Oxford Street district of London. Moreover a certain number of industries which make small items and require a small but specialised staff, often go unnoticed, as with electrical equipment or certain luxury industries.

On the other hand, textile industries create a cultural landscape of tall spinning mills with adjacent weaving-sheds, the latter with serrated roofs, glazed on the north-facing sides. Somewhat similar forms are found in engineering workshops. Metallurgical towns with their dominating blast furnaces and huge chimneys are obvious examples. Chemical industries are noteworthy for their huge buildings and tall chimneys emitting gases; Choisy-le-Roi can be recognised by the chimneys of its chemical industries. These industries are often consigned to out-of-town areas, to avoid smoke and fumes over populated areas, and also to take advantage of cheaper land. Sometimes however, the expanding town engulfs these establishments outside its limits and they become part of the agglomeration. So in planning a town, it is desirable to reserve some parts for industry, separate from the residential areas, and to place these near the docks or the station, or in valleys which act as transport arteries.

It is not enough to consider only industrial towns. Industrial complexes form a group of towns which together contribute to the industrial function formerly concentrated in one town. In this way the textile industry sprang up spontaneously around Manchester and Lyons. Throughout the Lyons region there are centres specialising in the spinning, throwing or weaving of silk; some weave raw thread and others thread that has been dyed. In Lancashire, the northern towns used to specialise in cotton weaving—with Burnley, for example dealing in coarser cloth, and Colne in finer qualities—whilst the towns around Manchester dealt mainly in spinning—with Oldham specialising in coarser counts and Bolton in finer ones.

Sometimes industries cluster together in order to rationalise their activities. The resulting concentration may present serious disadvantages: high-priced land, difficulty of traffic circulation, protracted journeys for commuters. So secondary industrial centres

[1] J. Klatzmann, *Le travail à domicile dans l'habillement*, Paris, 1957.

round large towns are planned, containing both factories and residential areas for the working population. The New Towns of the London basin (see p. 253 below) correspond to the Soviet satellite towns.

The result could be to empty an industrial town of its industry. Lyons is now no more than the headquarters, ordering the raw material, devising the patterns and taking charge of the preparatory dyeing and finishing of cloth, and then of its sale. The Lyons 'manufacturer' now no longer 'manufactures' anything. The centre of an industrial complex now tends to be only a business office.

There can also be dissociation between research departments, laboratories and commercial organisations on the one hand, and on the other, the manufacturing which requires huge premises and many workers. Some Parisian enterprises have kept their administrative services in Paris but their factories are in the south or east.

The Philips company owns forty establishments in the Netherlands apart from Eindhoven. They have kept the administrative services, the research offices and the manufacture of experimental items at Eindhoven, but 35,000 workers are employed in other centres.[1]

It thus becomes difficult to distinguish between the industrial and the commercial function; there are towns in which the only industrial function is the commercial service of an industry.

But the industrial function is completed by other functions that become integrated with it. There is the question of the recruitment of workpeople and managerial staff. The profession is concerned about the quality of its recruits, so an industrial centre is bound to develop facilities for technical training. In this context the Zurich Polytechnicum, the Chalmers School at Gothenburg, the Institut d'Electricité at Grenoble, and the Ecole des Industries Plastiques at Oyonnaux may be quoted as being among those whose reputations and area of recruitment go far beyond the framework of the city.

Cartography of industrial function (Fig. 8). The simplest method of representation is to show the proportion of the population of a given town that derives its livelihood from industry. This can be represented by an inner circle (or sphere), the remainder of the population forming a ring around it. But this is not enough. Differentiation of the town's various industries should also be attempted on the map. The different categories of industrial towns are in fact clearly definable: textiles, metallurgy, woodworking, cellulose production, chemical industries, etc. Towns may be differentiated by the numbers of workpeople employed in each industry. On W. William-Olsson's economic map of Europe, towns are represented by spheres, each of which is coloured

[1] P. George, *op. cit.*

to represent the industry employing most workers, and the little red spheres of the metallurgy towns express strikingly the intense concentration of metallurgy in Europe.

The ingenious way in which William-Olsson has solved the problem of indicating secondary activities is described on p. 116 above. This method only permits the indication of one form of industry, or two at the outside, but larger complexities must still be accounted for. In representation by circles, the circle can be divided into a number of sectors, each of which bears a symbol corresponding to an industry. In the case of representation by spheres, the sphere can be cut into slices, each one bearing the required symbol (Fig. 8.5). Other methods of representation have been adopted. They are not always successful.[1]

In some cases, the town name is underlined with lines each of which corresponds to a certain industry, and these lines can be placed in order of importance. But on looking at the map, as the names are not always the same length, an industry appears to be much more important when the name is longer (Fig. 8.6).

Neither is the system using a box placed next to the town very successful, for the symbol can extend far beyond the area occupied by the town on the map, out into the country or even into the sea (Fig. 8.7).

The choice of symbols also presents a problem. They can be abstract—coloured, shaded or dotted. However, more evocative symbols are preferred: a retort for chemical industries, a toothed wheel for metallurgy, and so on. This is acceptable as long as the symbols are few and very clearly identifiable. If it is necessary to refer to the key in order to understand what they mean, their advantage is lost (Fig. 8.8 and 9).

It can also be of interest to show not only the percentage of active industrial population and the way it is divided among the various industries; representation of the supply of and demand for employment can also be attempted as in the *Atlas of northern France*.

[1] Some interesting and clear variations on the circle-and-sector theme are to be found in the *Atlas of Britain*, Oxford, 1964.

CHAPTER 14

THE CULTURAL FUNCTION

RELIGIOUS ASSOCIATIONS

Not all urban functions are economic. There are towns with attractions of a spiritual nature, which have a religious function. This term must be defined. The practice of religion by the inhabitants, however intense, no more confers a religious function on the town than the presence of fortifications confers a military function. Nor does it suffice for the town to have been surrounded in earlier times by religious ceremonies, nor for the foundation to owe its being to religious connexions. Salt Lake City is a town of farmers and traders and Tel-Aviv is an industrial town.

Towns with a true religious function are those where people come from outside to practise their religion. The embryonic form of this is exemplified by the 'Sunday towns' which are found in countries where the population is widely scattered, like some of the South American countries or certain subarctic regions.[1] The isolated church is buried in the country; long journeys are therefore necessary to fulfil the demands of religious practice, and it is necessary to provide inns, stables and coach-houses near the church. Some of the faithful may even build themselves temporary lodgings which they use on Sundays; and there are small centres which come to life once a week. These can hardly be called towns, but very often in districts where there are no other nuclei, it is around these 'Sunday towns' that urban life is organised.[2]

Without actually owing their origin to it, many towns exercise a religious as well as a commercial function in relation to the neighbouring countryside. This does not apply where each district has its own church and religious services which the faithful can attend every Sunday. Moreover it is necessary for all the inhabitants to share the same religion. The town with churches of different denominations offers the scattered congregation of the region the possiblity of practising their own cult. And even in countries where the religion is the same, the lack of churches in rural areas may bring the faithful into the town. In Moscow churches can be seen crowded with people from the neighbouring countryside.

But some religious centres are the scene of much longer journeys

[1] P. Deffontaines, in *C. R. Congrès internat. de Géog.*, Amsterdam, 1938.
[2] G. Enequist, *Nedre Luledalens byar*, Uppsala, 1937.

and are the origin of the real religious towns; these are the places of pilgrimage which attract hundreds of thousands of the faithful.

The question of site is irrelevant. Such a town may be found in the middle of the desert, like Mecca; but the place has been invested with a sacred character. This may be attributed to a tomb as at Jerusalem, to a mysterious stone fallen from the heavens as at Mecca; to the relic of a saint; or to the evocation of a supernatural apparition. A mystical meaning may also be attributed to a natural phenomenon. Draughts from subterranean cavities made Delphi famous and some rivers, like the Ganges, have been invested with sacred virtues.

People go on pilgrimage to all these holy places. Pilgrimages are very diverse and they do not all take place in towns. Some, with a purely local attraction, animate a village from time to time. But there are some with an international reputation which attract the faithful from all parts of the world. The origin of the pilgrimage is sometimes lost in the mists of time, like the one which leads the Hindus to the sacred river Ganges; it may coincide with the actual origin of the religion, as in the case of Jerusalem and Mecca; or it may be stimulated by a phenomenon arising within a religious community, as at Lourdes or Lisieux. Some relics attracted crowds in the Middle Ages but are now practically forgotten. On the other hand there are places of pilgrimage which seem to enjoy a certain permanence. Le Puy was possibly a place of pilgrimage in pre-Christian times.

Forms of pilgrimage are also very varied. The large centres attract the faithful all the year round and only climatic considerations may emphasise the summer season, as at Lourdes. But there are also pilgrimages held on certain dates: people come to Le Puy three times a year and especially on 15 August, the Feast of the Assumption.

Some religions favour pilgrimages more than others; these are the religions in which the adherents are most frequently assembled, in which the external forms of the cult have more importance, and discipline is stricter. There are hardly any pilgrimages in the Protestant churches.

Naturally, the more adherents a religion possesses the more people join in the pilgrimages and the larger grows the town concerned. It is difficult to estimate the crowds which throng certain towns in this way. It is estimated that more than a million pilgrims come to Benares each year; at Mecca they estimate 500,000, at Lourdes 800,000. These figures certainly err on the conservative side.

The organisation of the pilgrimage depends first on the means of assembly. In the Middle Ages places of refuge were staked out along the road to St James of Compostella; a port was built on the Red

Sea at Jedda to receive the pilgrims coming to Mecca; special trains are laid on for the pilgrimages to Lourdes.

Whether the town has been specially built for the pilgrims like Le Puy, or has developed under their influence like Lisieux, it has to be organised to receive them. Everyone must be accommodated; as the pilgrims are of all sorts and conditions, modest inns and dormitory accommodation must be provided as well as luxury hotels. Attempts are made to regulate the flow and to avoid enormous influxes. The pilgrimages are spaced out over as long a period as possible. For Lourdes there is a secretariat which spaces the journeys out over five months, according to the dioceses. This afflux of population presents problems of hygiene which can be formidable when one is dealing with a poverty-stricken population, often attracted by the hope of miraculous cures. Hindu practices were particularly undesirable with mass bathing in the waters of the all-purifying Ganges; half-burnt corpses were flung in and the water was drunk unfiltered to avoid losing its holy properties. This led to terrible epidemics like the cholera outbreak at Hardwar in 1847. Quarantine periods for the pilgrims from Mecca were instituted in Egypt.

The town plan often reflects the pilgrimage's destination. Benares faces the river, which is reached by magnificent flights of steps, and there are no fewer than two thousand temples; at Lourdes a wide square allows space for processions on their way to the miraculous grotto.

But pilgrimage towns are not able to remain exclusively dedicated to the practice of religion. The sale of articles of devotion is an offshoot of the religious function. These articles are often manufactured on the spot: at Saint-Claude in the Jura mountains, rosaries in boxwood are turned; the shepherd girls of Velay used to thread rosary beads to be sold at Le Puy; today metal medallions are machine-stamped. In the same way Puy lace is manufactured for altar decoration.

This is not all. The pilgrims take the opportunity of making their purchases and shops abound to meet their needs. Mahomet actually foresaw that pilgrims would be able to bring with them articles for sale and thus earn the cost of their journey. Mecca has become one of the big trading centres of the East; it was the great coffee market of the eighteenth century, and Jedda, its port, is today the main trading centre of Saudi-Arabia, in the same way that Benares and Allahabad are great commercial centres in India.

Places of pilgrimage are often tourist centres, if the surroundings are suitable. Lourdes is situated in a highly picturesque position, and the pilgrims take the opportunity of visiting the neighbouring countryside. It is extremely difficult to distinguish among the visitors to Lourdes those who are there on a strictly religious basis and those who have come as tourists, taking advantage of the hotel and excursion facilities and regarding the religious ceremonies merely as interesting folklore.

Finally, in countries where education is in the hands of the clergy, these religious towns are often university towns. It is here that the scholars of the faith are moulded; Benares possesses a Hindu university, and at Kairouan, the centre of Tunisian pilgrimages, the great mosque is a kind of university.[1]

So gradually the facets of the religious towns are increased. These places of pilgrimage, which seem to be nothing but urban frames across which pass an endless and ever-renewed throng, house a permanent population which receives them. At Lourdes this population consists of 15,000 inhabitants in winter, outside the pilgrimage season. One cannot, of course, say the same of the 341,000 inhabitants of Benares, any more than it is possible to say which, among the 13,000 inhabitants of Lisieux or the 23,000 of Le Puy, sustain the religious function.[2]

UNIVERSITY TOWNS

The university function was in the past often combined with the religious function. The teaching of theology occupied an important place in the universities; the university of Paris is simply the descendant of the School of the Outer Sanctuary of Notre Dame, and until the Revolution it was the priests who provided this education. Higher Muslim education is organised around the mosques and many universities have been created in a spirit of religious controversy: the counterreformation was the parent of new universities. Uppsala, the most famous university of Sweden, is close beside the see of the Swedish archbishop. Finally, some universities, or some groups of faculties, in certain countries still depend on the ecclesiastical authorities.

This university function is exercised in the field of higher education, attracting many students, often from far afield. Secondary education, to which we shall refer again on p. 407, is mainly regional.

The university towns are, then, essentially towns of faculties and higher education. There are two kinds. First we have the towns which are dominated by the universities founded there, usually towards the end of the Middle Ages (Bologna, the oldest European university, dates from the twelfth century). These towns have remained faithful to their university and have not attempted to extend

[1] J. Despois, 'Kairouan', *Ann. de Géog.*, 1930.
[2] P. R. Gaussin, 'La ville du Puy-en-Velay et les pèlerinages', *Rev. Géog. de Lyon*, 1951.

their activities to other fields. To some extent their aristocratic style derives from this. People come from all over the world to lose themselves in an atmosphere untroubled by the fever of commerce and industry. And the students group themselves according to their country of origin. Uppsala still has its 'Nations' or provincial students' clubs. The list of small towns with great universities would be a long one: Heidelberg, Göttingen, Erlangen, Tubingen, Marburg in Germany; Bologna, Pisa in Italy; Salamanca in Spain; Coimbra in Portugal; Oxford, Cambridge in England; Lund, Uppsala in Sweden; Aarhus in Denmark; Louvain in Belgium; Princeton in the United States.

All these towns have a very special character: the university buildings—colleges, lecture rooms and libraries—are impressive. Moreover, university needs have increased, laboratories have multiplied and auxiliary schools for the training of engineers, technicians and nurses have been founded. The whole army of lecturers and students has to be housed and fed; hostels have been built, restaurants opened. Even the street names express the character of the town, recalling literary or scientific celebrities, or, like the Philosophenweg in Heidelberg, the preoccupations of the inhabitants.

The life of the town is completely regulated by university routine; the highlights are the university festivals when the examinations are over, or sporting events; the vacation is the dead season. The atmosphere of youth makes a deep impression on those who have experienced it; they remain proudly attached to their university in an almost patriotic sense. The attraction of these universities is such that, for their sake, students often forsake the university of their nearest large town: London for Oxford, Stockholm for Uppsala.

In contrast to these big universities in small towns are the universities in large towns which combine their university function with many others. It is indeed normal that a town should attempt to ensure that all stages of academic life, including the very highest, are available to its young people. This meets the democratic demands of a population which does not always possess the means of supporting its young for several years away from home. So now many large cities have their own university. One cannot imagine Moscow, Berlin, London, New York, Paris without one. Sometimes it goes a long way back: Copenhagen and Paris are among the oldest medieval universities. France, where the universities are, in general, to the scale of the towns that gives them shelter, has remained true to this pattern.

Sometimes a university forms a town within a town. In Paris the university was built on the slopes of Mount Sainte-Geneviève, inheriting a quarter with Roman and religious traditions. The Latin quarter has grown into a real small town while all around it commercial and industrial districts developed. Today university life still dominates the whole quarter.

The advantages and disadvantages of these two patterns have often been weighed. In the small university towns there is more community spirit among the students. Intercourse between the various disciplines, which is so beneficial, is easier. There is less difficulty in finding sites for new buildings, and the various institutes are often built in the middle of parks and gardens, in healthy and cheerful surroundings. They are in great contrast to the universities which, submerged in the big towns, try desperately to expand. There is no better example of this than the university of Paris, squeezed on to the site where Robert Sorbon built it. Buildings have been added where possible by means of eviction and expropriation right across the Latin quarter, overflowing into the Wine Market and finally, in part, moving out to the suburbs. It appears that the university function makes demands on the town in exactly the same way as certain industries.

Even so the university which is an integral part of the city and in keeping with its size does more than merely provide facilities for local students. University life is better integrated with national life and avoids the undesirable segregation of youth in a closed community dissociated from the problems of the town. Students in all countries now represent a force in politics which cannot be discounted. And it is a fact that new universities are being built in large towns, or at least in towns which seem destined to become the centre of an important region: Dakar, Oulu in Finland, Umeå in Sweden.[1]

But the contrast between these two patterns of university life is beginning to weaken. The pressure of urbanisation is tending to invade the most peaceful of towns, the oases of thought: Oxford and Uppsala have become towns which are semi-industrial, semi-academic. On the other hand, the university, increasingly involved in national life and more and more democratic in every country, refuses to be imprisoned in an ivory tower. Then again the speed of communications means that few towns remain in a state of isolated seclusion. The university of Provence was situated at Aix-en-Provence, twenty miles from Marseilles where the science and medical faculties were located; but there is no longer any question of rivalry, for the old faculties of literature and law are in effect within the suburbs of Marseilles, to

[1] In England this was so during the first half of the twentieth century, with universities in such towns as Nottingham, Leicester, Southampton and Hull since 1950 the tendency has been for new universities to be created on 'campuses' outside small towns—such as Keele (Newcastle-under-Lyme), Colchester, York and Lancaster. (Editor's note).

which they might in any case have emigrated had they been sited in the city.

Indeed in order to enlarge their buildings and laboratories, the universities are expanding outside town limits. University districts grow up becoming small subsidiary towns in embryo: Blindern near Oslo, Orsay near Paris, Fann near Dakar. The success of a large university planned at Orléans is considered to depend chiefly on the proximity of Paris. Berkeley, close to San Francisco, has the appearance of a town.

There are problems relating to university cities. For many years students have found accommodation as best they could according to their means, often in unhygienic conditions. The remedy has been, mainly in the twentieth century, the construction of hostels, halls of residence or 'cités universitaires'. Here the students are able to find decent living and working conditions at a reasonable cost. Such accommodation has been built in all university towns; and in response to great demand has been extended in order that it may be within the reach of as many students as possible. These districts sometimes have the appearance of small towns, often outside the town proper. They can be criticised as being segregated communities, of forcing the students to live apart. It is, however, certain that they represent an enormous progress in accommodating students of limited means and today they appear to be an essential auxiliary of any university.

One last problem is that of university concentration and specialisation. In former times it seemed necessary to build many universities in order to stimulate intellectual life in the various regions. In France, quite wrongly, scorn was sometimes poured on the small provincial universities with limited resources, providing public courses for the local population. But scientific and literary activities gathered round them and the students, living at home, were able to pursue their studies and make careers for themselves in their home town. Many celebrated scholars have pursued their early studies here who would never have been able to attend famous universities further away.

It seems ridiculous today not to revise the location of universities in every country in accordance with new transport facilities: on the other hand it becomes increasingly difficult to ensure that each centre possesses the complex equipment which is needed to meet the demands of teaching and research. However, in view of an increasing population it appears that it is possible, by means of specialisation, to maintain existing establishments. Indeed, in order to relieve overcrowded universities, and to facilitate more practical instruction, attempts are being made to create new institutions where first year students can receive instruction. Probably this will go even further and the university function will tend to overflow national frontiers. The reputation of this or that professor attracts students from all over the world; the exchange of teaching staff is increasing and more and more students go abroad for one or more terms, taking advantage of grants that are offered; similar degree courses are followed. We observe that the universities are thus recovering the international rôle which they often fulfilled at the time of their foundation.

Thus the function of the university has developed according to the degree by which civilisation has become imbued with science, and modern life has become dominated by technical advance. The towns exercising this function derive from it prestige and profit.

In conclusion we must not overlook education which is obtained neither from universities nor colleges but which nonetheless contributes greatly to the reputation of certain towns. It is still education that young artists seek in the vicinity of famous studios.

LITERARY AND ARTISTIC CENTRES

Culture does not stem only from the universities. A town has to provide a cultural centre for everyone. It does this by making available to all the intellectual or artistic riches which spring either from the internal life of the town or from its tourist attractions. This cultural function is exercised by books, newspapers, and works of art.

Artistic or literary production is not, of course, the prerogative of towns: masterpieces have emerged from the countryside; the fact remains that works of art are often inspired in a favourable urban atmosphere and, to some degree, exclusively in large cities. At one time a visit to Rome was an essential part of every artist's career. Almost all European painters at the turn of the last century passed through Montparnasse and it left its stamp upon them. To this day it is famous for its studios, its schools of art, its somewhat Bohemian cafés and its shop windows displaying paints and pictures. Chelsea is the London counterpart.

It is not only a question of moulding artists or men of letters; they have to be established. It is here that urban function is valuable. Printing can be done anywhere; publishing is undertaken only in large towns because only there the means of promotion, criticism and intelligent discussion are available. This is even more true of theatrical productions which call for collective appreciation and cannot exist without large and changing audiences.

Literary or artistic works are aimed at an élite. But magazines and newspapers have a large circulation. There is hardly a town which does not possess its own newspaper, weekly if the population is small, daily if it is larger. This is the normal channel by which the town influences its region and we shall return to it again. But there are also newspapers with a very large circulation and magazines which are read outside their own particular district. These are produced only in large centres; they are nationally or even internationally famous. A certain number of these such as learned or trade journals, are aimed at a specialised public. But many find their readers among all strata of the population.

Finally, we must not forget to mention scientific research which occupies an increasingly important place in the modern world.

There are, of course, experimental centres scattered across the countryside which were established there from the start. But it is in the towns that laboratories are normally found. They are in fact annexes either of universities or of industrial concerns; they work in collaboration with other laboratories, and one cannot imagine them other than in surroundings possessing adequate equipment. These centres do not necessarily correspond to the dimension of the towns where they are situated and some of them have made the reputation of small towns like Giessen, famous for the work of Liebig.

In the study of the humanities the library plays a part comparable with that of the laboratory. But while a laboratory is equipped with modern materials which rapidly become obsolete, a library acquires a background amassed over the years: the destruction of certain libraries during the war caused irreparable losses. A cultural function has long been exercised by the towns that possess these valuable libraries.

MUSEUM TOWNS

There are not only religious pilgrimages; it is possible to speak also of lay cults. Is it not such a cult which every week in Moscow attracts impressive queues of visitors to Lenin's tomb in Red Square?

There are other towns where people go to obtain aesthetic rather than spiritual satisfaction. Naturally every town has its museum, and it is extremely difficult to say how many of the tourists who go to Paris are attracted there by the Louvre, to London by the National Gallery or to Madrid by the Prado. This is all part of the outward appearance of the town; but there are towns the attractions of which derive more particularly from their art treasures. Visitors go to Florence and Dijon for their art galleries and museums, to Rheims and Canterbury for their cathedrals. The palaces of Versailles and Fontainebleau are their main attractions.

And what can be said of towns which are themselves museums,

8 Venice: the city rising from the waters; Lido in distance

9 Monte Carlo and the rocky promontory of Monaco

10 Rome: the Forum and Trajan's market; on the right, the Loggia of the Knights of Rhodes

11 Palermo: narrow street near the port

12 Rome: Piazza del Popolo from the Pincio gardens; in the distance the dome of St. Peter's

where visitors seek the charm of past splendours? Venice is one of the most important of these. The magnificence of the ancient city of the Doges, which can be seen along the whole length of the Grand Canal from St Mark's Square, is the major attraction, as well as the strangeness of this city where canals take the place of streets. There are a great many visitors and the trade of the town, as in the places of pilgrimage, depends mainly on the souvenir shops. On the other shore of the Adriatic, Dubrovnik too displays the charms of an ancient city, with its fortifications and its tiny alley ways leading off from the main avenue.

Again in the Spanish towns of Avila and Seville or in the Hanseatic town of Lübeck one breathes the air of other centuries and other civilisations. The Acropolis and Parthenon are relics where people come to meditate at the cradle of Greek thought.

Each town attempts to exploit to the utmost its advantages and to advertise itself by placards along the highways; at nightfall public monuments are floodlit, and festivals, enhanced by the beauty of their setting, are combined with the tourist season.

The function of the museum town leads to other requirements, and expansion means the building of motor roads, excessively high buildings, and residential blocks which disfigure the town; a protest was recently made for the protection of Venice.

FESTIVAL AND CONGRESS TOWNS

In the same way that there are temporary trade fairs, cultural life also has its temporary gatherings. At certain times some small towns assume the function of large cultural centres. This is the case with dramatic or musical festivals. Oberammergau in Bavaria is a small village which, at the time of the Passion play, is animated with the life of a town. Bayreuth, as a result of the Wagnerian cult, and Salzburg both hold famous musical festivals.

But the form of cultural event which is increasing more and more, is that of the congress. An increasing need is felt to meet and to discuss the great scientific, professional, social and economic problems. And these conferences often have an international character.

For a long time seaside resorts have, during their dead season, accommodated congresses. The number of members continues to grow and is further increased by secretaries, typists and interpreters and technical requirements grow more and more demanding, so the towns transform themselves into congressional centres in order to take advantage of these temporary invasions. They offer, apart from numerous hotels of every category, conference halls especially equipped with large amphitheatres capable of holding 1,500 or 2,000 people, equipment for simultaneous translation, committee rooms, restaurants.

It is, of course, possible to organise all this in the great capital cities, but to avoid the complicated life of these cities the profitable honour of being congressional cities is often accorded to medium sized towns in advantageous positions (for example, Cannes, Evian, Dijon, and Liège, or in England Brighton, Blackpool and Harrogate).

CHAPTER 15

THE TOWN AS A RESORT

It is a function of certain towns to provide the means of cure, rest or distraction. The essential factor is that the largest part of the population is temporary and possesses elsewhere—generally in another town—its permanent residence and is here without employment.

HOSPITAL TOWNS AND SPAS

Every town naturally possesses hospital accommodation and admits patients living in the district. But there are other towns which derive their function from the hospital services which are concentrated within them, notably when these services are specialised. The small town of Clermont de l'Oise houses the largest psychiatric hospital in France with four thousand patients and, out of a population of six thousand, two thousand are employees of the hospital.[1]

Sanatorium towns. On the other hand certain therapeutic treatments have been developed in towns far from any previous urban concentration. The cure of certain diseases requires sun, sea, or mountain air. At first sight it would seem that there is no reason for special urban concentration; the sun shines everywhere and if its rays are more beneficial at certain altitudes, accommodation can be found here and there in the mountains. In fact it is not the material conditions but the requirements of the treatment which necessitate urban facilities. This is particularly the case with tubercular conditions, whether of the lungs or of the bones. These are patients who come with the sole purpose of seeking a cure and who have to follow very strict régimes under close supervision by the medical staff. They are therefore grouped around centres where medical attention can be concentrated.

Thus towns for the sick have been created and reserved for them. In the case of tuberculous diseases risk of infection precludes other visitors and the town adapts itself to the requirements of the cure, for example by insisting on silence round the hospital area. For the treatment of pulmonary tuberculosis dry, well-sheltered and sunny plateaux are needed; these are numerous in Switzerland which is particularly suitable because of its altitude and railway communications. Since 1860 Davos has become a sanatorium town at an altitude of 5,100 feet; others have followed of which Leysin is the

[1] M. Ramponi, 'Clermont de l'Oise', *La vie urbaine*, 1961.

best known. For tuberculosis of the bones the seaside is indicated and in France an example of this is Berck on the shores of the English Channel.

All these treatments vary according to the progress of medical science. Residence in the mountains is no longer the only method of treating pulmonary tuberculosis, consequently towns brought into being to provide cures are falling into decay. Switzerland, which had become one vast sanatorium, has experienced very adverse effects. Fortunately other forms of the tourist industry are replacing medical treatment and the hotels of Leysin are being sold for conversion into holiday flats.

Spa towns. From time immemorial man has attributed therapeutic virtues to certain waters. These are generally springs gushing from the depths of the earth and containing radioactive particles, sulphur or other mineral substances. These were especially sought after when the pharmaceutical industry was less developed; the waters had to be drunk on the spot as they could not be transported in any great quantity. From times of antiquity some of these springs were famous and were responsible for the foundation of towns (the Roman Aquae). People have visited them ever since, often travelling long distances, and this frequenting of spa towns became a fashion not always justified on health grounds. Healing springs were sought in all directions. This at first was simply a matter of establishing exclusive institutions for high society. The democratisation of means of transport has increased the clientèle; modest hotels and guest houses are within the reach of all purses.

Easy access is of primary importance; Bad Ilitze, near Sarajevo in Jugoslavia, possesses water with a sulphur content, but it is too inaccessible to provide a spa. A pleasant environment is also necessary; Forges-les-Eaux, a spa near Paris was abandoned in the craze for the mountains.

But all that is of little importance compared with the proximity of a sufficiently large population with a high standard of living. There are practically no spas in underdeveloped countries. In Brazil there are luxury thermal health resorts but these have not brought towns into being. It is along the fissures of central and western Europe that spas are particularly favoured.

The success of the resort depends very much on the style in which it is launched; Vichy became very fashionable thanks to Prunelle, the medical inspector, and to Napoleon III.

We may find it surprising that American civilisation has not given rise to such resorts in spite of the density of a large population with a high standard of living and with easy transport facilities; but it was also an essential factor that these towns were developed at a time when thermal treatment was fashionable, when they benefited from a long tradition. Recent civilisations have usually attributed more therapeutic value to the sea or the mountains and there are few towns in the United States which, like Hot Springs, play this rôle.[1]

Towns of a very special nature are grouped round a thermal establishment. Luxurious hotels stand beside modest family guest houses. People come for treatment of a limited duration, but they are not patients in the ordinary sense of the word. The treatment only takes up a short part of the day and they expect holiday amenities as well. Parks are laid out; seats in the nearby woods are arranged at strategic points with a fine view; there are also golf courses; concerts and theatres are provided. Recreational facilities are a prerequisite of success and indeed often become the essential factor. In addition to those taking the cure are visitors who come to take advantage of these other amenities, making the latter all the more essential for the success of the centre. In the last resort the casino attracts more visitors than the thermal spring. It is significant that the company exploiting the springs often does not sell the water; this is offered free to all visitors, and profits are confined to the exploitation of hotels, casinos and race courses.

The most popular of the spa towns may accommodate more than 100,000 visitors; 131,000 visited Vichy annually before the Second World War. The permanent population is not only connected with the hotel industry; the exploitation of thermal springs has become an industry, for example, the bottling of thermal water for export and the manufacture of tablets at Vichy. Confectionery and pharmaceutical products seem to be auxiliary industries, and there are others more or less linked with the presence of a wealthy and invalid clientèle (rubber hot water bottles; spare parts for automobiles). To fulfil these requirements the agglomeration of Vichy possesses 44,000 inhabitants.

These spa towns have a somewhat precarious existence. They are scarcely frequented except in seasons when travelling is easy and pleasant, at the most for six months of the year. And it is difficult to prolong the season; well-known doctors, actors and musicians and high-class shops pursue their activities elsewhere during the rest of the year. Then there are fashions in cures and the clientèle has other engagements. Vichy, now forsaken by its former devotees from the erstwhile French territories in North Africa or the Near East, is seeking resources other than thermal springs; other forms of entertainment are being developed and it is attempting to become a sports town.

[1] David Lowenthal, 'Tourists and thermalists', *Geog. Rev.*, 1962.

Thus spa towns tend more and more to resemble holiday resorts in their way of life as in their problems.

HOLIDAY RESORTS

It seems paradoxical to see towns grow up which arise from the need to escape from the town; for it is a fact, and one most typical of twentieth-century civilisation, at any rate in the western hemisphere, that town dwellers periodically feel the need to escape from their urban environment.[1] But although many attempt to disperse in the mountains or in the country, many too, while escaping from small flats, find that during the hot summer months they miss the social life, the parties and entertainments to which they are accustomed. Has it not been said that man is an individualist provided he can be one within society? Some entertainments can only be organised around large centres. Golf courses and race meetings are hardly possible as auxiliaries to a village inn. Those staying in furnished accommodation and even in guest houses expect the facility of being able to buy supplies outside. Stocking up with foodstuffs at reasonable prices is only possible in centres already possessing an urban form. Hence the growth of towns with no other function than *Fremdenverkehr*, the supplying of visitors who have come from other towns. Of course these do not represent all holiday resorts; outside this classification there is a large number of isolated hotels, villas and resorts scattered across the countryside and the mountains.

To some extent these holiday towns replace other forms of urban life; they are particularly numerous in countries where there is a densely populated urban network, large agglomerations and a high standard of living. They were mainly developed during the nineteenth century; in earlier centuries privileged members of society were content to go off to their country estates. They are rare in countries which have remained essentially rural in character; in developing countries they hardly exist except in places accessible from the large towns where there are many foreigners.

Holiday towns can be placed in three categories, according to the desire for sea, mountains or sun. Within the limits of the financial resources available the individual has a free choice. There is also an element of fashion.

Seaside resorts. As early as 1823 a hotel was established at Arcachon for the use of the inhabitants of the Bordeaux region; however it was not until the period of the Second Empire that sea bathing became fashionable. One can, of course, bathe almost anywhere along the

[1] Georges Chabot, 'L'évasion urbaine', *La vie urbaine*, 1957.

coast, but people have crowded round places which are specially favoured by the character of the coast. The beach must be large enough to allow for a considerable number of bathers, and yet not too long, to allow a degree of concentration. Fine sand is preferable, as pebble beaches are not attractive to most people (though thousands sit on the rough flints of Brighton because there is nothing else to sit on). Favourable climatic conditions are important; warmth is necessary—resorts are more numerous in the south of every country. A microclimate is desirable which offers beaches exposed to the sun and sheltered from the wind. The hinterland too is important, forests are popular, as at Le Touquet, or picturesque landscapes, as at Biarritz. In order to attract large numbers of people to take advantage of these conditions, the places must be easily accessible. Every great city has its favourite beaches and these give rise to towns: Brighton, which has a population of 162,000, began as a beach on the south coast used by Londoners[1]; Atlantic City is the beach of Philadelphia; Bordeaux has Royan and Arcachon; Montpellier has Palavas[2]; the inhabitants of all the Lancashire towns congregate at Blackpool. But there are also beaches which attract visitors from far away—the inhabitants of Megalopolis meet in Miami, and the magnificent sandy beaches of northern Poland—as at Swinioujscie and Kolobrzeg—attract holiday-makers from Sweden, where there are no such beaches.

The way in which these towns originated is also worth noting. Le Touquet—Paris-Plage, where the first villas were built in 1882, made a slow start until, in 1902, an English company took the business in hand and turned the place into a luxury resort which now receives 30,000 summer visitors, 42 per cent of whom are foreign, and which has a permanent population of only 3,600.

A beach has sometimes saved a small port from decay, as in the case of Les Sables-d'Olonne which was ruined by the decline of the cod-fishing industry.[3]

All sea-bathing resorts are planned in relation to the sea front; the best houses and largest hotels are near the sea, on the road running along the coast; the casino is as near as possible to the beach. Behind these stand the more modest hotels, boarding houses and shops. Sports grounds, golf courses and even race tracks develop around the town.

Nevertheless these seaside towns differ widely. There is more rigid segregation than in any other type. There are expensive resorts with

[1] E. W. Gilbert, 'Inland and seaside health resorts in England', *Scot. Geog. Mag.*, 1939.
[2] F. Doumenge, 'Un type méditerranéan de colonisation côtière: Palavas' *Bull. Soc. languedocienne de Géog.*, 1951.
[3] A. Huetz de Lemps, *'Les Sables-d'Olonne'*, 1951.

splendid villas and high-class hotels, and others used by families of more modest means.[1] In some cases there are twin towns. Deauville, built complete at the mouth of the river Dives, was launched under the Second Empire as a fashionable resort; but the old fishing-port of Trouville became the more humble resort with no four-star hotels; near Le Touquet, Stella-Plage was devised to take the bathers who were unwelcome at Le Touquet. On the French Riviera, there is one resort after another, each having a well-defined class.

Beaches are becoming increasingly democratic. At Mers, on a beach 1,200 metres long, an aristocratic resort was set up. During the Second World War the casinos and large sea-front hotels were destroyed. These were not rebuilt, and today the visitors are not so wealthy. Nowadays campers are among the customers of a resort. They should not perhaps be counted as part of the town, and yet the town's activities could not be considered complete without them. At Valras in the Languedoc they represent a tenth of the population of 30,000 summer visitors.

Mountain resorts. Recreation in the mountains lends itself much less to urban concentration. It extends over vast expanses of territory, instead of being linear as in the case of beaches. However, certain forms of mountain tourism do lend themselves to concentration. This applies especially to winter sports for which special equipment is needed, such as chair-lifts, ski-lifts, etc., which can only be afforded by a few. Furthermore, the short duration of the winter days necessitates the organisation of communal amusements, and this has caused the formation of centres like Megève (pop. 3,600) and Garmisch-Partenkirchen.

Moreover, the tourist industry in the mountains revolves around certain centres, places supplying provisions to people holidaying in isolated spots, departure points for races, gathering centres for guides and instructors, like Interlaken, Innsbruck, Chamonix and Zakopane.

Sunshine towns. The sea and the mountains are not the only things which attract visitors. The sun makes longer and more rewarding holidays for all. Inhabitants of northern countries feel a real longing for it, and this is the reason for the success of the towns of the Riviera, the towns which spread over the foothills of the Italian Alps, the towns of the Crimea and of Florida. The combination of sun and sea increases the attraction of these places. Monaco receives 100,000 visitors a year, with a population of 22,000, of whom 3,000 are nationals. In the past these towns were sought after chiefly in winter. Sunbathing now guarantees them a double season. People are moving

[1] A. Amouroux, 'Valras', *Méditerranée*, 1960.

further and further south in search of the sun, and with this in mind, towns on the edge of the Sahara, like Biskra, are being prepared.

The problems of holiday towns. These towns equipped with facilities for the amusement of townspeople from elsewhere have grown up in many different ways. Some developed round an ancient nucleus, a mountain village or small fishing town. In such places, the nucleus has usually remained intact, only a few additions—balconies, verandas and garages to the houses which were sold to outsiders—having been made; new villas have been arranged around the nucleus, in a more or less geometric pattern. Sometimes, when the resort has been very successful, as in the case of Biarritz, the old nucleus has been submerged or destroyed.

On the other hand, there have been towns built almost complete, following a pre-established plan—rectangular or grid-iron in the nineteenth century, more skilfully curved in the twentieth. In both varieties, life—very worldly—revolves around a centre surrounded by hotels, cafés and cinemas. The town then expands according to its rôle; along the coast in seaside resorts, fanwise in mountain towns.

All these towns have problems in common. They have small permanent populations which may increase tenfold at certain seasons. They are at the mercy of fashion. Such and such a beach which was once all the rage may see its reputation declining. The clientèle is always fluctuating. People move around very easily; customers who come by car often leave after only a few days. Fortunately the owners or tenants of villas remain, if not as permanent residents, at least as a seasonal population.

Lastly, ease of travel does not favour urban concentration; it is no longer necessary to live in the town itself to take advantage of its facilities. People settle in the surrounding district, because, as the development of towns creates noise, they cease to be places of repose. Chamonix was once a place for country holidays; now it depends increasingly on commerce, augmented by a clientèle of tourists from the surrounding area. The town is becoming a sort of urbanised whole, very loosely knit, grouped round a communal centre.

This is especially true of the *condominios*, quoted by Pierre Deffontaines, which are being created in Brazil, centres around which isolated *sitios* for recreation are grouped to maintain a communal social life; these are clubs offering facilities to their members—restaurants, meeting rooms and swimming pools.[1]

However, the principal problem in all these towns is the short duration of the season, as in health resorts. Enough money must be

[1] P. Deffontaines, 'Du patrimonio au condominio. Contribution à l'étude de la géographie urbaine du Brasil', *Cahiers d'Outre-Mer*, 1961.

made during the short summer season to live on for the rest of the year. The further north one goes, the shorter the season, for in September the weather begins to deteriorate—at least in temperature. Efforts of various kinds are being made to remedy the situation. Staggered holidays would lessen the disadvantage, so pressure has been put on public authorities to alter the dates of school vacations and on factories and businesses to spread their holidays; reductions are made on hotel bills at off-peak times. It is more advantageous to have two seasons. Mountain resorts equip themselves as ski stations. The Riviera, once in great favour in winter, is now more affluent in summer. But off-seasons when tariffs must be reduced are inevitable. Hotels recruit waiters and chambermaids for the season only, from the surrounding country. The pity—in some countries at any rate, though this would hardly apply to Britain—is that they are taken just when they are needed to work in the fields. This is the advantage of winter sports resorts which use this labour force while it is idle.[1]

Now these seasonal visitors have deserted other towns. This coming and going is turned to good account. Employees are recruited for the summer season, from the staffs of large restaurants and expensive shops, and artistes who are liable to be disengaged in the large cities at that time of year. Some business houses have two shops which take turns in this way. So reception towns are formed, superfluous, luxury towns which have become the annexes of big, working towns. They might almost be described as 'reserve' towns, and they are, like watering-places, valuable on many occasions. They can house congresses and international conferences without difficulty. Their names are landmarks in diplomatic history. During the Second World War, Vichy, one of these resorts, was able to receive the entire French administration. And after the war, Bad-Godesberg, a town with hot springs and green parks, housed many diplomats when the federal capital of Bonn was set up six miles away.

It has sometimes been asked whether these really are towns at all. Certain American geographers have denied Miami this name, although it has a population of nearly 300,000. It is true that the town does no work destined for the service of areas outside it, which is the usual characteristic of the urban function; but all the elements described in the definition of towns are to be found there. Supplying opportunities for rest is a contribution to work, and in the study of a town parks and cinemas are not ignored.

It could be asked whether this function of recreation is tending more towards an urban form, or whether townspeople are going to prefer to scatter over the countryside which is becoming increasingly

[1] G. Veyret-Verner, 'Le tourisme au secours de la montagne', *Rev. Géog. Alpine*, 1956.

comfortable and convenient. It must be remembered that up to now the function of recreation has appealed almost exclusively to the inhabitants of towns. The rural world has been practically excluded from it. It is desirable that rural recreation should follow. As tourism is based essentially upon people's desire to move into a different environment, it may be supposed that inhabitants of the country will, as tourists, prefer to visit towns.

The study of recreation towns is by no means easy. The census is generally taken in winter, in order to avoid the disadvantages of population movement, and only gives information about the permanent population, without, incidentally, any exact indication of professions or their relations with the tourist trade. Information bureaux and chambers of commerce give the number of beds at the disposal of visitors, that is, the reception capacity at peak periods. Bus companies only give the number of their services, and railway tickets are seldom issued in the town's station. Therefore indirect means must be resorted to; the supplies of flour to bakers, for example.[1]

TOWNS FOR THE RETIRED

Up to now, under the heading of resorts, or the function of reception, we have only been considering 'temporary' or seasonal towns. But there are also towns which form permanent places of refuge where people go for material or social advantages. These are towns for the unoccupied, for people who have accumulated money elsewhere and who have given up all activity: towns for the retired.

These towns are selected from among those with a pleasant climate which offer the diversions of a tourist town or which are near to a large town without being overcrowded. Thus all round Paris there are small towns where the retired people of the capital are happy to take refuge, returning to the capital occasionally for a break in their tranquil lives. In England, places on the more relaxing parts of the south coast, such as Worthing and Bournemouth—have a similar function.

In France the most typical town in this category is Nice, which numbers 136,000 inactive inhabitants against 100,000 who are active. Individuals over sixty-five years old represent 12 per cent of the population. The census of 1954 found 26,000 retired people from

[1] E. Juillard, 'La Côte des Maures', *Rev. de Géog. alpine*, 1957; M. Boyer, 'Le tourisme dans le sud-est Méditerranéen français', *Bull. de la Section de Géographie du Comité des Travaux Hist. et Scientifiques*, 1958; F. Cribier, 'Variations de consommation de farine et migration touristique d'été en France', *Bull. Ass. Géog. français*, 1961.

public administration and private industry—the former managers and directors of industry and commerce.[1] The town combines provision for retirement with that for visits of varying lengths by the leisured rich. High-class businesses and banks are particularly well represented. There are 549 hairdressing salons employing 1,000 people, one doctor for every 500 inhabitants, and a dental surgeon for every 1,500. The town is kept scrupulously clean, with many gardens and trees; it is well-lit, and constantly arranges amusements of all kinds, concerts, festivals and exhibitions.

[1] R. Blanchard, *Le Comté de Nice*, Paris, 1960; cf. also 'La Haye, ville de retraités', P. George, *Bull. de la Section de Géographie du Comité des Travaux Hist. et Scientifiques*, 1958.

Chapter 16

THE ADMINISTRATIVE AND POLITICAL FUNCTION

THE ADMINISTRATIVE RÔLE

All the functions we have studied, commercial, industrial and cultural, form and develop more or less spontaneously. Even the regulation of markets by the authorities only offered a town something that it could exploit and incidentally this was usually a confirmation of an existing function. This is not the case with the administrative function. By conferring this function on a town, public authorities give it a certain primacy to which other towns are obliged to submit, and they guarantee the proper exercise of the machinery of administration. So a certain activity is, *ipso facto*, maintained, whatever the circumstances.

The administrative function does not apply only in towns, but communal or district administration is too restricted to be considered as an urban function. At higher levels administration probably coincides with other functions, but the rôle of a capital, and the court of justice which accompanies it, is far from being negligible. This is especially true for the chief town of a division (district, county, etc. according to the country). The various branches of administration unite in these towns: judiciary, university and ecclesiastical authorities have them as their centre. The increasing complication of administration and the expansion of economic undertakings multiply their services.

Care is usually taken to establish this administrative function in the principal town of the region. This has sometimes been a delicate question, where several towns have competed for primacy; as happened in France when the departments were being established. In the Jura, three rivals towns, Lons-le-Saunier, Dôle and Saint-Claude, stood as candidates; an itinerant prefecture was even considered and while Lons-le-Saunier was decided upon as being the most central, the bishopric was centred on Saint-Claude, by way of compensation.[1]

This function of capital enjoys a certain stability. A town's way of life is not suddenly upset, nor are its prerogatives needlessly removed, if it has been carrying out its activity to the best of its ability. So some towns retain administrative supremacy in spite of

[1] J. Brelot, *Le Jura pendant cent ans*, Lons-le-Saunier, 1953.

the growth of its neighbours, which remain subordinate. Quimper has a population of 19,000, while the subprefecture of Brest has 110,000; Châlons has 37,000, while the subprefecture of Reims has 121,000.

NATIONAL CAPITALS

Location. The administrative function can be seen fully developed when coupled with a political function: it is no longer a question of being the centre of a region, but of being the supreme centre of a State. The function of national capital is among those best adapted to the development of large towns.

These capitals are of course very varied. They depend in the first place upon the dimensions of the State. Some of them were identified in the past with the State itself, which was scarcely bigger than the town: the towns of ancient Greece and medieval Italy, for example. Without even mentioning the tiny principalities, the tasks of Reykjavik, capital of Iceland, whose population is 185,000, and of London, capital of a country with a population of over 50 million, cannot be compared. The right of peoples to run their own affairs brought about a great increase in the numbers of states, and therefore of capitals, after the First World War, as did independence in Africa after the Second.

The rôle of the capital varies from country to country. There are cases where the capital takes over all administrative and political tasks; this happens in centralised States. On the other hand, some States assume the federal form; these consist of a union of states, each having its own capital; the federal capital only takes charge of communal matters; a large proportion of the administrative work falls to the states; and this can cause conflict, as has been seen with the measures taken against racial discrimination in the United States. Often care is taken not to make one town both federal capital and capital of a state, for its influence would then be dangerously increased. Rio de Janeiro used to be federal capital of Brazil but Niteroi was made capital of the State. Federal districts were formed for Washington, Canberra and Brasilia; the relationship between Moscow and the Soviet Socialist Republics is more complicated still.

If the capital can only be studied in relation to the State, the converse is also true, that the future of the State is inseparable from that of its capital. To foreigners the capital is the symbol of the State. In time of war a country seldom continues to resist after the surrender of its capital; the armistice of Leoben was signed at the gates of Vienna, and the armistice of San Stefano at the gates of Constantinople.

A country's activities are directed, whether it likes it or not, towards its capital. In France people from north and south alike all refer to 'going up' to Paris. This is where a country's history is made, and there is more and more pressure towards centralisation. For the last 150 years, French political systems have been made and unmade in Paris. Paris was the scene of the uprisings of the Revolution, and the Girondins' scheme of reducing the influence of the capital to an eighty-third part was only a whim.

The choice of site for a capital must therefore not be haphazard. The location has been imposed in some countries by the presence of a primate town; sometimes it has been the result of a sovereign's influence. Addis-Ababa was built round the imperial palace, to be the capital of Abyssinia; Madrid was a royal residence to which Philip II brought his administration, forsaking the ancient capitals of Burgos, Valladolid and Toledo.

The problem of the geographical situation of capitals has often been studied;[1] sometimes with hindsight to discover the advantages and disadvantages of a capital city; and sometimes with the planning of a new capital in mind, as in the case of Canberra and Brasilia.

The question of relations with the rest of the country is of primary importance. A central position would therefore seem the obvious choice. Madrid is in the centre of Spain, Rome in the centre of Italy, Moscow of European Russia and Ankara of Turkey. Situated in the middle of a country, the capital is less exposed to assault from an enemy in case of land invasion; Belgrade was forced to contrive a fortified slope for its protection on the left bank of the Danube. It has been calculated that Brussels stands at the centre of gravity (according to centrographic formulae) of Belgium's urban population.[2]

But the geometrical centre is no more the vital factor than the centre of gravity; everything depends on facilities of communication and the position of valleys. When a country's main artery is a river, the capital can only be placed on the banks of that river: Warsaw on the Vistula, Vienna on the Danube. Railways have altered the possibilities. In order to place Antananarivo in the centre of the country, a railway was hurriedly built to link it with the coast.

A country's centre of gravity cannot be indicated solely by equidistant places and isochrones; it depends upon the distribution of the population, and the wealth of certain provinces. At the time when Washington was founded in the United States, there could have been no question of placing it anywhere but in the north-east. The

[1] V. Cornish, *The great capitals*, London, 1922.
[2] L. Dethier, 'Contribution à l'étude du réseau urbain de la Belgique', *Bull. Soc. belge d'Etudes géog.*, 1962.

Australian capital could only be in the south-east, where the rainfall guaranteed sufficient resources, and Antananarivo was in the centre of the most highly populated part of Madagascar.[1]

This centre of gravity has more than a merely economic significance. The districts or provinces which have played an important part in the formation of the country must be taken into account. The Electors of Brandenburg, who became Kings of Prussia and then Emperors of Germany, brought with them the fortune of their capital, Berlin. It could equally be said that France was formed round the Duchy of France, the capital of which gradually extended its ascendancy over the whole kingdom. If Ankara was chosen as capital of the new Turkey, it was probably partly because Istanbul seemed too far away from the centre, but also because Ataturk found his greatest support in that province, and the town was the seat of the National Committee.[2]

It sometimes happens that a whole State is reduced to being the hinterland to its capital. Denmark is scarcely more than the hinterland of Copenhagen, and, especially in the past, England was the hinterland of London. In this way macrocephalous States have been formed, in which the lack of balance between the capital city and the rest of the country presents very difficult problems.

This is not all: a capital expresses the whole aspect of its country. Colonies of necessity had their capitals on the coast of the sea joining them to their metropolis. Rio de Janeiro, Montevideo and Buenos Aires were established on the coasts which received ships from Europe. Calcutta was chosen as capital of British India. When Finland was attached to Sweden it had as its capital Turku on the west of the Gulf of Bothnia; when conquered by Russia at the beginning of the nineteenth century it had to transfer its capital to Helsinki on the Gulf of Finland. Norway had Trondheim as its capital, a port of easy access joined to the south-eastern regions by the great valleys of Gudbrandsdal and Österdal, while Oslo at the end of its fiord shaped like the finger of a glove, was hardly accessible by sailing ships; but Norway was joined to Denmark and then to Sweden, and whereas Trondheim was too far from the other capitals, Oslo was nearer to its Scandinavian neighbours, and became the capital even before steamships and railways released it from its difficulties and made it the centre of Norway. China had six successive capitals, following the vicissitudes of politics.[3]

Nothing is more significant than the movements of the Russian capital. Moscow was at the heart of the self-centred empire of the Czars. When Peter the Great wanted to open it to Europe, he built

[1] C. Robequain, 'Tananarive', *Rev. de Géog. Alpine*, 1949.
[2] Akçura, 'Ankara et ses fonctions urbaines', *La vie urbaine*, 1960.
[3] Chiao-Min Hsich, *Congrès int. de Géog.*, Stockholm, 1960.

St Petersburg at great expense in the marshes, and for two centuries this town was the Russian capital. When Russia became the Union of Soviet Socialist Republics she disowned the Czarist heritage, and Moscow recovered the rôle of capital.

Some capitals are of recent creation, decided upon after long deliberations; the problem of location is seen here at full strength. Canberra was officially founded on 1 January 1911 in a district specially created for it, and parliament met there in 1927.[1] The town was chosen at the limits of the temperate and subtropical zones in New South Wales, the richest and busiest State in Australia, between Sydney and Melbourne, although somewhat off the route joining those two cities. More recently public opinion has been aroused by the problem of capitals with the creation of Brasilia.[2] Canberra was chosen in the wealthiest province, near to two towns with populations of about two million. In Brazil, on the other hand, Brasilia, inaugurated in 1960, was deliberately placed far from the two great rival cities of Rio de Janeiro and São Paulo, and, by a decision reminiscent of that of Peter the Great, President Kubitschek sited the capital in the poorest region. This region cannot support its existing population, thousands of workers leaving it every year for the towns of the south; but it is a region with a stake in the future, particularly hydro-electric power from rivers like the São Francisco. It was nonetheless a gamble to found a town in a semi-arid region far from large towns and for which completely new roads and railways had to be built. This foundation has been described as an act of faith: enthusiasts[3] already see businesses multiplying round the capital and irrigation bringing about a flourishing agricultural activity, while pessimists calculate the initial outlay, the cost of its operation, and the illusory nature of the agricultural transformation.

It is only through use that it can be seen whether such a capital was well or badly sited. It has been said[4] that Madrid was badly sited to be the capital of an overseas empire, that Seville was more worthy, as was Athens in ancient times, and Lisbon in more recent times, and that this unfortunate position may have precipitated the downfall of the Spanish empire. There are also capitals which have difficulty in making themselves accepted. São Paulo has never lowered its flag to Rio de Janeiro, and has sometimes come out in open revolt against Rio, and this was a factor in the decision to found a new capital.

[1] E. Reiner, 'Canberra', *Geog. Rundschau*, 1961.
[2] P. James and S. Faissol, 'The problem of Brazil's capital city', *Geog. Rev.*, 1956.
[3] J. Boudeville, 'Brasilia', *Industrie*, 1960.
[4] Cornish, *op. cit.*

Growth and survival of capitals. In spite of a few recalcitrant rivals, every capital takes on an almost sacred character; it becomes a country's heritage. Time strengthens this prestige, which remains attached to it even if the capital itself is dethroned. Istanbul is still the leading city of Turkey, even though the capital has been transferred to Ankara. It often happens that these unattached capitals tend to recover their original rôle. Delhi, the former religious and political capital, had its functions restored to it in 1912; Moscow, which remained the holy city, regained its place as capital in the twentieth century, and in reunified Italy, Rome seemed the symbolic capital.

This is because in a state, the tasks of a capital grow continually. Government and administration require an increasingly large personnel, and the functions of the state have multiplied in modern times. Government machinery becomes more cumbersome. Ministries proliferate, requiring an ever larger civil service. Members of parliament, foreign ambassadors and consuls settle in the capital; and the desire to be at the central point near the 'control room' concentrates many businesses in the capital. Large finance companies and banks have their head offices there, and we must not omit theatres, museums maintained by the State, large newspaper offices, publishing houses, and the luxury industries attracted by the presence of a rich clientèle.

The capital becomes a tourist attraction. Four million tourists (not counting businessmen) visited Washington in 1951 and the tourist function in this capital is second only to the administrative function. So all capital cities tend to grow automatically and quickly: Brasilia, only founded in 1956, acquired a population of 65,000 in less than ten years. And old towns, once promoted to the rank of capital, grow at a prodigious rate. Ankara, which in 1923 was a small town with a population of 30,000, had 453,000 by 1955; Bucharest had a population of 100,000 when it became capital of the principalities of Romania in 1859; a century later it had a population of 1,300,000. The population of Moscow, which stood at one million at the time of the First World War, increased sixfold forty years later, after its capital status was restored.

So a capital naturally tends to become the first town of a country, if it is not so already. Rome has overtaken Milan and Naples, which had larger populations at the beginning of the century. All the capitals of Europe, excluding the very special cases of Bonn and Berlin, are pre-eminent towns, except for Brussels which cannot compete with the port of Antwerp, and Berne, which is the capital of a federal State. In federal States, in fact, concentration towards the capital often happens within the associated States, and this is why Ottawa, Washington and Delhi, not to mention Brasilia, are all equally exceptional to the general rule.

This rapid growth is, however, in no way comparable to that of an ordinary town; it depends on the rôle of the capital and happens in relation to that function. The capital always seems something of a luxury town. Houses of parliament and embassies occupy an important place, as do ministries and public buildings. Broad vistas are contrived wherever possible to form an impression of splendour, and State revenues are spent freely on the embellishment of the capital. Successive French governments have worked towards the formation of that architectural harmony which, starting from the Louvre, is composed of the Tuileries, the Champs-Élysées, the Arc de Triomphe, the Place de l'Étoile, and goes as far as the Rond-Point de la Défense. The plan tends to be centred on the palace of the sovereign: Karlsruhe is built in a semicircle round the grand-ducal palace, Dijon round the Place d'Armes, opposite the ducal palace; Versailles was built in relation to the palace. At Nancy King Stanislas planned the grandiose display of gardens which run from the governmental palace to the town hall. Washington is built round two poles, the White House and the Capitol; at Canberra the buildings are arranged concentrically around Capitol Hill. In Ankara, it is striking to see the development, adjacent to the old quarter swarming round its mosques and open air markets, of a quite recent capital district, separated from the old part by the railway, and aligned majestically along the Ataturk boulevard, with ministries, embassies, villas, and the palace of the National Assembly. Even if it is less distinct than at Ankara, a kind of segregation often arises in capital cities, with a district for ministries, and a district for embassies. It should be noted, however, that capital cities do not have the monopoly of all branches of activity. New York, Zurich, Calcutta and Istanbul all exercise cultural and economic pre-eminence over their respective capitals of Washington, Berne, New Delhi and Ankara.

A sort of proliferation of capitals has been referred to. In Germany, following on the special and peculiar status of Berlin, Bonn became the political capital, but Hamburg remains the great centre of commerce, and Frankfurt the centre of finance. In South Africa, Pretoria has the executive power, Cape Town the legislative power and Bloemfontein the judiciary power. In Britain and in France there is talk of decentralising the government by evacuating certain ministries to major provincial cities. There was a proposal in 1963 to send the French parliament to Orléans, the Ministry of Finance to Lyons, and the Ministry of Justice to Bordeaux; and in England the Ministry of Social Security (formerly Pensions and National Insurance) already has its seat in Newcastle-upon-Tyne.

SUPERCAPITALS

Over and above States and their capitals there are more comprehensive organisations which pay no regard to frontiers, and they too have their capitals. These could be called supercapitals. The colonial empires had capitals which ruled over local ones, and London remains the centre of the British Commonwealth.

There are also religious capitals for the faithful all over the world. The Vatican City is one. This city has formed an enclave within the Italian state since 1871, and its independence was recognised by the Lateran Treaty in 1929. This is not strictly speaking a State, because its citizenship depends upon the employment offered; nor is it really a town, because the City, with its population of about a thousand, receives from the Roman services its water, gas and food supplies. But it is from there that mandates for every part of the world are sent out; States have their authorised diplomats there and, in short, its authority extends over more than 400 million subjects. Supercapitals also exist in the sphere of supranational lay organisations. It seems that this has been the counterpart of the decline of the empires. Groups of nations form, needing a permanent meeting-place: the European Coal and Steel Community in Luxembourg, Euratom in Brussels, O.E.E.C. in Paris. Several towns canvas for the honour of being capital of Europe—in central and western Europe at least: Paris, Nice, Strasbourg, and Luxembourg. The choice is made difficult by rivalry between states: but this rivalry itself proves how much this title is sought after for the profit and prestige the winning town would receive.

Still broader organisations have been set up, extending to the whole world. The first was the League of Nations; it had its seat at Geneva which thus became a sort of world capital, and a palace was built on the hills overlooking the lake with the International Labour Office as a neighbour. The United Nations Organisation, which took over from the League of Nations after the Second World War, was transferred to New York and housed in a huge palace with walls of blue-tinted glass; this forms an enclave with extraterritorial rights. The Organisation has not yet reached the stage where an island has been reserved for it, belonging to no nation in particular.

CHAPTER 17

VARIATIONS AND PLURALITY OF FUNCTIONS

HOW FUNCTIONS CHANGE

We have examined the various urban functions which promote the founding and development of towns, and we have tried to give examples by choosing the most typical cases, where the town has remained faithful to its original function.

But very often this is not the case. This or that town may have been founded to fulfil certain needs under certain conditions which have altered with time, so that the town withers or disappears, loses its urban quality and becomes little more than a village. The English rotten boroughs, which were once accepted towns, were eclipsed and ruined by the development of other towns in the neighbourhood and lost the markets for which they were founded. Many towns of ancient times are dead today—Troy, Sparta, Mycenae, and Babylon. Ports which become silted up lose their commercial function, and wither, like Aigues-Mortes, Le Crotoy and Ravenna. Mining towns are ruined when the mineral is exhausted, like the 'ghost' towns of California and Nevada. Cassel and Bergues, once fortresses on the Flemish frontier, have fallen close to the minimum required by French statistics for urban qualification. And certain commercial centres in Poland became villages after the extermination of the Jews who ran the businesses.

Often, however, towns subsist in spite of setbacks. The connexions formed with the region remain, and uphold commercial activities. We have observed that the industrial function tends to subsist by replacing old industries with new ones, and other functions appear to guarantee perpetuity or even a more brilliant future. Then graphs show the population curve, which was tending to level off, starting to rise in a parabola.

The new function is often connected with the original one. Fortress towns were placed on invasion routes, in corridors between the mountains and at the entrances of the great passes. But the routes taken by warring invaders are the same as those used by traders in peacetime. The fortress can even sometimes initiate a commercial function by changing the direction of a route. The crossing of the route which follows the northern edge of the limestone plateaux with the route from Paris to the north has been fixed at the bottom of the hill at Laon.

The function of the naval base is no longer uniquely military. Brest has opened its docks to heavy merchant shipping, and Le Havre, a former military port, has become commercially important.

Sometimes one activity is substituted for another by chance. Douai has thus passed through a series of vicissitudes.[1] Originally a small fortress, it benefited from being the upper navigation limit of the River Scarpe, and from its position on the route between Flanders and Champagne; in the twelfth and thirteenth centuries it was an important commercial centre; trade brought a cloth-making industry, and in the sixteenth century a university was founded there. The cloth industry was in danger of fading away, but in the seventeenth century a fortress was established to bar the route to Flanders, because of the rivalry of France and Spain. This made it a garrison town. Finally the exploitation of the subsoil in the nineteenth century turned it into a coal-mining town.

Politics also can bring about changes in function; as at Beirut, which was a Roman colony, an Arabian port, then the centre of Ottoman culture, and finally was made capital of the Lebanon in 1920.[2] Belgrade usurped the rôle of capital of Serbia from Kragujevac, which then became a small agricultural centre until it was revived recently by the growth of industry.[3] Helsinki was once only a fortress coupled with a small agricultural market; out of a population of 9,000 in 1905, 3,000 were military personnel. On its jagged peninsula, well protected by rocky islands, the town had the task of watching over the Gulf of Finland, opposite Tallinn. A large part of the town was destroyed by fire in 1808; this was at the time when Czar Alexander I was looking for a capital for the Grand Duchy of Finland; it seemed that the reconstruction of Helsinki offered an opportunity to build a large administrative and residential city, and the capital (which until then had been at Åbo (Turku)) was transferred there. The town became fashionable; the aristocracy of Russia and the Baltic countries met there for sea bathing; it became a commercial centre in largely rural Finland. Finally when Finland was liberated, the rôle of capital grew to give the town a population of 700,000.[4]

Usually, however, the town itself, uninfluenced by any external causes, attracts new activities by its own existence, and these contain the elements of new functions. In fact every town whatever its function represents a capital investment which cannot remain unproductive. Capital in terms of people, first of all. The inhabitants of a town

[1] J. R. Leborgne, 'Le site et l'évolution urbaine de Douai', *Ann. de Géog.*, 1950.
[2] S. Chehabe-ed-Dine, *Géographie humaine de Beyrouth*, Beyrouth, 1960.
[3] M. Popovic, *Kragujevac et sa région économique*, Belgrade, 1955.
[4] S. E. Äström, *Samhällsplanering och Regionsbildning i Kejarstidens Helsingfors*, Helsingfors, 1957.

cannot always be fully employed by a single function, even if the internal services of the town are included. The male labour force of a factory leaves available a female labour force which can be employed in another industry but which can equally well find its work in the tertiary sector. Conversely, a commercial place, employing a numerous female population, can attract mechanical industries for the employment of men.

The infrastructure also represents capital. Means of communication —roads, railways and stations to serve the town—have been established. These must be used to the fullest possible extent. The railways serving an administrative or university town can be used to bring raw materials and to take away manufactured products. The university towns of Oxford in England and Uppsala in Sweden have thus become industrial centres without giving up their original function. And, chiefly for financial reasons, Monaco, which once catered exclusively for the luxury tourist trade, now has light industry which employs 5,000 people and has all the financial and banking facilities of the Principality at its disposal.

This material capital can be represented by a building. The College of Arts and Crafts which is the livelihood of Cluny, was established there because it contained a large group of buildings inherited from the ancient abbey. Saint-Nazaire was a large shipbuilding seaport; rebuilt after the war, it became a port with many different industries.[1]

This tendency to diversify is expressed today by the efforts many commercial towns are making to create industrial zones. A large car-building factory has been built at Rennes, formerly a commercial and administrative town.

Similarly, many towns try to take advantage of the tourist industry. Every fishing port, given fairly favourable conditions, can become a seaside resort. Trouville began as a place for country holidays, then the beach became the vital factor when trawling ruined it as a fishing port, and Deauville was created. In many a small Cornish town fishing, formerly the mainstay of the population, has ceased to be a commercial operation and remains only as a sport, part of the holiday industry.

This is not always successful. Sometimes certain functions cannot easily be carried on side by side. The tourist trade suffers if mixed with an industrial function. Smoke and noise are annoying to visitors; watering places would lose their clientèle if they encouraged heavy industries. Annecy had this problem, and was obliged to consign its industries to the perimeter of the town. University towns, isolated in

[1] G. Le Guen, 'Aspects économiques et urbains de la reconstruction de Saint-Nazaire', *Ann. de Bretagne*, 1954.

their ivory towers, where people come to take refuge from the big cities, are running the risk of losing their essential characteristic of seclusion. Uppsala has carefully segregated the university town from the industrial town, one on either side of the River Fyris: while in Oxford the university has been described as 'the "latin quarter" of Cowley'!

Capital and initiative are needed for the development of a new urban function. Now capital and initiative are often absorbed by the town's original function. In some commercial towns the development of industry is hindered by the exclusive place occupied by commerce. Chambéry has remained a commercial town, while industry has established itself more easily in the less business-minded town of Annecy. This is perhaps less true today, when capital and technicians are more mobile.

All towns have a tendency to snowball and to develop several forms of activity. In most cases this is limited to certain secondary activities, but in favourable circumstances, these grow in importance. They multiply, and the moment comes when it would be difficult to label the town as having a particular function.

THE METROPOLIS

This gives rise to the metropolitan city.[1]

The metropolis appears today in every study of urban geography as forming a special category, differing more from the small town than the latter does from the village.[2] It is a recent development, occurring in countries with a slow population increase as well as in those with high birth rates. It is the triumph, but at the same time the despair, of the urban phenomenon.

The large town is represented by the number of functions it collects rather than by the number of thousands of inhabitants. In the Middle Ages, Paris, which then had a population of under 100,000 was a very large town. Nowadays a larger population is demanded, and no town with a population under several hundred thousand (some even say a million) would be considered large. We are presented with the same problem as with the definition of a town. The criterion of function must be accompanied by one of size.

In metropolitan cities each function has benefited from the conditions which brought about the other functions and has found

---

[1] The German term is *Grossstadt*, the English *metropolitan city*. The French expression *grande ville* is already obsolete. It is a pity perhaps that the term *Megalopolis* has already been used in a different sense. Perhaps a better French term would be *villes géantes*.

[2] Max Sorre, *Fondements de la géographie humaine*, 1952, vol. 3.

reasons for developing there itself. The town acts as a magnet, and the bigger the town, the more powerful the magnet.[1] In every country there is a town the success of which for various reasons, is particularly striking, the 'primate' town which, without necessarily being a capital, stands out clearly from the mass formed by the country's other towns. Mark Jefferson has calculated that this primate town, in twenty-eight countries out of fifty-one countries under consideration, has a population at least twice the size of the next largest town to it, and in eighteen countries a population at least three times the size.[2]

The power of attraction is likely to be particularly strong in capital cities. The best example is Paris, because of French centralisation. Almost every urban function is represented there, the rôle of capital being conclusive. But it would be difficult to say whether London owes to its commercial or political rôle its industrial, university and tourist functions. Once the movement has started, it does not stop, even if the town is divested of its political attributes. Istanbul hardly suffered at all from the transfer of the capital to Ankara.

But capitals are not alone in concentrating all their functions together. So-called industrial civilisation is well adapted to the migration of functions, for it allows them to flourish in places not particularly suitable to them.

Commerce was already represented in most towns, but all economic evolution tends to increase the fluidity of capital; banks play an increasingly important part, and every town is the banking centre for its area; the town handles and delivers more and more goods, and controls the movement of those it does not handle itself. Similarly large-scale industry can now be attached to any town. In the past handicraft industry could only be superseded at a heavy cost in transport. Nowadays raw materials can be carried more easily, and in some cases the cost of transport has little effect on the value of the product. All towns are supplied with electricity. Light manufacturing industries which can be established everywhere have multiplied the fastest. Electrical equipment and precision tools can be manufactured without moving heavy materials and these are the kinds of industry chiefly found developing around towns on their way to industrialisation.

Moreover, apart from the millions of emigrants who crossed the seas in the past, a labour force is easily attracted from the country to the town, ready to carry out the new functions. So every town tends to become industrial. And today we see industries springing up in the former colonial towns.[3]

[1] Jean Chardonnet, *Métropoles économiques*, Paris, 1959.
[2] M. Jefferson, 'The law of the primate city', *Geog. Rev.*, 1939.
[3] R. Douessin, 'Les industries de Tananarive', *Madagascar*, 1962.

It should be added that most towns today have some pretensions towards a tourist function, founded on their artistic wealth, their site, the curiosities of the surrounding area, or the fact that they form stopping places on busy routes; sometimes they are content merely to attract customers by gastronomic temptations.

Finally, with education today being widespread, the need for specialised scientific institutions brings about the creation of educational centres in every large town. So there is a tendency towards plurality of functions. All these functions help one another. Commercial life is grafted on to production centres; industry goes to the consumer centres. In this way large, multi-functional towns with high populations develop.

An analysis of the conditions of this development may be attempted. It should be observed first of all that, if only towns with populations over 1 million are considered,[1] 50 per cent of them are sea ports.[2] In fact ports lend themselves particularly well to the concentration of functions: on the one hand, they are by definition commercial places, and on the other they attract manufacturing industries by the materials of all sorts which they receive.

This was true even in the days of the ancient Mediterranean civilisations.[3] It proved even more true in the colonial era. The great commercial connexions were then maritime links with the metropolis. Hence the great ports on either side of the Atlantic in America and Europe. The development of maritime trade enormously increased the rôle of large ports as well as the size of the towns they supported.

Among the 'millionaire' cities in the interior of continents, capitals should first be considered separately; these represent 25 per cent of such cities; there are in fact eighteen, apart from those already counted as ports, which number six. It should be noted, incidentally, that in Europe the capitals are usually larger than the ports.

The other inland cities, seventeen in number, are either old capitals which have kept their prestige and have remained the central point of a large area (Milan, Leningrad, Kiev, Munich), capitals of federated States (São Paulo), sometimes industrial towns which have benefited from the riches of the earth (Birmingham, Johannesburg), or else (and these are the majority) towns of India and the Far East where, with extraordinarily dense populations, every agglomeration grows automatically, and reaches enormous figures.

Some of these metropolitan cities have the advantage of particularly favourable conditions and acquire a sort of permanence which allows them to triumph over time. They are their own heirs. Carthage was

---

[1] There are seventy-one such 'millionaire' cities in the world today.
[2] Together with Chicago and Detroit—which are ports on an inland 'seaway'.
[3] But although Athens, Tyre and Carthage were ports, Rome was not.

destroyed, but its resurrection was assured; at Istanbul every hole made in the ground reveals the remains of an ancient city, and is full of history. In China, Pearl Buck has written of towns where man built on top of man, and where no house has earth foundations.[1]

These gigantic towns have a strange power, a sort of predestination, defying time, and finding in every age and in every civilisation a new reason to prosper. They seem compelled to a development which mankind started but now finds difficult to master and control.

[1] *Pavilion of Women*, London, 1947.

PART THREE

THE PLAN AND EXTENT OF TOWNS

Chapter 18

THE PART PLAYED BY THE SITE

The requirements of its function govern the position of a town. Sometimes they cause a town to grow in particularly uncomfortable conditions. In order to exploit mines, towns have of necessity been established in the middle of deserts, like Kalgoorlie, or in frozen wastes, like Kiruna or Schefferville. But men have generally managed to find a site which, while meeting functional needs, nonetheless allows them to live in reasonable comfort.

The choice of a site, being imposed by functional requirements as well as by geographical position, is sometimes very limited. When an estuary port has to be placed as far upstream as possible, at the high tide limit, the only choice is between the two banks. A mining town is established in the immediate neighbourhood of the mineshaft.

The advantages of the geographical situation often limit the choice of site to watercourses. The most obvious place for taking advantage of river traffic is the point where this traffic ceases, where the watercourse is no longer navigable and where transshipment must be effected. Similarly, a waterfall offers a marvellous opportunity to channel off energy, which will drive tilt-hammers and textile machinery and supply electricity. Many sites are therefore influenced by the proximity of a waterfall, as in the case of Tampere, situated between two lakes of different altitude in Finland.

It is rare, however, for the situation to determine the site absolutely; several neighbouring situations may be equally suitable, and the one which appears most favourable, according to local conditions of relief and drainage, is selected.

The relief is primarily a defence against water, all the more vital because water is often the deciding factor in the choice of position. Plains which attract trade and valleys which favour traffic circulation are defenceless against flooding. Along a river, the water-soaked alluvial soil does not lend itself to the laying of firm foundations, and floods are often disastrous. By climbing up a few feet, these disadvantages may be avoided without losing the advantages of the situation. The concave bank of a meander is preferred to the often marshy convex bank, as at Rouen and Prague. Wide glacial valleys furnished with alluvium through which a river wanders lazily are particularly to be avoided; but rock outcrops present ideal situations, which, moreover, are easily fortified. A small hillock gave Graz its site in the Mur valley in Austria. Sion was sited on a rock-bar in the upper Rhône valley.

The valleys of young mountain ranges, high up and well watered, are drained by swollen rivers, the violence of which is greatly increased by melting snow in the spring; but the tributaries joining these rivers build up alluvial cones where they may easily be artificially channelled, thus making extensive terraces available.

At the foot of the mountain ranges where the plains begin, the towns carry on an exchange of complementary commodities. These towns will be established on the well-drained first slopes of the hills. Piedmont slopes, erosion surfaces, and terraces at the foot of mountain ranges can all be sites for towns, or a small hill detached from the range may be chosen, as at Salzburg. It is even better if an indentation in the range allows easier communications, and guarantees a sheltered site: this is the case in the deep, high Jura valleys with towns like Salins, Arbois, Poligny and Lons-le-Saunier.

Across the plains it was not so easy in ancient times to guard against the ever-present possibility of flooding, or to guarantee defences against invasion. Every hillock becomes precious in these circumstances. Such hills may be relics of former erosion surfaces, such as those that provide sites for towns in Flanders (e.g. Cassel). Or they may be former islands in the marshes, now joined to the shore again, like the spur on which Sète is situated. Within the Lika *polje*, in western Yugoslavia, Gospić, its principal town, is built on a hum. In the glaciated plains of northern Europe drumlins and eskers play similar parts. Dublin is built on a drumlin, Stockholm on an esker.

The relief of the ground does more than protect against flooding. It allows better advantage to be taken of sunlight, by spacing the houses out espalier fashion, along the contours; and this is true not only of health resorts. In all towns in the temperate zones and in arctic regions the residential districts spread up slopes with a southern aspect.

We have regarded water hitherto as the principal danger to man and one which he has tried to avoid when choosing a site, but it is also a precious and much sought after advantage. It can be a defensive element. A water-course is a natural moat. What better protection could one have than the loop of a river? Besançon is situated on a loop of the Doubs, Semur on a loop of the Serein, and Durham on a loop of the Wear. Even marshes can extend their natural protection all round a town. In Picardy peaceful rivers drain the broad valleys cut into the chalk, and are seldom subjected to flooding. The valley bottoms are boggy, and Amiens found an element of security in the marshes of the Somme. But above all water represents one of the conditions of human life, and is as indispensable as air. It has been

13 Moscow: the Moskva river from the Gorki park

14 Tashkent: a high school amidst mud huts

15 New York by night

brought to deserts at great expense through pipes many hundreds of miles long. This is a luxury which can only be resorted to in exceptional circumstances, such as the lure of gold mines.

The first concern of the small group which forms the embryo of a town is therefore its water supply. Rivers assure the supply of water necessary for animals and washing. They can, if necessary, supply human needs too, but these, of course, require purer water. This can usually be found near these water-courses. In alluvial valleys the water-table is high and can be reached easily by wells. The same applies to most plains. Springs rise on the sides of valleys, tributaries of the river, fed by the rains of the plateau; they may be small springs or great underground rivers bursting out from the limestone. Lines of springs accompany alternations of porous and non-porous rocks, and towns are often established on slopes where sand meets clay. Such alternating layers are sometimes to be found in synclinal folds; in these conditions, the water trapped in a sandy layer between two non-porous layers permits the sinking of artesian wells, as in Paris; in London (where the fountains in Trafalgar Square originally flowed naturally under artesian conditions) the water contained in the chalk is imprisoned between the Gault clay beneath and the London clay above.

Springs occupied such an important place in the life of ancient towns that divine origins were often attributed to them. Dijon took its name from the holy spring (Divio) near which it was established.

Of course, at the time of its foundation a town's needs were small, and the choice of site only took such needs into account. Populations were small and people, in those days, relatively undemanding. We are surprised now to see large towns situated on tiny rivulets. In the course of a town's development, this contradiction has presented extremely serious problems, which have not all been solved. At Ankara water is sold by the litre, and is brought every morning on little donkeys from the spring, where people are obliged to queue. We shall return to this matter when dealing with the provisioning of towns.

The site of a town is, as we have seen, the one which seemed, at the time it originated, the most convenient to those who were to carry out the town's function. This site may retain its advantages if it was chosen from the first with the intention of expansive development, as for example, in Canberra and Brasilia. But more often, as the town develops the site no longer fulfils the conditions which it originally offered.

It can sometimes adapt itself quite well when there is an abundant water supply and if there is an extensive plain suitable for new buildings. It can even find new outlets on the old site,[1] but this is by no means always the case. Fortunately, technical progress has kept pace with urban growth. A river is no longer an obstacle: Rouen, established on the concave bank of the loop, is expanding on to the convex bank. The two banks of the Elbe at Hamburg are joined by a tunnel. To allow for the expansion of the city of Caracas, hills have been cleared away and small valleys filled in.[2] Few cities have suffered more from their site than New York, squeezed into the narrow Manhattan peninsula. Large numbers of bridges, tunnels and ferries over the Hudson and the East River have had to be provided as the agglomeration has grown and the cost in wasted time, tolls and extra transport charges of this initial error of site-selection for the ten million inhabitants has been enormous.[3] Elsewhere the upper and lower parts of a town are joined by funicular, or by tram. This often applies to towns built by the sea (San Francisco, Salvador) or by a lake (Lausanne), which have to climb in order to accommodate their increasing populations, or to towns built inconveniently on a rise which have to extend over the plain, like Lyons.

There are plenty of ways of overcoming obstacles. Soft, marshy ground is overcome by building on pile foundations (as at Amsterdam) or by laying down concrete rafts; the latter solution has allowed the construction of a new quarter of Belgrade on the left bank of the Save. The new towns of the Soviet Union bear witness to the supremacy of technical methods.

It could be argued that in this case man could adapt himself to any site at all. He has freed himself from the conditions imposed on him by nature. 'Towns are becoming more and more removed from nature, and are sometimes quite contrary to it', as Pierre Lavedan has written.[4] In fact, as the old adage says, *natura non nisi parendo vincitur*. If the site is no longer imposed so strictly by natural conditions, it always reflects them, and the urban landscape, like the problems of daily life, bears witness to them.

[1] H. Baulig, 'Le site de Strasbourg', *Univ. de Strasbourg*, 1946.
[2] A. Journaux, 'Problèmes de site et d'extension à Caracas, Bogota et Quito', *Inf. Geog.*, 1960.
[3] E. Huntington,' The water barriers of New York', *Geog. Rev.*, 1916.
[4] *Géographie des villes*, 1959, p. 7.

CHAPTER 19

THE TOWN PLAN [1]

The needs of the urban functions make themselves felt, and daily life is organised, within the framework established by the town plan.

Many towns were built casually. In order to carry out a new function, people settled as best they could, sometimes making use of

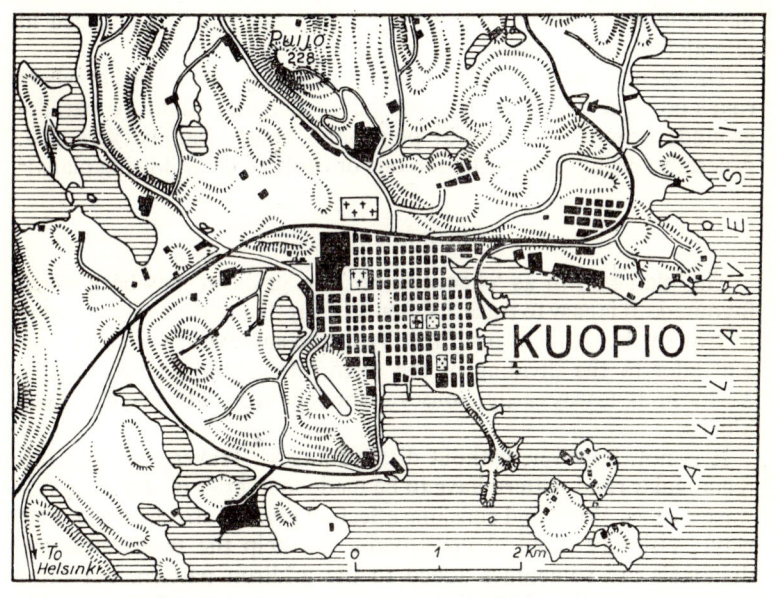

Fig. 10. Plan of gridiron town: Kuopio, in Finland (pop. 40,000)
From Georges Chabot, *L'Europe du Nord et du Nord-Est*, vol. 2: *La Finlande et les pays scandinaves*, Paris, 1958, p. 113.

an already existing nucleus,[2] crowding into a fortress, or spreading out along a trade route. People are sometimes surprised by the maze of streets crossing one another which is so characteristic of the ancient quarters of European towns, or of present day Arab towns; whilst the dead-end alleys of oriental towns like Antioch and Damascus defy all attempts at rational explanation.

[1] Pierre Lavedan, *Histoire de l'urbanisme*, 3 vols., Paris, 1926–52.
[2] Gunnar Lering, 'Om Stadsplan och Byplan: det centrala Västergötland', *Gothia*, 1961.

For a long time, however, men have nurtured the idea of the town they would like to build. Sometimes they saw it as similar to their villages: the circular villages of the Slavs seem to have served as models for many towns. The Hittites adopted the circular form and the Egyptians a quadrangular one, for religious reasons. The circle appeared, too, to be the image of perfection, and Jerusalem is represented thus in the Middle Ages. But in every age the most convenient plan was also sought. Renaissance architects enjoyed developing these theories. Modern town planners have taken up this subject again, seeking the ideal plan. Two or three geometric plans may be distinguished.[1] These have inspired, usually intentionally, the building of towns, and have guided town planners engaged in remodelling these towns.

The plan is often arranged round a central point: a market where roads cross, a cathedral, a manorial castle, or a military parade-ground; around this central point, the houses can be arranged, generally speaking, according to one or other of the following plans:[2]

THE GRIDIRON PLAN

The plan known as 'chequerboard' or 'gridiron' has been much favoured. It is easy to trace.[3] Often it is foreshadowed by the crossing of two roads. It lends itself readily to the subdivision of properties and later to the dividing of the town into administrative sectors. Traffic finds its way easily; houses are built conveniently in rectangular blocks.

It is not, however, without disadvantages. Traverses across this rectangular road pattern have to be made in a series of steps and this wastes time. Wind and sunshine affect all parallel streets in the same way. Visibility is nil at crossroads. To help the traffic situation diagonals are sometimes added to the gridiron pattern, but these cause some blocks of buildings to end in points and to be awkwardly triangular.

The gridiron was chosen by the Romans, who modelled their colonial towns on their military camps. It took its bearings from two wide, perpendicular roads which crossed at a central point: the *cardo* and the *decumanus*. The remains of these can still be seen in many towns on the Roman frontiers. Turin is the best example of a Roman

---

[1] J. Gottmann, 'Plans de villes des deux côtés de l'Atlantique', in *Mélanges géographiques canadiens offerts à Raoul Blanchard*, Quebec, 1959.

[2] R. E. Dickinson, 'The town plans of East Anglia: a study in urban morphology', *Geography*, 1934.

[3] R. E. Dickinson, 'Le développement et la distribution du plan médiéval en échiquier dans le sud de la France et l'est de Allemagne', *La vie urbaine*, 1938.

town which has maintained its original form. The medieval hill-top towns of the south of France followed the military heritage of the Roman world.

Most town-builders were attracted by this geometrical simplicity. Naucratis and Alexandria followed a gridiron pattern, and this

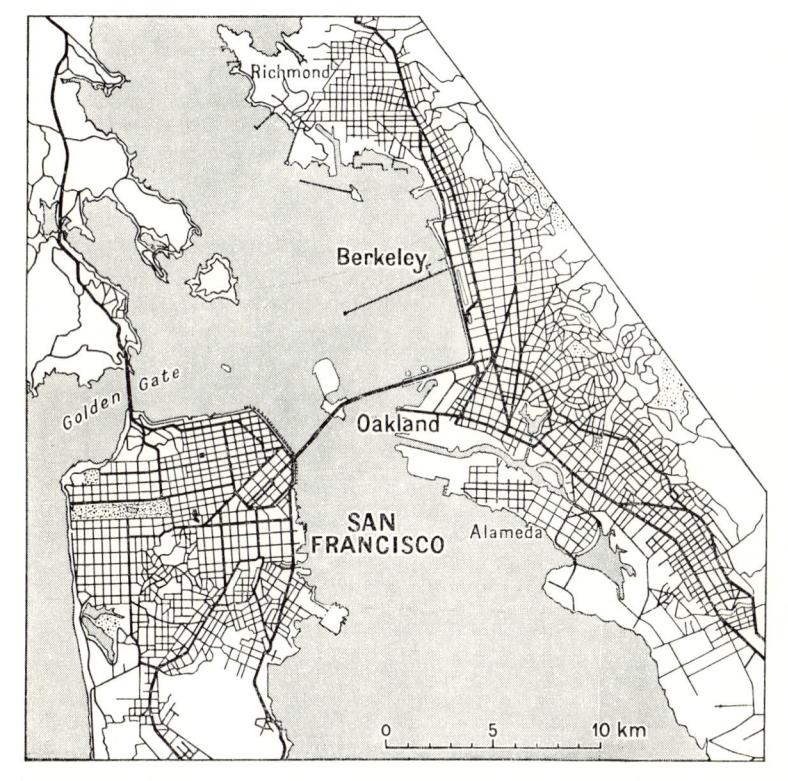

Fig. 11. San Francisco: Gridiron plan to the west of the Bay

convenient scheme can be seen throughout the ages. Western Europe returned to it after the twelfth century. Quadrangular towns like Vitry-le-François seem the most perfect expression of it; but the outside limits can be elliptical, or circular, as at Riom, and still maintain the interior squared pattern. This was the pattern of the towns built by Vauban (Neuf Brisach for example) which dissociate the interior plan (orthogonal) and the outside limits; Sarrelouis is a gridiron inside a polygon.

Many American towns have adopted this plan. One is Chicago, and on the long Manhattan peninsula parallel avenues are intersected by roads at right-angles.

The town most often quoted in this respect is Mannheim, rebuilt on a gridiron plan. The chessboard pattern is made even more effective by the numbering of the roads by figures in one direction and letters of the alphabet in the other.

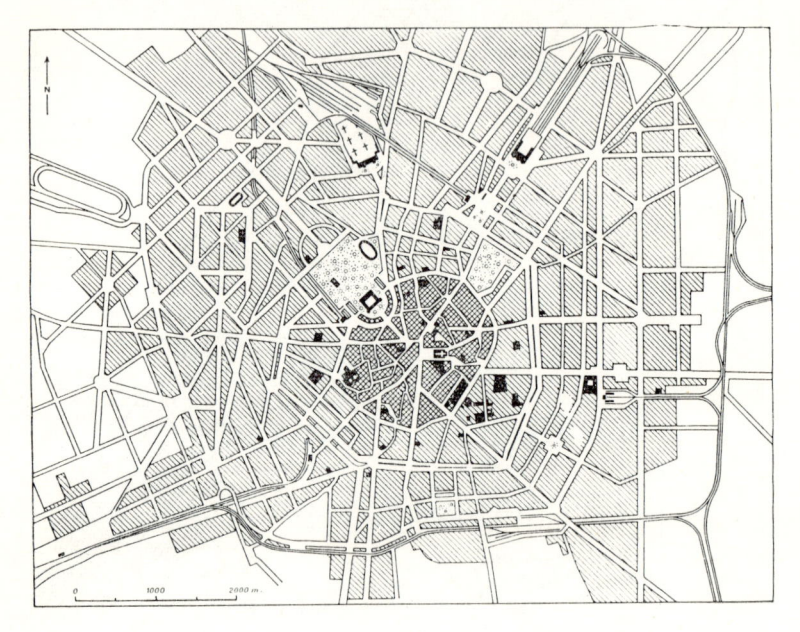

Fig. 12. An example of a radioconcentric plan; Milan

THE RADIOCONCENTRIC PLAN

The town arranged in *concentric rings* places the town harmoniously round the centre. Arteries crossing the circular roads leave this centre in a star pattern, hence the term *radioconcentric*. The centre, that is, the business and public service districts, may be reached easily from all points. Traffic, however, which mostly uses the arcs of the circles, is slowed down and suffers from poor visibility at the intersections of the arcs and the radii.

Some of these faults can be corrected by substituting concentric hexagons for the circles. Distances on the outside are shortened, and visibility is better than with the gridiron plan, with angles of 120° instead of 90°.

THE TOWN PLAN

The circular plan was adopted in the orient in ancient times, and was in favour in the Middle Ages in imitation of the walls of a fortress. It can be seen preserved in small towns like Brive. One would hesitate to find an echo of this in completely newly planned towns like Brasilia. The plan seems to suggest a bird with semi-circular wings, but instead

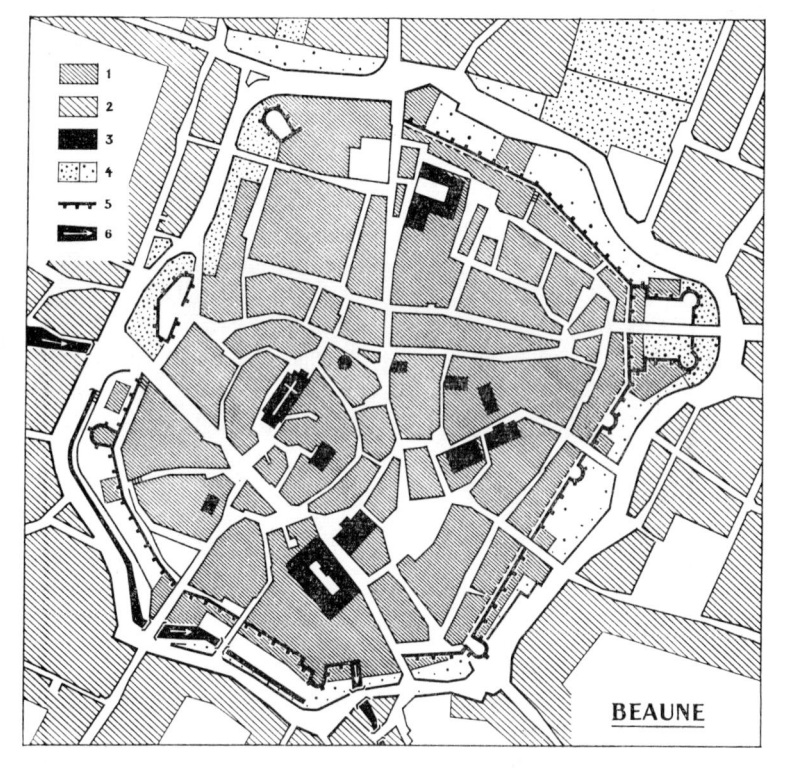

Fig. 13. An example of a radioconcentric plan: Beaune

1. Old nucleus. **2.** Recent additions. **3.** Principal public buildings. **4.** Green spaces (parks, walks, etc.). **5.** Ramparts. **6.** Course of River Bouzaise (partially underground).

of coinciding with the geometrical centre, the control centre, Three Powers Square, is to be found on the outside.

Throughout history the quadrangular plan and the concentric plan have vied with one another. Nevertheless, other formulae have appeared. The star-shaped arrangement was adopted usually when the town expanded, and we shall study this in a later chapter.

THE LINEAR TOWN

The linear town is now in favour. Formerly it only seemed acceptable in the case of a small town; it was a restriction imposed by the relief of the land which was accepted reluctantly. Nonetheless it was applied to Madrid in 1894 by Arturo Soria; Soviet town planners made use of it at Stalingrad.

Sven Dahl[1] has devised an ideal town with houses arranged over a

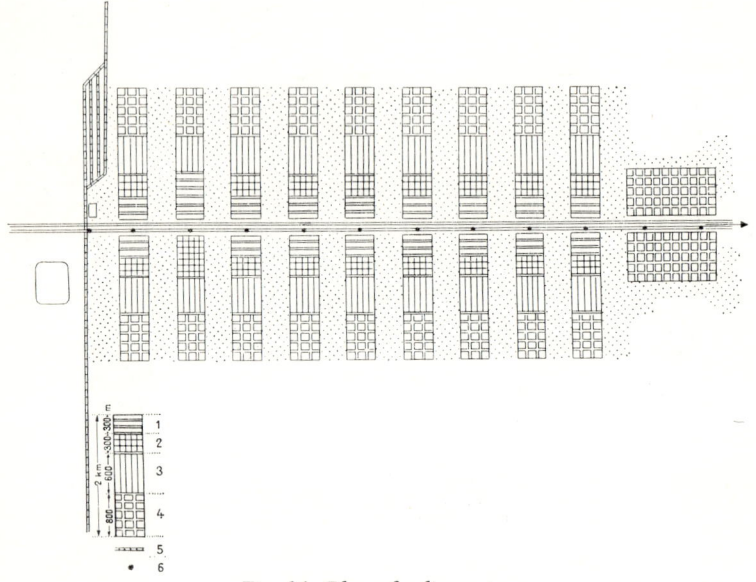

Fig. 14. Plan of a linear town

After Sven Dahl, *En ny stadstyp in Stade Några stadsbygd Problem*, Gothenburg, 1955, p. 31.

1. Workplaces. **2.** Large blocks (each housing about 1,700 people). **3.** Rows of houses (each block housing about 1,700 people). **4.** Villas (each block housing about 1,700 people). **5.** Railway. **6.** Underground railway station.

distance of 2 kilometres forming rectangular blocks on either side of a busy road or railway about 10 kilometres long. The administrative, business and industrial centres would be situated along this road or railway which would have a stopping-place every kilometre. In this way an elongated town would be obtained where, for about 20,000 inhabitants, life could be carried on with the maximum simplicity. It would be a sort of very elongated gridiron (Fig. 14.).

[1] S. Dahl, *En ny stadstyp in Stade Några stadsbygd Problem*, Gothenburg, 1955.

Some public health enthusiasts see as the only solution the expansion of towns in such a way as to guarantee every inhabitant the proximity of green spaces. It has even been suggested that Paris should be replaced by buildings stretching the length of the Seine as far as Rouen.

ADAPTATION TO OBSTACLES

These plans have usually had to adapt themselves to the physical conditions imposed upon them. A town can adopt the shape of the hillock on which it is built. After the founding of Montauban, which amazed its twelfth-century contemporaries, the shape of the promontory on which the town was built was copied; the ramparts followed the form of the ground, and the roads ran more or less parallel.

At Morez-du-Jura, the town can only stretch along the Bienne river across the narrow steep sided transverse valley, and the roads run parallel to the river. Many similar examples can be found in the valleys of South Wales and the Appalachians. In the case of Besançon the town stands on the loop of a meander the curve of which has been the deciding factor in the direction of its main axis.[1] Physical restrictions are particularly stringent when composed of two elements: Trieste is forced to adopt the linear form between mountain and sea.

Circumstances can truncate the plan. In port towns the radioconcentric plan can only develop in a semicircle the centre being obstructed on the sea or river side, as in the case of Bordeaux, Amsterdam and Cologne. At Pointe Noire in the Congo, the African town is enclosed in the loop of a marshy river and has developed fanwise from the point which joins it to the European town.[2]

The gridiron too is sometimes broken up by force of circumstance. Chicago and Los Angeles are composed of juxtaposed gridirons. At Helsinki, on the triangular lake-studded peninsula, the town has developed in two gridirons, one going north-westward and the other north-eastward. At Canberra several radioconcentric fragments are assembled to form the town.

Ultimately the adaptation of geometrically conceived plans to natural conditions is not enough. People have grown tired of their disadvantages and banality, and have tried to give them some sort of character. As a contrast, what Pierre Lavedan has called 'the flexible plan' has been devised, giving currency to all kinds of attractive and fantastic ideas. This has happened in the new towns of the Tennessee valley, such as those created by Norris in 1935.

[1] J. Soyer, 'Evolution d'un site urbain: le méandre, Niort et Cahors', *La vie urbaine*, 1961.
[2] J. Denis, 'Pointe Noire', *Les Cahiers d'Outre-mer*, 1955.

Thus plans follow very different formulae. It is often difficult to pick out the original nucleus in a town plan. Circular avenues give a concentric appearance to towns which have in fact only increased their exterior circumference several times. Plans as they appear today are seldom anything more than the ancient heritage remodelled, and are hard to unravel.

REMODELLING THE PLAN

The plan of a town, as first established and as fashioned by its growth, is not modified only along its original fringes. The ancient nucleus must be adapted to new conditions created by the activities of a population increased tenfold and sometimes a hundredfold. The old quarters are often unhealthy; moreover the life of the perimeter reacts on the centre, which continues to be the administrative and business quarter. As the traffic converges on them the narrow winding streets of the nucleus act like a cork. As early as the Roman era, people were laughing at ancient Athens with its narrow streets.

Such adaptation to new conditions is the most difficult of all the problems of town planning. The appearance of the old quarters must be respected as far as possible, for they are a valued heritage. They are the town's charm, and in an age when tourism is so profitable they are an attraction to visitors, so they are often protected by being classified as buildings of historic interest or architectural merit.

Nevertheless, where conditions are unhealthy, it is tempting to demolish rickety buildings; but how can one create new blocks to harmonise with the ancient scene? How is the Balance district of Avignon to be rebuilt without injuring the Papal Palace which dominates it? Town planners, like Marcel Poète, have stressed the necessity of a town's fidelity to its past.[1]

This problem, unfortunately, has been solved in many cases by catastrophic destruction.[2] Wars and fires have wiped out subservience to the past. Scarcely any of those northern towns, which were built of wood, escaped being burnt several times, right up to the nineteenth century.

Earthquakes destroyed San Francisco in 1906, Messina in 1908 and Tokyo in 1923. Wars have caused even greater havoc. Such destruction occurred in ancient times; Troy and Carthage perished at the hands of their conquerors. The power of modern inventions has multiplied these catastrophies. During the Second World War, many

[1] M. Poète, 'L'évolution des villes', *La vie urbaine*, 1930.
[2] It is from the cellars alone that it has been possible to reconstruct the plan of certain towns, cf. E. Lambert, *Bull. Ass. Géog. Français*, 1946.

European towns were almost completely destroyed. Their reconstruction gave an opportunity for adaptation, and destruction was in some cases the basis of beneficial remodelling. As P. Zaremba puts it, reconstruction does not mean the perpetuation of errors.[1] When roads had to be traced through the ruins by bulldozer, it was easy to give them the necessary width and direction. In this way the great east–west artery in Hamburg was created, facilitating transport parallel to the Elbe.

Reconstruction has sometimes cleared perspectives, allowing certain views to be appreciated to the full. At the foot of the terrace on which Warsaw stands, houses had crowded right up to the banks of the Vistula. By not reconstructing them, the town was given back its original appearance of a balcony overlooking the river. At Szczecin, similarly, a wide promenade was created below the town, along the Oder.[2]

Reconstruction has not always been so revolutionary. Landowners sometimes continued to be masters of their properties and expropriation is expensive. Sewer systems and foundations remained intact, and when the inhabitants returned, they wanted to resume their former way of life. Even the most ruined cities retained a certain fidelity to the past. In Warsaw which was almost totally destroyed, the old square was rebuilt almost exactly as before. The same was done at Danzig. At Reims, after the First World War, innovations were limited to the widening of the old roads.

The problem is much harder to solve when it is not simply a matter of reconstruction after a disaster and the town still retains the full weight of its past. In these circumstances rebuilding must be particularly sensitive. The widening of a road by the demolition of façades, as has been done recently in Warsaw (Jerozolimskie street), is normally a difficult operation. It is often more convenient to make a new road crossing yards and gardens, and to leave untouched the imposing façades of the old streets.

The most classic example of this type of rebuilding is the work carried out by Haussmann under the French Second Empire. In Paris, nothing had hindered the continuous development of the town since the romano-gallic era. The great east-west and north-south traffic arteries had become quite insufficient. In the north-south direction the Rue Saint-Jacques was retained; since Roman times this

---

[1] P. Zaremba, *The spatial development of Szczecin in 1945–1961*, Poznan, 1962.
[2] G. Chabot, 'Hambourg: le port et la ville', *Inf. Géog.*, 1952; cf. P. Lavedan, 'La reconstruction en Allemagne', *Urbanisme et habitation*, 1954; R. Hulot, 'Saint-Malo et sa reconstruction', *La vie urbaine*, 1956; P. George, 'Problèmes géographiques de la reconstruction des villes en Europe Occidentale depuis 1945', *Ann. de Géog.*, 1960.

road had continued southward the route coming from the north, across the Ile de la Cité. Parallel with it, however, the Boulevard Saint-Michel was created, making use of the gardens which climbed in terraces up the slope of the Latin quarter. The east-west artery, the need for which had been felt for centuries, was more difficult to create: indeed, it was never completely created, though the Rue de Rivoli carried it for quite a distance.[1]

In the face of these difficulties, some have been tempted to abandon this 'patching up', and to build a new and independent town alongside the old; Haussmann proposed the building of a new town in the Roman countryside distinct from ancient Rome and Rome of the Popes. Such a course would be a death blow to ancient towns, which would never accept reduction to the status of museums. The 'sliding' theory is more flexible; a centre is built close to the old town, and gradually the town's life slides towards this centre: this may be what was intended in Paris around the Rond-Point de la Défense, but such sliding does not happen by itself, and even if it does obey the command of its promoters it is likely to be a very slow process.

Often then we must be content with 'patching up' in order to adapt old towns to the needs of modern life, to make traffic movement possible and to get rid of the old, insanitary districts. To do this, the original plan must be cut up, unpicked and patched.

However, all these remodelling problems concern principally the central nucleus around which a town has developed. The appearance of a town depends more upon its new quarters and its extension beyond the small, central island.

[1] P. Lavedan, *Les villes françaises*, 1960 and *Histoire de l'urbanisme*, 1952, vol. 3; P. Limouzin, 'Le boulevard Haussmann', *La vie urbaine*, 1957.

CHAPTER 20

THE EXPANSION OF TOWNS

All towns have a tendency to grow.[1] The function for which the town was founded takes an increasingly important place, and other functions are added to it. Each function demands the services of more and more people; immigration stimulates the growth both of functions and of personnel. Then, the raising of the standard of living requires more numerous and more spacious houses. Growth has often entailed increasing congestion; this was inevitable when it was not possible to spread beyond the city walls or the limits of the urban territory.[2] But demographic pressure bursts these boundaries, and in countries where there were no restrictions of this kind, it was more convenient to spread outwards from the beginning. Often, and almost invariably for a long time, this expansion was spontaneous, even anarchic. It happened as and when it suited the convenience and means of the inhabitants. Two forms of such growth may be distinguished.

ACCRETION

The simplest of these is accretion; inside a town all space is built on, or land is too expensive for housing; people build at the approach to the town, and as near to it as possible. This is what happened in the past, when the town's dimensions were limited, and when people went to work on foot: they settled beyond the city precincts outside the protection guaranteed by the walls.

In order to protect this area it was necessary to extend the fortifications outwards and to demolish the old walls. From this kind of action resulted the series of circular boulevards which, in many towns, as in Paris, repeat the original plan in concentric rings. This was particularly true of circular towns; it was important to be as close as possible to the centre, and all directions were of equal value where the contours of the land permitted. The development of Moscow is a typical example. The successive belts are more or less circular, the Kremlin being in the centre: Bjely Gorod, Semljanoj Gorod, the Kammer-Kolleshkij belt and the line of the circular railway.[3]

[1] P. Geddes, *Cities in Evolution*, 2nd edn., London, 1949.
[2] See for example G. Pinchemel, 'Les cours et courettes lilloises', *Urbanisme et Habitation*, 1954.
[3] J.-L. Saushkin, *Moskva*, quoted by J. F. Gilbert in 'Dṛas geographische Milieu und die räumliche Entwicklung von Moskau', *Urania*, 1958.

This concentric expansion was the starting point of E. W. Burgess's theory, in which he defined five concentric zones. The transition zone, the one which corresponds to current growth, forms a ring surrounding the whole town.[1] But new districts do not necessarily correspond to the original plan. Sometimes a systematic gridiron pattern is built round an old nucleus of twisting streets. Sometimes, on the other hand, it is the peripheral districts which present a fanciful layout as an expression of their dislike of the monotony of a systematically built nucleus.

Expansion depends to a great extent upon the facilities available and upon individual initiative. For building purposes, parcels of fields were bought, thereby increasing their value enormously; city properties were then copied on the rural lots. The town, inheritor of the countryside, has maintained the ways of the country. On the slopes overlooking Nancy, vineyards were bought in lots, and gave rise to districts composed of small villas.[2] At Bordeaux the country roads, trapped between the grounds of ancient castles, have imposed their sinuous form on the expansion of the agglomeration.[3] Marseilles presents an example of this scattered and disorganised speculation which took place in intermittent bursts throughout the nineteenth century and which is responsible for a large share of the town's expansion, especially in the direction of Notre-Dame-de-la Garde.[4] These city villas crystallise the former state of things, and today form an obstacle to the great schemes of town-planners. At Stockholm they have blocked expansion on the sea side, forcing the town to develop towards the west.

In African countries, round the newly formed European quarters, indigenous dwellings are to be found in great numbers.[5] Near to Brazzaville, Moukondji-Ngouaka is growing thus by accretion; the town, which was founded in 1931, was a continuous built-up area by 1959.[6]

[1] 'The growth of the city', in *The City Chicago*, 1925.
[2] C. Prêcheur, 'Nancy, Rapport de l'actuelle structure urbaine et de l'ancienne structure agraire', *Bull. Ass. Géog. français*, 1953; cf. L. Dethier, 'L'influence de la structure foncière et du dessin parcellaire sur le développement urbain. Le quartier de Salzinnes (Namur)', *Bull. Soc. belge d'Etudes géog.*, 1962.
[3] P. Barrière, 'Les quartiers de Bordeaux', *Rev. géog. des Pyrénées et du sud-ouest*, 1956.
[4] M. Roncayolo, 'Marseilles. Plan de la ville et spéculation', *Bull. section de Géog. du Comité des Trav. hist. et scientif.*, 1958.
[5] P. George, 'Les problèmes posés par l'accroissement urbain spontané dans les pays en cours de développement', *Publications de l'Institut d'Etude de développement économique et sociale*, n.d.; cf. G. Lasserre, 'Le paysage urbain des Librevilles noires', *Cahiers d'Outre-mer*, 1956.
[6] P. Vasselier, 'Brazzaville', *Bull. de l'Institut d'Etudes centrafricaines*. 1960.

This sort of expansion naturally gives rise in a free country to all kinds of speculation. Around Rio de Janeiro, within a radius of fifty miles or more, land has been bought in anticipation of its future increase in value. Landowners wait for their lots to become ripe for development and refuse to let them for agriculture.

This growth does not necessarily resemble the regular growth-rings of a tree trunk. It must, even more than the urban nucleus, bow to the conditions imposed upon it. In fact the nucleus being more compact escapes more easily from topographic influences, but wherever the covering of a wider space is concerned, they cannot be treated with indifference.[1] The site of Perm, as shown by N. Stepanov,[2] was already expanding along the river Kama; it is now growing both upstream and downstream. Beauvais was cupped in the bottom of the Thérain valley, but four tributary valleys which meet at that point decided the directions in which the town should expand. Being at the foot of a mountain, Rio de Janeiro could only extend along the sea, leap-frogging the hillocks strung along the coast.

The proximity of high ground is not always an obstacle. If the slope is gentle and communication with the town is easy, it attracts houses to take advantage of drier ground and a sunnier outlook. If the slope is too steep funiculars give access to the higher levels.[3]

On the other hand, the obstacle of a river, especially a violent and irregular one, has for a long time forced the towns built on its banks to develop in semicircles. Blois and Tours used to be on one side of the Loire only. Lyons was situated on the right bank of the Rhône; between the twelfth to the sixteenth century, half a dozen attempts to build a bridge ended in failure, and it was only from 1581 that they had a real bridge and that the town could comfortably expand across the river.[4] Marshes have played a similar part, for example at Hanover.

The development of island or peninsula towns has, of course, been a long struggle against the water. When Stockholm grew too big for its small island—Staden—it joined itself to the mainland in the north (Norrmalm, Östermalm), then to the neighbouring island of Kungsholmen, but for a long time the need to maintain a passage with locks between lake Malar and the Baltic prevented its development

---

[1] P. Plavinet, 'Le quartier d'Amérique: un quartier périphérique parisienne profondément marqué par les influences du sous-sol', *Bull. Soc. Etudes hist. géog. et scientifiques de la région parisienne*, 1955.

[2] *Soviet Geography. Review and Translation*, 1962.

[3] P. Monbeig, 'La croissance de la ville de São Paulo', *Rev. de Géog. alpine*, 1953, describes the way in which the anarchic development of São Paulo has accentuated the fragmentation latent in the topography.

[4] S. Charléty, *Histoire de Lyon*, 1902; cf. Russo and Audin, 'Le site de Lyon', *Rev. Géog. de Lyon*, 1961.

southward (Södermalm).[1] At Abidjan the town has expanded as best it can over the ground through which the Ebrié lagoon threads its way. The plateau which forms its business centre is a narrow peninsula. Future growth will only be possible for this rapidly expanding town if several bridges are built.[2] So the growth of towns on peninsulas often takes on the shape of a comet's tail. Helsinki, blocked on three sides by the sea, could only extend northwards. Conakry, on the tiny island of Tumbo, at the very end of the Kaloum peninsula to which it is joined by a causeway, may be considered as a peninsula town. It extends along the Kaloum in two rows of suburbs, the northern side for European inhabitants, and the southern side for Africans, with the railway in between.[3]

In the case of a great metropolis, the problem becomes quite dramatic. San Francisco, at the extreme tip of its fingerlike peninsula spills over the narrow Golden Gate to the opposite shore, strides across the Bay Bridge towards Oakland and Berkeley, and, occupying all the edges of the bay, ends by making it a sort of urban lake—the modern urban counterpart of the medieval 'green village' arranged around its pond.

The most outstanding example of this struggle against the waters is that of Manhattan which, in order to become the New York we know today, had to cross the Hudson and East Rivers, and link the neighbouring islands to it by means of ferries, bridges and tunnels.

Sometimes the water has been treated as just an extension of the land. In certain Chinese towns part of the population lives aboard junks. At Salvador de Bahia, in Brazil, a whole district is composed of huts built on wooden piles to avoid climbing the heights—and also to avoid paying rent!—and to take advantage of the fishing. But the accumulation of rubbish results in the joining of the original lake town to the mainland. It is, to quote Pierre Monbeig, a human polder which is gradually increasing in value, driving away the poverty-stricken outcasts who were the original inhabitants.[4]

The site of the enlarged town is not always the same as the original site. Saint-Étienne used to be a street town running along the little river Furens, a tributary of the Gier, but the valley of the tributary was too confined and the town extended perpendicular to the original axis, making use of the Gier valley; there was thus a complete change of direction.[5]

[1] W. William-Olsson, 'Stockholm: its structure and development', *Geog. Rev.*, 1940.
[2] R. Blanc, 'Abidjan et son nouveau pont', *La vie urbaine*, 1958.
[3] O. Dollfus, *Conakry en 1951–52*, 1953.
[4] P. Monbeig, 'Salvador de Bahia', *Industrie*, 1960.
[5] M. Perrin, *Saint-Étienne et sa région économique*, Paris, 1937.

Similarly, towns perched on defensive hillocks could only develop by letting the new districts swarm round the foot of the hill (e.g. Laon, Belfort).

Not all obstacles are physical. While some towns benefit from being close to a frontier, the development of others is inhibited by it, as there is a tendency for twin towns to grow, one on each side of the frontier.

Towns produce their own obstacles too. Smoke is troublesome, and residential districts avoid the side of town affected by the prevailing wind; this is why the smart districts of both London and Paris have developed to the west. The railway, originally at a tangent to the town, can sometimes be a major obstacle. It can be as difficult to cross as a river. Up to 1914 the growth of Argenteuil was blocked by the railway.[1] When the railway is build as a viaduct, however, the roads can pass underneath it in tunnels as at Lyons and Dijon. Once urban development has crossed the obstacle, incidentally, the railway can be a terrible nuisance inside a town: as for example in Ottawa, which is cut in pieces by railways. At Ankara the railway separates two towns, the ancient town, and the one created by the capital.

THE STAR

One form of accretion produces a star-shaped town, for while some directions may be advantageous, others are more difficult. The network of roads and railways becomes the dominating factor in expansion from the moment when it is no longer possible to get to work on foot. The extension does not fulfil the same needs as the original town.

Sometimes the method of transport is individual: there are countries where everyone gets about on bicycles, as in the Netherlands and Denmark. Adding a motor to the two wheels increases the speed and therefore the capacity for distance. The private car allows its owner to come regularly from even greater distances and the park beside American factories for the workers' cars is most impressive.

Star-shaped expansion is favoured particularly by public transport. In fact private transport, even though of importance to the development of American towns, is inconvenient and expensive. The true revolution has generally been brought about by public transport. This brings to the centre large numbers of people who have found pleasanter living conditions and lower rents out of town.

Some of this traffic uses the roads. This includes the suburban tramlines laid along the streets in the nineteenth century and the buses and coaches which have multiplied in the twentieth century.

[1] P. Leheu, 'Argenteuil', *La vie urbaine*, 1954; cf. M. Th. Ménétrier, 'L'extension de Nanterre au-delà de la voie ferrée,' *La vie urbaine*, 1958.

Possibilities for expansion are in these cases linked to the roads. Houses which prolong the town are aligned along these roads, leaving wide empty spaces between them. The roads represent the arms of the star. Of course these arms increase in number, and the spaces between them are gradually filled, but the principal arteries grow longer and the new plan remains similar to the old.

The railways carry a still greater number of passengers. Very few lines—except around great cities such as London—were created for this purpose, but in the neighbourhood of towns the stations collect the local residents to take them to their work every day (see p. 417, commuting). The railways were the instigators of residential suburbs. The electrification of the Sceaux line in Paris in 1938 gave the urbanisation of the southern suburb its final incentive, and the traffic carried by the line increased from 8 million passengers in 1937 to 47 million in 1960.[1] The pattern of suburban railways also helps bring about the star-shaped plan. This can be seen clearly in Tokyo.[2]

Thus growth depends on the means of transport which the various social classes have at their disposal. The Faubourg Saint-Germain in the south of Paris was the home of people who owned private horse-drawn carriages and it was taken over by the working classes during the era of public transport; however, the motor car has created a new residential area for the privileged beyond this.

Expansion also depends upon the form of civilisation. In ill-equipped countries, everyone strives to remain as near as possible to his place of work. Towns pile up just as medieval ones used to. In other civilisations the road is the commanding factor. In the towns of the American West the private car has allowed and indeed caused immense expansion. Los Angeles stretches for nearly sixty miles with hardly any public transport at all.[3] The contrast can be seen clearly in erstwhile colonial towns. The indigenous quarters grew by accretion whereas the European residential quarters tend to be scattered around the outside. This can be seen at Madras, Abidjan and Libreville.

ABSORPTION OF VILLAGES

Towns do not always extend into open country. The tide of new buildings reaches out to existing peripheral villages and engulfs them.

[1] J. Chauvet, 'La ligne de Sceaux et le développement d'une banlieue parisienne', *La vie urbaine*, 1950; J. Lapébie, in *L'année ferroviaire*, 1962; cf. R. Clozier, *La Gare du Nord*, Paris, 1940. The case of London is similar; see, for example, H. Rees, 'A growth map for north-east London during the railway age', *Geog. Rev.*, 1945; E. Course, *London Railways*, London, 1962, chapter 10.

[2] S. Kiuchi, 'Problems of comparative urban Geography', in *Geog. Studies* presented to Prof. Taro Tsujimina, 1961.

[3] M. Tabuteau, 'Los Angeles', *Cahiers d'Outre-mer*, 1953.

These villages keep their old way of life for a long time. The villagers live alongside the new city-dwellers until the day when all rural life becomes impossible and the fields are sold by the square yard to receive new buildings. In 1920 the fusion of Berlin with twenty-three neighbouring localities, seven of which were towns, was decided by law.[1] But the former look of the village does not always completely disappear. When crossing Berlin from north to south in the direction of Potsdam, one is struck by the old country churches lining the streets, each one the only remaining evidence of an old village. Wissous, at the entrance to Paris, has remained a village to this day, and many old rural houses stand as witnesses of the past. Many are the villages that have been absorbed in the growth of London, and some of them still retain something of their original nucleus with cottages, an inn and a parish church—like Edgware and Willesden, for example.

These rural centres cannot be absorbed without bringing up a number of administrative questions; the rural commune remains autonomous and until it is attached to the town administratively this presents delicate problems. Generally speaking, two sorts of suburbs may be distinguished: those which have the amenities of a town, and where people often go to escape the town, and those which lack the amenities and appear to be proletarian.

SHANTY TOWNS

The most lamentable form of uncontrolled expansion by accretion is to be seen in shanty towns (French *bidonvilles*, literally 'towns built of tin-cans'). When all that is needed is a tin roof and a few planks, they are all too easy to build at the approach to a town. Many large modern towns unfortunately have districts formed in this way. The *favellas* of Rio de Janeiro are simply an expression of the working-class districts climbing up the unwanted spaces all round the town: no water is piped to these steep slopes, and their inhabitants have to go down daily for supplies (see below, p. 308).

These miserable huts clinging to the walls of the town are also to be found surrounding African towns. Their inhabitants live in the most deplorable state of hygiene.[2] The great cities of Europe are not always exempt: in the fortifications district of Paris where permanent buildings were not allowed, huts of this sort were to be seen, and this state of affairs was worsened by the arrival of workers from North Africa. Shanty towns result from the influx of people driven from

---

[1] M. Halbwachs, 'Gross Berlin: grande agglomération ou Grande Ville', *Ann. Hist. Ec. et Soc.*, 1934.
[2] J. Pelletier, 'Un aspect de l'habitat à Alger: les bidonvilles', *La vie urbaine*, 1951; P. Mas, 'L'urbanisation actuelle du Maroc: les bidonvilles', *La vie urbaine*, 1951.

their native place by hunger, and for whom the city offers at least the possibility of earning a living.

Fortunately, accretion does not always occur in this dismal guise. It can happen in different areas at different times, thus revealing the architectural style of several periods. As one leaves the centre, one sees successively the tall, old houses which are so difficult to heat, with their narrow windows, modern blocks with broad spaces between them, and then huge sky-scrapers. Where expansion has happened suddenly, the contrast between the old nucleus and the new quarters is startling.

POLYNUCLEAR EXPANSION

In contrast to accretion is polynuclear expansion. At first this consists only of nuclei at the approaches to a town of which they more or less form a part. The *faubourg* or inner suburb which is attached to a town today was once the 'borough' at the gates of the town, *foris burgus*. For want of space it was built outside the town, it was the commercial centre, whose growth was continuous; and Henri Pirenne has demonstrated how the *faubourg* became far more important than the administrative and military borough in the formation of the town.

The simplest case of polynuclear expansion is the placing of a new town beside an old one. Very often in the Middle Ages a completely independent town sprang up beside the feudal town; sometimes this was the bishop's town, or a centre formed round an abbey. At Dijon the Abbey of Saint-Bénigne caused the formation, in the ninth century, of a new town opposite the *castrum*, on the other bank of the Suzon, reached by the *Rue du Bourg* which has kept its name.[1] Nancy increased in size in the sixteenth century when a new town was founded beside the old one. When Rome became the capital of reunited Italy, a town was built to the east of medieval Rome to fill this new rôle. At Erlangen in the seventeenth century a new town was built to house the Huguenots, and is recognisable by its quadrangular plan.[2]

New towns founded beside old ones sometimes fulfilled a desire for segregation. This might be social segregation, as in the case of Zagreb where the bourgeois class wanted their own quarter; or professional segregation, as with the merchants of Istanbul who had their own town called Galata. Or it might be racial, for foreigners do not like integrating; European building plots were established in China beside the ancient Chinese towns.

[1] M. Oursel, *Les Origines de la commune de Dijon*, Dijon, 1939; cf. A. Durrand, *Aurillac*, Aurillac, 1946.
[2] J. Bluthgen, *Erlangen*, Erlangen, 1961.

This juxtaposition was especially the rule in colonial settlements, where new arrivals, either for comfort or security reasons, settled away from existing towns. This may be seen in almost all African towns: Rabat, built for the French administration on the opposite side of the river mouth from Salé, is a good example. Similarly, at Bandung in the island of Java, the Chinese, Europeans and Indonesians have their separate quarters. In Algiers the Europeans settled on both sides of the Casbah, either in the direction of Mustapha and Belcourt, or to the west in the direction of Bab el Oued.[1]

In such circumstances it becomes difficult to distinguish expansion by accretion from polynuclear expansion, for these are certainly new urban nuclei each distinct from the other but situated close to the old town.[2] Colonial juxtaposition can take very different forms. In countries where urban life already existed, a white town was sometimes founded beside the indigenous town. But this juxtaposition could also develop later, after the foundation of a white town, when the old villages adopted the pace of modern life. Médina, near Dakar, was founded in 1914 to regroup certain African villages. At Bamako the original town was in fact only a large village.

The white man brought town life to the Congo, commercial towns for the products of trading or for banking operations, and administrative or military towns. And the 'white man's work' caused an influx of rural inhabitants, for the white man needed labourers, workers, clerks and 'boys' of all sorts.[3] In this way two towns develop side by side, each with its own style, the whites living in villas and having cars to go from one place to another. It is significant incidentally that the indigenous town continues to be referred to as the village or the villages, for it often consists of several villages joined together.[4] Negro towns are only towns because of the mass and density of their populations and to the extent that they are dependancies of a white town.[5] In all these towns, the formation of different nuclei

[1] J. Pelletier, 'Alger 1955', *Cahiers de Géog. de Besançon*, 1959.
[2] Cf. J. Richard-Molard, 'Villes d'Afrique Noire', *France Outre-mer*, 1950; Olivier Dollfus, 'Conakry en 1951–52', 1953; Soret, 'Démographie et problèmes urbains en A.E.F.', *Pub. de l'Institut d'Etudes Centrafricaines*, 1954; Lombard, 'Cotonou, ville africaine', *Centre I.F.A.N.*, 1953 and 1954; P. Brasseur-Marion, 'Cotonou, porte du Dahomey', *Cahiers d'Outre-mer*, 1953; C. Savonnet, 'La ville de Thiès', *Centre I.F.A.N.*, 1955; J. Denis, 'Léopoldville', *Zaire*, 1956; A. Seck, 'Dakar', *Cahiers d'Outre-mer*, 1961; Xavier de Planhol, 'La formation de la population musulmane à Blida', *Rev. de Géog. Lyon*, 1961; A. Mabogunje, 'The growth of residential districts in Ibadan', *Geog. Rev.*, 1962.
[3] Jean Dresch, 'Villes d'Afrique occidentale', *Cahiers d'Outre-mer*, 1951, and 'Villes congolaises', *Rev. de Géog. Humaine et d'éthnologie*, 1948; P. Vannetier, 'Banlieue noire de Brazzaville', *Cahiers d'Outre-mer*, 1957.
[4] P. Lasserre, *Libreville*, Paris, 1958.
[5] J. Dresch, 'Villes d'Afrique occidentale', *loc. cit.*

corresponds with a desire for segregation: separation of the lay and religious powers, ethnic and political segregation.

However, the appearance of an urban centre distinct from but attached to the original town has often happened spontaneously, and in various ways. Thus bridge towns usually grew up on one bank only; but on the other side of the bridge the bridgehead constituted a small, distinct centre, sometimes defended by some kind of fortification, like the Châtelet to the north of the Cité in Paris. Merchants used to stop there before crossing the bridge.

Railways were usually designed to avoid crossing the town centre, taking advantage of cheaper land. Businesses immediately grew up around the station and gave rise to a district which in some cases became the busiest and which was linked to the town by an avenue that soon became populated and full of life. The difficulties encountered when the line was established sometimes meant that the station had to be some distance away: outside the great river meander at Besançon, over a mile from the town at the mouth of the deep Poligny valley, and at Laon in the plain at the foot of the hill.[1]

When large-scale industry replaced craftsmen, space was required. This was hard to find inside towns, so industries moved out. Outside the walls they had the advantage of cheaper land and goods could be collected and delivered more easily. Generally factories were sited in valleys where traffic circulation was easy, and advantage was also taken of the proximity of railways. The expansion of ports allowed industries to make use of new docks. Consideration of health and safety obliged the oil industries to be set up away from towns. Thus industrial centres were created as annexes to the town. The reasons for this were generally valid for several types of industry. The more a factory is modernised and 'automated', the more space it requires.

However, it was not simply a question of finding room for factories and workshops. The numerous working population required by each centre did not usually wish to live in the town, where rents were higher, and so settled by the factory. The business enterprise often took the initiative and built housing estates which attached the workman to his job. In this way Bournville grew around Cadbury's chocolate factory, and Port Sunlight around Lever's soap works.

At Kansas City the factories of North Kansas City[2] were built on cheaper land across the river Missouri. At Gothenburg the naval shipyards on the right bank of the Götaälv gave rise on the other side of the estuary to a centre which was for a long time badly served by a ferry and a footbridge.[3] Industrial areas often began as independent

[1] Cf. R. Crozet, 'Le problème de la gare de Tours', *La vie urbaine*, 1952.
[2] J. Adams, 'The North Kansas City Urban District', *Econ. Geog.*, 1932.
[3] G. Chabot, 'Göteborg', *Bull. Ass. Géog. français* 1949.

centres formed around a factory or a group of factories: Siemensstadt, to the west of Berlin, was the town of the Siemens enterprises. The smelting town enclosed in the Huerta de Sagunto in Spain could also be quoted[1] in this context, or the industrial plant of Dalmaniagar near Dehri (Bihar State, in India).[2] These industrial tentacles often extend for long distances. The towns which extend the agglomeration of Lyons to the south have been described as a rosary.[3]

Some of these centres are not connected with a particular factory. They simply answer the population's need to find cheaper and more spacious housing near the town. This has caused the phenomenon of the suburb, which we shall examine later on (p. 238). Almost all the localities around Paris have served as settlement centres in this way. Between the census of 1954 and that of 1962, about ten localities more than doubled their populations. Villeneuve-la-Garenne grew by 233 per cent (from 4,000 to 13,400), and Sarcelles by 326 per cent (from 8,400 to 35,800). In contrast to Burgess's concentric theory, Hoyt puts forward the idea of sectors where townspeople settle according to their preferences and to the convenience of the district, and which form the nuclei around which the town grows.[4]

So a town develops little centres at its approaches which increase its size without being really separate from it. These outposts are soon absorbed as the irresistible pressure of the town fills up the green spaces and welds them together. The process continues. Nevertheless, contemporary forms of expansion are very different from traditional methods. Let us consider, for example, the development of Vienna as shown on the two maps in the Austrian Atlas. The contrast becomes all the more striking because after 1918 Vienna underwent a period of stagnation, having lost three-quarters of the empire of which it was capital. When expansion began again, it was at a different pace. The neighbouring villages lost their rural character and became urbanised up to a distance of about six miles. Little houses hitherto outside the town and used as weekend cottages, became permanent homes; the Schrebergärten, small part-time agricultural holdings, now house families all the year round (Fig. 15).

Polynuclear expansion has in fact developed at a quite different rate since the increase of rapid transport. The influence of town life has spread much more widely and has enlivened distant centres. There was no need to remain at the entrance to the town. It was often

[1] A. G. Crispo, *C. R. du Congrès Géog. int. Rio de Janeiro,* 1956.
[2] R. L. Singh and S. K. N. Singh, *ibid.*
[3] *Bull. du comité pour l'aménagement et l'expansion économique de la région lyonnaise,* 1955.
[4] The structure of Americans cities in the past-war era; *Amer. Jour Sociology,* 1943.

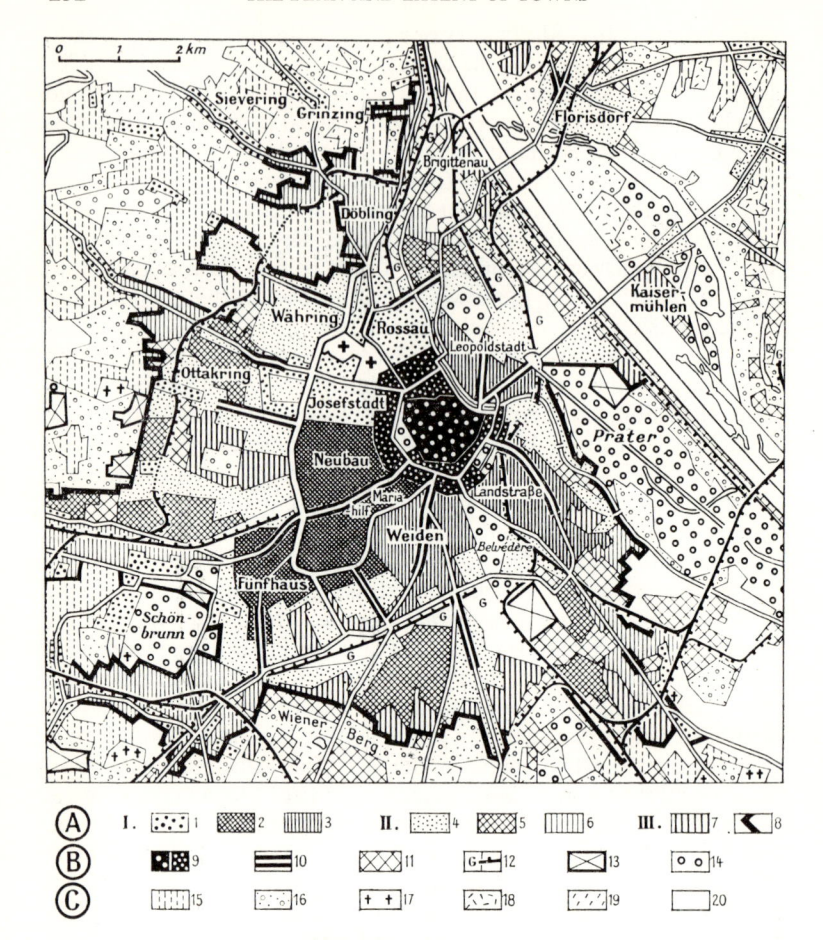

Fig. 15. The expansion of Vienna

After Elizabeth Lichtenberger, Austrian Atlas, and *Geographische Rundschau*, June 1962; maps vi, 10 and vi, 11 (H. Bobek and E. Lichtenberger).

A. Urban nucleus.

I. Areas built before 1840 (old town, suburbs, outskirts): **1.** Principal residential areas. **2.** Districts with back-yard industries. **3.** Suburbs on exit roads.

II. Areas built between 1840 and 1918: **4.** Principal residential areas. **5.** Districts with industrial establishments. **6.** Starting points for growth of various kinds.

THE EXPANSION OF TOWNS 233

III. Between the wars and postwar periods: 7. Areas of marginal growth, using open spaces. 8. Limit between closely built urban nucleus and spaced out peripheral area.
B. Separate elements.
9. 'City' structure. 10. Principal trading arteries. 11. Industrial establishments. 12. Stations and railways. 13. Public places. 14. Green spaces.
C. Peripheral areas.
15. Residential areas. 16. Areas undergoing transition (Schrebergärten, temporary housing, etc.). 17. Cemeteries. 18. Wasteland. 19. Vineyards. 20. Other cultivated land.

simpler to take advantage of nearby housing, to start new activities and to settle there. To some extent, a town lays claim to everything surrounding it. Tunis has grown through the surrounding localities, Carthage, Bizerta and La Goulette, as much as it has in itself.

We may therefore ask whether it is still appropriate to speak of town expansion. Formerly, external nuclei of crystallisation were not far distant; they became suburbs without difficulty. Today it is hard to see how all the external nuclei dependent upon one town could become annexed to it. It is not so much a question of expansion as of the formation of a new urban mass or 'city region'.[1]

PLANNED EXPANSION

Towns in many cases owe their origin to an act of civil or religious authority (see p. 168). But the growth of many more was held back by those in power; military, administrative and fiscal requirements were strengthened by religious restraints. 'In the social conception of the Middle Ages everything had to take place inside the ramparts.... The building of suburbs, which might threaten the security of a town, was forbidden.... When land became too scarce and the population was almost suffocating, when every space was occupied by houses, gardens, courtyards and even bridges, then the ramparts were extended'.[2]

In modern times these restraints might seem to have been shaken off, leaving everyone free to choose his own land to build on as he liked. But the community life of towns makes its demands, and individuals found themselves facing problems which were too big for them.

Only public authorities could undertake certain forms of land improvement, so they controlled the expansion. From the seventeenth century on, in Amsterdam, the municipality took charge of the

[1] M. J. Wise, 'The City Region', *Adv. of Sci.*, 1966, pp. 571–88.
[2] G. Des Marez, *Le développement territorial de Bruxelles au Moyen Age*, 1935.

expropriation of polders and their parcelling-out for building purposes.[1] The building of roads, and the establishment of tramlines and bus routes, which were subsidised by municipalities, guided the directions of urban expansion.

Moreover, free, uncontrolled expansion has many disadvantages. Doubtless, everyone could be left to build his own house and to organise his own individual water supply and cesspool, as happens sometimes in rural districts. This was in fact often the case. However, there was seldom complete liberty, and in recent times building has almost always been subject to prior authorisation. 'Every independent action in matters of town planning is fraught with consequences for collectivity', as William Oualid declared.[2]

It became evident, too, that all below-ground construction, water and gas mains, and sewage disposal, had to precede building. The public authority often took charge of this, not only in socialist countries, but also in countries with liberal economies. Municipalities prepared the ground before placing it at the disposal of new inhabitants.

The burden was a heavy one, and involved the running of considerable risks. Mass expropriation of the whole neighbourhood could not be undertaken. The choice of land was hazardous and presented endless complications. Except in a few special cases, companies were allowed to act at will, the authorities being content to control them and to impose certain restrictions upon them.

These companies would seek out suitable land, bargain over the price and re-sell it in plots when they were not building on it themselves. This division of land into plots was often open to abuse; new landowners sometimes found themselves in a muddy wasteland without any guarantee of drainage. Strict ruling was introduced in every country to impose certain obligations on land agents and to protect future purchasers. Unfortunately these laws were often made too late, when a great deal of harm had already been done.

Real estate companies calculated where a town's activities were leading, and the subsequent form communications would take. The speculative element inevitably played a part. The societies' acquisitions sometimes provided them with enormous profits, but also left numerous disappointments.[3]

The fact remains that this distribution of land in lots allowed many small purchasers to procure land and build their own house, which is the wish of the majority of people in every country.

[1] F. Verger, 'La morphologie urbaine d'Amsterdam', *Norois*, 1961.
[2] *La vie urbaine*, 1936.
[3] J. D. Feldmann, 'Pre-building growth patterns in Chicago', *Ann. Ass. Amer. Geog.*, 1957.

These purchasers usually belong to the middle class. They are people who already have a little capital and, certain of their jobs as artisans or white-collar workers, can obtain loans through the facilities of building societies.

In the Paris region many parks belonging to old estates were divided into lots in this way. The Ormesson estate on a loop of the Seine remained intact for a long time, although at the gates of Paris. It became the object of successive divisions into lots, uncontrolled up to 1924 but more disciplined and expensive later.[1]

Whether by public authorities or private companies, the expansion of residential areas is being increasingly controlled and directed. The expansion of industrial areas is even more rigidly controlled. Certain factories making toxic or dangerous products are not allowed in built-up areas. Moreover nowadays every country sets aside special areas for industrial establishments.

But planned expansion has not been content merely with the selection of sites. The siting of public buildings may be the origin of new centres. Administrative areas have been created, around which housing has developed. Among the glaciated rocks of Blindern, Oslo has planted a university, and a radio station, around which a new district is being built.

This is not all. Attempts are made to house inhabitants who have not the means to do it themselves. This is the rule in socialist-run countries. Most new buildings are erected by the State which, as landowner, supplies housing. Buildings are often standardised, so to avoid monotony, exteriors are painted different colours.

Even in non-socialist countries, public authorities are interested in housing, and do not leave it entirely to individual initiative. It has been necessary to help immigrants to find homes, to avoid the multiplication of shanty towns, and to rehouse the occupants of decaying or unhealthy dwellings. Moreover, the fixing of rents has often stopped the construction of buildings to let. It became important to put decent housing at the disposal of people who needed it at a lower price than that demanded by profit-making private enterprise. The state and the municipality had to become landowners too, and at a loss. Very large sums were set aside for the construction of public housing almost everywhere. Advantage was sometimes taken of this to give social benefits, and to facilitate house-purchase after a certain number of years. Large firms adopted similar policies to give their workpeople comparable benefits.

This was done in several ways. The garden city formula was launched in England, near Liverpool, in 1887, and copied in Germany,

[1] P. Ruy-Cogordan, 'Trois communes de la banlieue est de Paris', *La vie urbaine*, 1953.

on the Taunus slopes near Cologne, in 1895.[1] It was a great success; it provided small, individual houses, sited among trees and grass, with public services, and answered a certain residential ideal.

It seems, however, that this formula no longer meets with the same approval. Individual houses take up too much space; it seems preferable to build higher blocks of flats or apartments, with courtyards and larger gardens where children can play. The famous Karl Marx City at Vienna was built between the wars according to this scheme. But even for this, space was often lacking. Municipalities were urged to build as much housing as possible at the lowest possible price. There was no longer any question of leaving ground unused. New housing is often in huge, multistorey blocks, built very close to one another. These are built to provide not less than 1,000 homes, and often over 10,000. They are the work of public authorities and big firms, and today are transforming the approaches to towns in practically every country—even to a certain extent in Great Britain, the traditional home of terraced houses and 'semi-detached'.

In developing countries the rapid growth of towns and the impossibility for the inhabitants of finding housing for themselves has put urban expansion into the hands of building societies, usually under the control of public authorities. This has happened particularly in Central African towns.[2]

Public authorities have gone even further. They are no longer content to make regulations concerning the activities of individuals, or to build flats to let at reduced rents here and there. The only way to stop uncontrolled and unlimited expansion of huge towns was to build new towns surrounding them, rather as fires are lit in a wood to check a conflagration from spreading.

Whole new townships have sprung up in this way, in districts surrounded by trees and grass, like La Dame Blanche, at Garges-lès-Gonesse, to the north of Paris, intended for a population of 21,000, or Wythenshawe, six miles or so from the centre of Manchester, which was a 'green-field' site in 1926 and by 1958 housed 85,000 people. Building has been started in the suburbs of Berlin, at Berlin-Buckow-Rudow, for a city for 45,000 people. Near Montbéliard a town is planned for a population of 50,000 to house the workers in the automobile industry. These are often dormitory towns. But it is hoped to do more than this, and create self-governing towns to relieve the congestion of huge cities. Attempts have been made at this in almost every country. Satellite towns are no longer strictly speaking

---

[1] Kampfmeyer, 'La cité-jardin de Gronauerwald à Bergisch Gladbach', *La vie urbaine*, 1928.

[2] J. Denis, *Léopoldville*, 1956, and *Le phénomène urbain en Afrique centrale*, Brussels, 1958.

the expansion of a town around its original plan, and are considered further in Chapter 23.

Thus public authorities, municipalities and the State have often taken the expansion of towns in hand, to avoid the confusion of a too free expansion. In this way they have generally—though not always—been able to preserve the green spaces and open land. Such spaces are not always just grass. It was feared that the market-gardening suburbs of Nantes were going to disappear. The town obtained its food from these and they could not be simply pushed further out, so building was prohibited in certain parts of these suburbs.

At Ottawa the Federal Government has reserved a territory twenty miles by forty miles which will form a green belt, but in which the Government may install certain administrative services and sports grounds.[1]

In most cases it was a question of preserving the parks around the large houses of the landed gentry or the upper middle classes in the neighbourhood or towns, where these were in danger of being submerged. Many landowners could no longer afford the upkeep of such establishments, and their estates began to acquire such a high value that they willingly gave up at least part of them. This often caused division into lots by private companies, but often too the operation was carried out by the municipality, anxious to retain green spaces.

Nowadays the expansion of towns is left less and less to the whims of their inhabitants or the work of speculators. Pierre Lavedan was able to write in *La Géographie des Villes* (1959): 'The tide of control is such that urban geography will soon become simply a section of the administration'.

[1] Jacques Greber, *La Vie urbaine*, 1959.

CHAPTER 21

THE SUBURBS

We have shown that towns tend to grow continually, and this has become more apparent since the industrial era. But this growth was only possible if the town spread outwards from its original centre. The multiplication of industries and the separation of home and workplace helped by the development of means of communication created an urbanised area round the town, which is no longer the town in the sense in which it was formerly understood. If the chief characteristic of the nineteenth century was the concentrated growth of towns, that of the twentieth has been chiefly suburbanisation, that is to say, the development of the outskirts or urban fringe (in French *banlieue*, in German *Vorort*). Between 1940 and 1950 in the United States the suburban population grew by 35 per cent and that within towns by only 13 per cent.[1]

The French *banlieue* was in the Middle Ages the space under the jurisdiction of a town proclaimed by the *Ban*.

For a long time these suburbs were limited to the immediate neighbourhood of a town. Often they were represented only by a strip of market-gardening land. The growth of towns and the development of means of communication imposed an increasingly great complexity on the suburb. It became an aureole round the town, the *Umkreis* of C. Ipsen.[2] Raoul Blanchard speaks of the 'formidable problem of the suburbs' and Pierre George notes how necessary is 'caution in the application of this term, and even greater caution in its generalisation'.

Sometimes the zone simply consists of dwellings scattered about the countryside. Nevertheless, all suburban life tends to become concentrated at points where communications are easiest; sometimes these centres grow spontaneously, often they use neighbouring villages or small towns as a support.

---

[1] E. L. Ullmann, 'New York in Regional Growth', *Proc. of Western area development conf.*, 1954.

[2] Paper read at the Akademie für Städtebau, und Landplannung, Dusseldorf, 8 July 1958.

THE VARIOUS TYPES[1]

Various types of suburb can be distinguished thus:

The market-garden suburb. This develops around every town to supply its population with fresh fruit and vegetables. The ease with which provisions can be obtained from great distances by rapid transport has not cancelled the advantage of foodstuffs gathered the night before, or that morning, brought fresh to market with but little cost of transport. So outside large towns there is usually an area, preferably on low-lying alluvial ground, where the soil is prepared artificially to provide for this intensive cultivation. And with the help of sludge from its sewage works, the town creates its own market-garden suburb.[2]

Allotments which indicate the approaches of a big town are another aspect of this market-garden suburb.

The dormitory suburb. The dormitory suburb comprises localities where town-dwellers come in search of cheaper housing, a few trees, and a small garden. These suburbs are responsible for the daily flow of commuters. Some townspeople consider it as a kind of exile, far from the amusements of social life. L. Rosenwayr noted that in Vienna 82 per cent of the people his survey reached wanted to remain in the town centre.[3] But Vienna is a great cultural centre; it is extremely doubtful whether anything like so high a percentage of the population of the average British industrial city would prefer to remain in the centre—from which they have been fleeing in vast numbers for half a century.[4]

The industrial suburb. As we have seen, factories can benefit from using the less expensive land in the suburbs; this applies not only to factories built at the approach to the town and still part of it, but also to those established some distance away; they too benefit from links with the town where the workers and managers often live; they are near the social centre, but they have much more space at their disposal, and sometimes even a local labour force. This is a phenomenon which operates around the town on a regional scale, while decentralisation is taking place on the national scale.

---

[1] On suburbs see R. Clozier, 'Essai sur la banlieue', *La Pensée*, 1945; P. George, 'La banlieue, une forme moderne de développement urbain', in *Etudes sur la banlieue de Paris*, 1950; G. Friedmann, 'Villes et campagnes', 1953; G. Chabot, 'Faubourg, banlieue, zone d'influence', *Urbanisme et Habitation*, 1954; C. A. Wissink, *American Cities in Perspective*, 1962.
[2] See below, p. 338.
[3] L. Rosenwayr, in *Wien: Die Stadt und ihr Umland*, 1956.
[4] On commuting, see p. 417.

In 1943 Chauncy D. Harris in his study, *Suburbs*, distinguished six types: 1. industrial suburbs where townspeople come to work in the factories; 2. industrial suburbs where people have their jobs on the spot (what we should probably call satellite towns today): 3 and 4. semi-residential, semi-industrial suburbs; 5. dormitory suburbs; 6. mining and industrial suburbs in mining areas. More recently C. A. Moser and W. Scott, in England, have distinguished upper-class residential suburbs, middle-class residential suburbs, suburbs with light industry, with old industries or with modern industries.[1]

There are smart, residential suburbs, often dating from before the introduction of modern transport, poor residential suburbs where people took refuge from high rents in the town; industrial suburbs where the workers' houses are crowded round the factory, suburbs where individuals have built their own houses at weekends, and which sometimes look like shanty towns. There are well-equipped suburbs with well-cared-for roads, and marshy suburbs where you flounder through the mud.

THE DEVELOPMENT OF THE SUBURBS

Suburbs are the expression of the life and functions of a town. Towns without suburbs are rare, and are only found in exceptional circumstances. It has been remarked that Rome was once a city without suburbs.[2] It was almost non-industrial and grew all at once; the neighbouring countryside, devastated by malaria, offered no rural centres from which suburbs might have grown. In order that suburbs may form, a certain expansion of urban life is necessary.

The suburb, as we have already seen, is hard to define. In fact it is a phenomenon rather than an area. The suburbs may be said to begin where the continuous built-up town ends. First there is the built-up area of houses with small gardens, forming dormitory communities from which more than half the active population works in the town; these suburbs may be considered as being still part of the town, in a broad sense, and for this reason may be described as inner suburbs. Secondly, there are less immediate, or outer suburbs; these are residential, industrial or commercial areas which rely on the town for their livelihood. They too are largely dormitories, though they have industries which rely heavily on the town, and they may also be characterised by the presence of weekend houses. Townspeople and country folk live here side by side.

What might be described as 'suburban features', with their characteristic building styles and use of land, thin out the further

[1] *British Towns*, 1962.
[2] A. M. Seronde, 'Les quartiers méridionaux de Rome', *Bull. Ass. Géog. français*, 1956.

16 Sydney: the geometry of suburbia

17 Noumea (New Caledonia): gridiron pattern of town centre, with dispersed housing on the hills

18 Salvador (Bahia): the core of the old city

19 Salvador (Bahia): housing on stilts in the lagoon

one gets from the town. So distant suburbs are spoken of; but this concept can become confused with the *umland* or zone of influence (see below, p. 425).

Suburbs depend upon the forms of transport available. In the days when only the aristocracy could easily afford to travel in carriages, they were glad to live in the suburbs. Railways put rapid travel at the disposal of all, without distinction, and the less fortunate hurried to the cheaper housing of the suburbs; however, in countries with scattered populations railways were not sufficiently numerous, and the transformation has been effected by road.[1] The private car has again given the advantage to the well-to-do.[2] In flat countries like Denmark and the Netherlands the bicycle has been a major contribution to the growth of suburbs; whilst of late the motorised two-wheeled vehicles has played a similar rôle.

There is, therefore, a migration towards the suburbs. The prime factor is the facility of communications with the town. Just as the towns of a conurbation merge together, so the town and the country confront each other. A certain village, perhaps several hours' journey from a city, was leading a more or less independent life, regarding the town only as a distant outlet: then townspeople reached it by train, and eventually faster train or bus services made it into a dormitory community.

The island of Amager to the south of Copenhagen is a case in point. Whereas the northern part of the island is an integral part of the city, the rest of the island for a distance of half a dozen miles or so is scarcely more than suburban, its old villages developing into different types of suburb. Some continue in their rural vocation, occupying themselves with market-gardening; others, like Kastrup, house industries; Tårnby is a dormitory village; Dragör, a former fishing village, has become a place for weekends and holidays.

The suburbs, then, represent the urbanisation of the countryside. In some of the highly urbanised parts of Europe and North America the entire countryside has adopted an urban look; certain types of farming have developed specially for towns, and the raising of dairy

[1] H. F. Wilson, 'The roads of Windsor', *Geog. Rev.*, 1931.
[2] A. E. Smailes, 'Some reflections on the geographical description and analyses of townscapes'. *Trans. Inst. Brit. Geog.*, 1955. Whilst this statement may be true as a generalisation, it is not universally applicable. In London for example, the increasing affluence of the so-called 'working classes' has resulted in roads in the older inner suburbs being used as unofficial car parks for the residents, many of whom are coloured immigrants. Every house in what a few years ago would have been regarded as 'poor' districts now has its car standing outside. On the other hand, the residents of the wealthy areas of inner London, where car-parking is severely restricted, either have to do without cars and rely on taxis, or pay very high rents for the few converted coach stables or modern garages that are available. (Editor's note.)

cattle has taken precedence. These regions look like huge suburbs. Jean Gottmann has called this the suburbanised zone. Here, less than 25 per cent of the active population works in agriculture. William A. Robson may well deplore these landscapes surrounding towns which are neither rural nor urban and which have suffered urban contamination.[1] Perhaps, indeed, 'subtopia' is not an inappropriate term to apply to them.

Will the future see this suburbanisation of the countryside within a megalopolis, or on the other hand the ruralisation of towns, making the countryside an integral part of the mother-town, which is the ideal of the school of architecture of Le Corbusier? Whatever the answer, town and country are coming closer to one another. A suburb is a town coming to grips with the country surrounding it.

[1] *Great Cities of the World*, London, 1954, p. 104.

Chapter 22

TOWN CLUSTERS AND CONURBATIONS

An urban mass does more than spread across the countryside. There may be other towns in the neighbourhood with which it comes in contact. Dietrich Bartels[1] has correctly stressed the usefulness of the study of interurban relations, so often neglected. For him, two neighbouring towns are towns at a distance of under twenty-two miles apart, from centre to centre. Some of these urban growths touch each other, others are separated by a completely rural area. Interrelationships are sometimes very close, while elsewhere there is no contact between, say, one bank of a river and the other, or across a frontier (as in the two Berlins). So in some cases towns complement one another (Elberfeld and Barmen), and in others compete with one another (Mainz and Wiesbaden).

Under the heading of neighbouring towns, conurbations occupy a separate place, and this must be stressed because the term has led to many misunderstandings, with the result that neighbourhood and conurbation have been confused.

CONURBATIONS: DEFINITION

The term 'conurbation' was introduced by P. Geddes who also suggested 'conglomeration'. It has been discussed and interpreted in various ways. Some geographers reject it, preferring to use the term 'agglomeration'. C. B. Fawcett in fact gives a definition very similar to that of agglomeration, since it required an uninterrupted built-up area.[2]

In his work on the conurbations of Great Britain,[3] T. W. Freeman declares that a conurbation consists of an accumulation of industrial towns where, incidentally, places of commercial or residential origin may be found. These towns may have kept a definite character, but Freeman—and with him the British census—considers that a town with its satellites or dormitories can form a conurbation; London is therefore considered as Britain's largest conurbation. Pierre George

---

[1] *Nachbarstädte*, 1960.
[2] C. B. Fawcett, 'Distribution of the urban population in Great Britain, 1931', *Geog. Journ.*, 1932, pp. 100–13.
[3] *Conurbations of Great Britain*, Manchester, 1959 and 'The Manchester conurbation', in *Manchester and its Region*, British Association Handbook, 1962.

has noted quite rightly that London is much more a case of an agglomeration.[1]

The terms agglomeration and conurbation should not be confused. They are two different ideas which may coincide, but do not necessarily do so. An agglomeration presupposes more centralisation; in a conurbation towns remain distinct while still forming part of a large group. On the other hand some authors define a conurbation as simply a cluster of towns close to one another.

It would be useful to make the definition more precise. A conurbation can in fact be defined by two conditions. The first is genetic. An agglomeration exists only when the whole was formed under the influence of one town. The proliferation of industrial or residential suburbs gives rise to an agglomeration. On the other hand, the juxtaposition of towns which have expanded separately gives rise to a conurbation, even if one of these towns wins in the end.[2]

The simplest form of conurbation is that of twin towns, of which classic examples are St Paul and Minneapolis.[3] Similarly, Ludwigshafen and Mannheim,[4] on either side of the Rhine, form a conurbation; a century ago Mannheim had a population of 22,000 and Ludwigshafen one of 2,000; Ludwigshafen did not develop under Mannheim's influence, but because of the establishment of a chemical products factory, the Badische Anilin und Soda Fabrik. Thus there can easily be conurbations of two towns, each one developing on its own account. The same applies to Liverpool and Birkenhead.

Besides double conurbations, there are multiple conurbations. The most often quoted French conurbation is that of Lille, Roubaix and Tourcoing.

Conurbations are often due to the presence of mineral beds which have given rise to several neighbouring towns: this is the case of St Etienne and the towns of the Gier valley. The towns of the Ruhr were all founded on the exploitation of coal and the industries or commerce arising from it. They represent today the most classic example of a conurbation, extending from Krefeld to Dortmund, with a population of over four million. Two other and somewhat similar coal-mining conurbations are those of the Donbass, where five towns have populations of over 500,000 each, and of Upper Silesia in Poland.

[1] *Ann. de Géog.*, 1960, p. 312.
[2] This is the meaning ascribed by A. Sestini ('Qualche osservatione Geografico—Statistica sulle conurbazioni italieni' in *Studi Geograf.* Florence, 1958) and by C. A. Wissink, *American Cities in Perspective*, Assen, 1962.
[3] H. Hartshorne, 'The twin city district', *Geog. Rev.*, 1932; H. W. H. King, 'The Canberra–Queenbeyan symbiosis, a study of urban mutualism', *Geog. Rev.*, 1954.
[4] A. Traband, 'Structure actuelle du complexe économique de Mannheim-Ludwigshafen', *Rev. Géog. de l'Est*, 1962.

Of Lancashire it could be said that all its towns depend to some extent upon Manchester, for which they work, some spinning fine thread, some coarser thread, and some weaving cloth. But although they have grown round Manchester, they once had their own life, and have kept it, like Oldham. Similarly in the west Midlands, with Birmingham, Walsall, Dudley and Wolverhampton.

A multiple conurbation has developed at the confluence of the Main and the Rhine: Frankfurt, Höchst, Offenbach, Hanau, Aschaffenburg, Russelheim, Darmstadt, Wiesbaden, Mainz. This consists of towns, which, in the middle of the nineteenth century, were still independent and, incidentally, did not carry out the same functions; today they form a continuous mass where the population density never falls below 750 people per square mile, and which contains a population of over 2 million.[1]

Some conurbations develop under peculiar conditions, across a strait. Sometimes this is only a case of two navigation terminals, reminiscent of the bridgeheads which occur so frequently; for example the conurbation of Messina and Reggio, which, including the suburbs, housed a population of 350,000 in 1951.[2] But a conurbation can spread even further, defying political boundaries: Aage Aagesen has indicated one, developing either side of the Sund; Helsingör and Hälsingborg are separated by a sea journey of only twenty minutes and Danes travel daily to work in Sweden and Swedes to work in Denmark. There are others. People from all over Swedish Scania go to do their shopping in Copenhagen, which is like the capital of the region. Danes and Swedes alike foresee an urban mass on the Öresund which would allow for two bridges across the Sund: one from Helsingör to Hälsingborg and the other from Copenhagen to Malmö; the Danish island of Salthom would be its airport and the Swedish island of Ven in the middle of the Sund would be the playground for the group. In Japan towns crowd around the strait of Shimonoseki (Shimonoseki, Wakamatsu, Moji, Yawata, Tabata).

The second condition to be satisfied in the definition of a conurbation is urban density, a density expressed in the number of urban units and the population of these units. This density involves the problems set by juxtaposition, and this is one of the characteristics of conurbations. Small towns which are quite well spaced, even if they form an urbanised area, can never form a conurbation, for they do not have a common problem to solve. The small textile towns of the Beaujolais district cannot be considered as conurbations although they are numerous and fairly close together.

[1] A. Krenzlin, 'Werden und Gefüge der Rhein—Mainischen Verstädterungsgebietes', *Frankfurter geographische Hefte*, 1961.
[2] L. Gambi, *Congrès int. de Géog.*, Stockholm, 1960.

The problems are primarily connected with transport, for this is what joins a conurbation and any improvement in transport helps to strengthen the link. It is also useful to have communal water, gas and electricity supplies. For fire-fighting, it is essential to have effective methods available which can be immediately directed to the scene of any disaster. The police cannot be effective unless their activities in the various parts of the urbanised area are coordinated. Finally, it is in the interests of the different elements of the conurbations to respect the green spaces which still remain between them. Better still, they can have parks and public amenities in common.

The need is often felt for intercommunal or supracommunal organisations to solve common problems. After 1920 the Ruhr was obliged to institute an authority over a certain group of administrative services, and almost everywhere the tendency is to form districts which include the towns of a conurbation and which are armed with special powers.

The group of towns in the Swedish central lowland is sometimes considered to be a conurbation. The towns of Nyköping, Oxelösund, Norrköping, Söderköping, Linköping, Skänninge, Motala, and Vadstena form a line, nowhere further than thirty miles apart. Each of these towns was founded on the case of communications in this plain where the lakes are all linked to one another (in the same way that elsewhere a conurbation is founded on coal); each one has grown independently of the others. If, of necessity, they are in close contact with one another, even though they may have had to specialise because of their proximity, there are scarcely any common problems of town planning or administration amongst them.

'INTERURBATIONS'

A conurbation is often the cause of complementary towns, which Nils Björsjö calls 'interurbations'; these no longer come into the category of towns which have developed side by side near one another and which have grown together. They are towns which, not having all services at their disposal, are mutually indispensable. Usually it is a question of a town which has grown up quickly beside an old one, and which forms a pair with it. A town which has formed round a factory can thus remain almost exclusively industrial if nearby there is another town which plays the part of protector, carrying out the administration and offering all commercial facilities. It is quicker to erect an industrial group than to organise a commercial complex. Skutskär, a little town with a population of 3,806 between Uppsala and Gävle, is the seat of one of the most important timber industries (saw and pulp mills) of Sweden; founded in the open countryside, it dates from

1868, and since then the town has remained exclusively industrial. The people of Skutskär go shopping in Upplandsbodarna, less than two miles away, which has a population of only 1,891.[1] Similarly, the small factory town of Ugines has found what it needed in Albertville, while at the same time bringing new life to that old country town. Generally speaking, however, this situation is only a passing phase which lasts until the new place has equipped itself in turn.

THE DEVELOPMENT OF A CONURBATION

Conurbation represents a somewhat unstable state of equilibrium, and the term is often of only temporary significance. In the course of their development towns enter the conurbation stage. In the early stages two towns may have but distant connexions, and are too far apart to form a conurbation. But their growth and the speed of transport bring about their union. Such towns which appear to be merging together are hard pressed to keep their individuality. St Paul and Minneapolis now form a single town crossing the Mississippi. The line of houses from Lille to Roubaix and Tourcoing is now continuous. The conurbation stretches along the roads which join its component parts.

In the course of this advance one town often beats the others to such an extent that the others are subordinate to it and look like its suburbs. The winner is not always the town which seemed the most favoured at the outset. In the seventeenth century, Chandernagore was much bigger than Calcutta; the two separate trading stations grew side by side until the day when the growth of Calcutta reduced Chandernagore to the state of a suburb. The list of towns absorbed by a stronger neighbour in this way would be long. An excellent example of a town drawn into the orbit of another is provided by Hallstein Myklebost, that of Drammen and Oslo.[2]

The conurbation at the mouth of the Main shows the progressive predominance of the successful towns. The map of journeys to work indicates the attraction of Frankfurt against which the other towns can do little to protect themselves (with the exception, up to now, of Aschaffenburg).[3]

In the final stage of evolution it becomes impossible to distinguish the original towns. Montferrand is included in Clermont-Ferrand

[1] N. Björsjö, 'Skutskär. Bebyggelse och Centrumproblem', *Svensk Geog. Arsbok*, 1959. This author distinguishes *inurbations* where industry and services are both present in the town, *suburbations* in which the industrial centre is but an annexe adjacent to a large regional centre, and *interurbations* in which the industrial centre depends for certain services upon the neighbouring town. (*Congr. Int. de Géog.*, Stockholm, 1960.)

[2] H. Myklebost, *Drammen*, Drammen, 1949.   [3] A. Krenzlin, *loc. cit.*

and Saint-Pierre is a suburb of Calais; town boundaries do not prevent Villeurbanne from being part of Lyons.

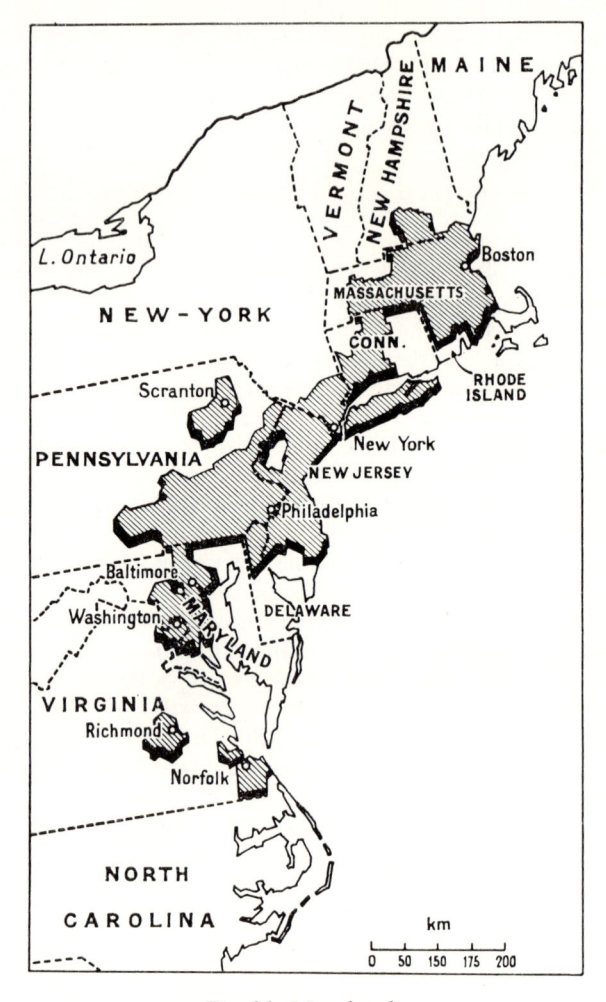

Fig. 16. Megalopolis

From Jean Gottmann, 'Megalopolis, or the urbanisation of the northeastern seaboard', *Econ. Geog.*, 1957, p. 191.

But conurbation does not of necessity mean merging. In the course of evolution, respective positions become more precise. From the outset, the towns of a conurbation may have taken on different functions. Nowa Huta near Krakow, and Ugines near Albertville were founded as industrial towns. Wiesbaden is still a spa. Often, too, this differentiation has increased in the course of development. Ludwigshafen, which is dominated by industry, has to a great extent (and this was especially true in the past) passed on its commercial function to Mannheim. The Ruhr towns also show a tendency to specialise in industry, commerce, or finance. Leeds has taken over the administrative functions while Bradford remains the manufacturing centre for wool. Often the old commercial town acts as a dormitory while the industrial function is taken over by a more go-ahead town, like The Hague and Rotterdam.

It is not always entirely pleasant for a town to be dragged as part of a conurbation into the orbit of a big city. Versailles was founded with certain connexions with Paris, but royalty built it in order to escape from Paris, and the town was independent of the capital. It was therefore a royal residence and almost a capital itself, numbering 70,000 inhabitants in 1789. The Revolution and the Empire appeared to have ruined it. In 1801 there was a population of only 27,000; plans were made to plant the park with fruit trees and vegetables. In the nineteenth century this decline turned it into a town for retired people. The two railway lines built in 1839 and 1840 added to this retirement and holiday town atmosphere for the inhabitants of Paris. Then communications became increasingly easy, and Versailles entered the twentieth century as part of the immediate suburbs. Now a dormitory town with a population of 84,000 (1954), it takes part in the function of capital of which it was once dispossessed.

'MEGALOPOLIS'

The most outstanding of conurbations, to which attention has recently been drawn,[1] is the one which has formed in the north-eastern United States. It has not generally been regarded as a conurbation. There are sixty miles between Philadelphia and New York, and 150 between Boston and New York, but these cities are joined by a whole chain of towns, several of which have populations of over 100,000. In this way an urbanised mass has formed, housing 38 million people, which Jean Gottmann has named Megalopolis. The industrial towns have kept their character and independence, but they are too close to ignore each other and all the space which separates them has been urbanised to some extent; this is what gives the area the character of a conurbation.

The urbanisation developed under very special, exceptionally favourable conditions. Many of the towns are ports. They have

[1] J. Gottmann, *Megalopolis*, New York, 1961.

benefited from a massive influx of immigrants from another continent; but they did not tie these people to the immediate vicinity; on the contrary they populated the surrounding countryside, and through them the entire country became colonised. Megalopolis arose from the conquest of the American continent, which developed under its direction. The power of America is concentrated there. The development of the whole country has benefited Megalopolis, in the course of 150 years of industrialisation, general prosperity and continuous expansion.

Even without such exceptional conditions, other conglomerations like Megalopolis seem already to be forming. They were not caused in the same way as the Megalopolis of America. In the Rhenish-Westphalian region it was the presence of coal which allowed the parallel development of several towns with populations of over 100,000. And the conurbation is becoming a megalopolis, with an uninterrupted series of towns from Krefeld to Hamm, north of the river Ruhr. Round the Pennine uplands, with Manchester, Birmingham and Leeds, there is a similar, U-shaped cluster of towns, which together contain over 13 million inhabitants and for which a national park is provided.[1] It is but a short distance from the outer fringe of the Birmingham conurbation to the agglomeration of London, which was formed under very different circumstances.

Commerce on the one hand and industry founded on coal on the other have given rise, from Flanders to the Netherlands, to groups of towns which are showing a tendency to join across frontiers and which are already extending tentacles towards the Ruhr and the agglomeration of Paris. Such a megalopolis disrupts the former division into States.

A similar group is taking shape round the Japanese Mediterranean in the Hanshin region from Osaka to Kobe where about thirty cities are grouped around the Bay of Osaka. Alongside the great capitals of Osaka (pop. 3,100,000) and Kobe (1,150,000) commercial and industrial towns like Sakai (330,000), Fuse, Amagasaki (400,000) and residential towns like Toyonaka and Nishinomiya (260,000) are to be found. This mass continues to extend towards Kyoto and the towns of the inland sea. We shall probably soon see a similar constellation in China at the mouth of the huge rivers.

The time has already been foreseen when Rio de Janeiro and Santos will be united by a series of towns which stand in line, like Volta Redonda and Taubate along the river Paraiba; the urban group in south-east Brazil will match the group in the north-eastern United States.

Almost everywhere towns are expanding and multiplying. They

[1] K. C. Edwards, 'Trends in urban expansion', *Advancement of Science*, 1959.

concentrate in places where they can take advantage of the same favourable conditions, and the area becomes organised as a polynuclear conurbation. A megalopolis is only a conurbation which has grown to giant proportions.

Is this the formula of tomorrow? Jean Gottmann seems to think so, and sees in it an irreversible movement, destined to transform certain areas of the world. Lucien Gachon sees in Megalopolis 'the inhabited world of tomorrow in its most favourable and most densely populated parts'.[1] Some town planners and geographers, on the other hand, think that Megalopolis is an exceptional occurrence and an undesirable one, and that large cities are condemned to decline.

In fact, the megalopolis of the future should not be considered as the aggravation of a town, as used to be thought, and is sometimes still so regarded. A new form of urban growth is now in existence, breaking the urban population up into several centres. The linear city, so dear to Le Corbusier is, in fact, a megalopolis. Conurbations dissolve into agglomerations.

[1] *Synthèses*, March 1960.

CHAPTER 23

SATELLITE TOWNS

In expanding, a town does more than urbanise the surrounding country or become part of a conurbation. It causes the formation of industrial or residential centres which depend on it, and without which its life and activities cannot be considered. Soviet geographers have called these localities 'Sputniks'. This word signifies a satellite, real or artificial, which revolves round a star.[1]

V. C. Davidovitch[2] defines a satellite by three characteristics: (i) People living in the satellite come to work in the central town. Satellites which possess this characteristic are called first class satellites; (ii) The central town guarantees a certain number of services, notably cultural services for the satellites; (iii) The satellites accommodate the town's population for holidays and relaxation. In any case communications between the town and its satellites must be easy and frequent. In certain cases (the single-centred agglomerations of V. C. Davidovitch) the central town is at least ten times bigger than the largest of its satellites.

Moscow has brought life to numerous satellites, with average populations of from 100,000 to 200,000. Round Leningrad are thirty-nine satellites scattered within a radius of forty miles.

In this interpretation, the term satellite signifies only a town which gravitates towards a major centre, and is particularly applicable to the dormitory towns of the perimeter. It is often interpreted in this way, and it should be remembered that its meaning is often as vague as is that of suburb. It is important therefore to define it precisely. In fact distinction should be made between what is sometimes called a consumer satellite (for example the dormitory town) and the production satellite. The term satellite should be reserved for the latter. The satellite town would then be the one which, while depending on the metropolis for all sorts of services, provides employment for its own inhabitants. This would correspond roughly with the second group of satellites of the Soviet geographers.

Great Britain has endeavoured to develop towns of this sort since the Second World War: these are the New Towns[3] (the term was preferred to that of satellite towns which was likewise suggested). It

---

[1] Cf. P. George, 'Gesto San Giovanni, banlieue ou satellite de Milan', *Bull. Ass. Géog. français*, 1956.
[2] According to *Soviet Geography: Review and Translation*, March, 1962.
[3] *Town and Country Planning*, London, H.M.S.O., 1959.

was principally a question of decongesting London, where population and industries were expanding uncontrolled, and of reorganisation after the economic disturbances caused by the depression of the 1930s and the physical disturbances caused by the war. The Greater London Plan of 1944, drawn up by Sir Patrick Abercrombie, suggested the establishment of a series of new towns within 60 miles of London, to which population and industry from the overcrowded boroughs of the metropolis could be moved. The New Towns Act of 1946 made the creation of such towns possible, and eight New Towns were developed around Greater London. The idea was also extended to the 'Development Areas' of Great Britain, where urban renewal and the provision of more diverisified industries were both necessary—two New Towns were created in Durham County, for example (and two more were designated later) four in Central Scotland, and one in South Wales. But these are hardly 'satellites' (except perhaps for East Kilbride near Glasgow). At the outset, an optimum population of 60,000 was envisaged, but this figure was later felt to be too low, and 100,000 is a more likely target for some of the New Towns (see below p. 349-51).

The undoubted success of the British New Towns has been due in large measure to the careful coordination by the Development Corporations of the parallel growth of residence and industrial employment. The one New Town which is different from all the others is Corby, where New Town status was given to a township that was already growing very fast around its great steelworks, and was in some danger of becoming a 'company town'.

Earlier examples of towns grouped round an industrial establishment come to mind: Le Creusot round the Schneider factories, Eindhoven built round Philips, Rochester round Kodak. But these come into a different category. The monopolisation of a town by a single company presents many disadvantages. In a satellite town efforts are made to organise industry so that the labour force may have greater freedom of choice. Moreover, a satellite gravitates towards the town and is to some extent its emanation; business for the most part continues to be carried on in the town where the firms' head offices often remain: the umbilical cord is not severed. This is in no way comparable to the moving of individual factories into the country; it is a coordinated effort the aims of which are at the same time economic, social and demographic, and the State alone is in a position to undertake such an operation.

These towns in fact have the great advantage of avoiding mass movements of workers, and of guaranteeing these workers better housing conditions than those in the metropolis, while at the same time maintaining the advantages of commercial, financial and cultural links with the metropolis. The irreversible movement which draws people towards the metropolis is thus channelled and humanised.

This formula is increasingly successful. Satellites are now being planned in the Rhenish-Westphalian region. Five have already been built to relieve existing towns: Sennestadt near Bielefeld, Espelkamp near Luebekke, Wulfen near Recklinghausen, and Hochdahl and Garath near Dusseldorf; another is being planned near Cologne, and yet another (at Meckenheim-Merl) near Bonn.

Vällingby in Sweden represents a similar attempt. A completely new town has been built 10 miles from Stockholm, according to the most advanced ideas of town-planning. The public services are grouped in the centre surrounded by living accommodation in tall blocks; further away stand individual houses. Parks are planned, and the town is divided into districts, each to house 3,000 people, and each enclosing sports grounds and swimming pools. Beside the town a strip of land is reserved for industry.[1] The success of the formula brought about, in 1962, the building of a new satellite to house about 10,000 people, Hägernästaden, to the north of Stockholm.

The creation of satellite towns is an interesting experiment, but it is still too recent to be judged. It should be noticed that direction and finance have come from public authorities, both in communist countries and in those with a liberal economy. The question is to what extent state support and control are indispensable to the continuance of these satellites. There is in fact the risk of seeing them become merely extra dormitories built at great expense.

In this respect one wonders what is the optimum distance of these satellites from their mother-town. If they are too near, they will eventually join it; if they are too far, they will no longer fulfil their rôle, which was to relieve the town while benefiting from its activities. Round London, such towns have been placed at a minimum distance of 20 miles from the centre of the city.

In reality, not all satellites are created *ex nihilo*; it is simply a case of rationalising a phenomenon which exists everywhere, in some confusion, usually round large industrial centres. There are, in fact, towns near a metropolis which remain under its direction, while still keeping a life of their own, and providing local work for their populations. These towns may be considered as satellites if they are integrated in the economic whole directed by the centre. The towns depending upon the textile centre of Lyons, towards which most of their activities are directed, could be considered as satellites of Lyons. The same could be said of the metallurgical towns of the lower Loire, around Nantes.

[1] J. Hugueney, 'Vällingby', *La vie urbaine*, 1957.

Sometimes the organisation of a group stems from the satellites themselves. V. C. Davidovitch quotes the case of the oil towns founded on the Apsheron peninsula, which found their centre of gravity in Baku. Łódź in Poland was for a long time only a village and in 1831 had only 5,000 inhabitants, while in neighbouring valleys little towns like Zgierz had populations twice the size; but this string of small industrial towns needed a centre, and in the space of a century the population figure of the centre reached 700,000.[1] The outskirts created the town.

Efforts are made to spread these satellites widely to avoid the congestion of large cities. In France, as we have seen, these artificially created towns are dormitories; attempts are now being made to bring new life to towns already existing around Paris but out of reach of commuters, and to make them into satellites. Meaux, Melun and Montereau could fill this rôle.

This term should not be wrongly applied just because it is fashionable. It is often used for any town situated within the zone of influence of a large centre: this is only justified when the town really gravitates towards the centre, but is not too far from it; it will then form part of the agglomeration.

[1] J. Dylik, *The development of settlement in the Łódź region*, London, 1948; L. Straszewicz, 'The Łódź industrial district', *Przeglad Geograficzny*, 1959.

CHAPTER 24

AGGLOMERATIONS

Surrounded by its satellites, an integrated part of a conurbation, enlarged by its places of recreation, a town is the centre of an urban system and to a certain extent its support. The word 'town' presupposes a continuous built-up area to which spaces are added and reserved for the daily life of the population. But it may well happen that the population of the fringes, depending heavily upon the town, may become more numerous than that of the town itself.

It is difficult to fix the limits of this urban mass, which is more easily defined by its centre, and by its communications pattern than by its actual limits. Thus, because of the general use of motor cars, Los Angeles is expanding immoderately, and 'the distinction between the city and its suburbs is one of the most artificial possible'.[1]

We thus arrive at a definition of an agglomeration: a concept more extensive than that of a town and more precise than that of a suburb. It is freely admitted nowadays that it is the basic concept, one which gives each town its personality and which we attempt to define by statistics.[2]

Attempts have been made to define the limit of an agglomeration by the outside limits of daily commuting.[3] It is important to be still more precise, for this presents the risk of over-enlarging the agglomeration: it should be limited at least to the communities where more than half the active population, for example; work in the town. But this takes into account only one element.

Abel Chatelain proposed a demographic criterion. An agglomeration should extend as far as the enclosed communities in which the population had ceased to decline.[4]

The university of Berkeley (see p. 30) has adopted, for its statistical definition of a 'metropolitan area' (which is in fact an agglomeration), norms which seem reasonable: a group of at least 100,000 people containing at least one town (continuous urban area) with a population of at least 50,000, plus the adjoining administrative divisions, which present similar characteristics and where, in particular, over 65 per

[1] M. Tabuteau, 'Los Angeles', *Cahiers d'Outre-Mer*, 1953.
[2] M.-F. Rouge, 'Définition des agglomérations', *Urbanisme*, no. 60; J. Coppolari, 'De quelques notions fondamentales et définitions en Géographie urbaine', *La vie urbaine*, 1960.
[3] P. H. Chombard de Lauwe, *Paris et l'agglomération parisienne*, 1952.
[4] *Etudes rhodaniennes*, 1946.

cent of the population are engaged in non-agricultural activities. A difficulty still remains, however, as regards the very variable extent of the administrative divisions. The metropolitan area thus defined represents a population which, in certain cases, is eight times bigger than that of the town (Brussels), or even seventeen times bigger (Charleroi). The definition depends upon relations with the central nucleus. Several zones must therefore be distinguished inside the agglomeration. Support for this has sometimes been found in the figures for population density.

Aage Aagesen[1] has distinguished five zones in Copenhagen:

1. The city, which corresponds to the town of 1850 and still retains the function of a 'city': the density here is 38,000 per square mile (falling in some parts as low as 14,000).

2. Around the city, districts dating from the end of the nineteenth century with a density of 33,000 to 65,000 per square mile.

These first two parts of the town show a decreasing population.

3. The outer urban districts where the population density is from 15,000 to 34,000 per square mile.

4. The inner suburban fringe (density 8,000 to 20,000) with rapid growth, often more than 20 per cent per annum.

5. The outer suburbs where density is below 20,000, with moderate growth.

S. Korzybski[2] divides an agglomeration into centre, inner suburbs and outer suburbs, according to the population density figures. These zones take the place today of the former zones which gave rise to a somewhat different interpretation. The Paris of yesterday (eighteenth and early nineteenth centuries) which was then distinct from the *faubourgs* or inner suburbs is now combined with them, the population densities being similar, to give the *town* which is distinguished from the suburbs by differences in the population density. And, with the weakening of administrative limits, we are proceeding towards a situation in which the term 'town' no longer means very much and should be replaced by that of 'centre'.

It seems reasonable, nonetheless, to take into account also the activities of the population.

Maurice-François Rouge[3] makes these distinctions:

1. The actual zone of agglomeration, which encloses the neighbouring communities, these being interconnected by the continuity of the urban zone (which we should readily consider as the town).

[1] *Geog. Tidsskrift*, 1960.
[2] 'Le profil de densité de population dans l'étude des zones urbaines de Londres et de Paris', *Urbanisme et Habitation*, 1954.
[3] 'Définition des agglomérations', *Urbanisme*, 1958.

2. The zone of interdependence where, without there being continuity of construction, activities of an urban nature provide characteristics in common.

3. The marginal zone where the majority of the population follows an activity or a way of life of an urban nature, and must go for work, shopping or recreation to neighbouring communities.

It should however be borne in mind that the marginal zone may expand (perimeter of shopping area for example) well beyond the limits normally ascribed to the agglomeration.

P. H. Chombard de Lauwe[1] distinguishes seven zones in the agglomeration of Paris: (i) the nucleus; (ii) acclimatisation zone, embracing populations of various origins; (iii) inner residential zone; (iv) industrial-residential zone, mainly occupied by the working-classes; (v) mixed residential zone; (vi) zone in the process of being developed for housing; (vii) marginal zone, outside the agglomeration proper.

A more recent study[2] has resulted in the division of Paris into the following zones:

1. The enclosed agglomeration, which includes: (i) the town of Paris; (ii) an urban ring which is itself divided into two: (*a*) a zone of flats and tenements having a high population density and almost continuous buildings; (*b*) a mixed housing zone having a high population density with individual houses and close continuity of buildings.

2. Suburban ring with low proportion of population engaged in agriculture, large number of daily commuters and widely-spaced buildings.

3. Amenity zone or neighbouring, non-agglomerated zone with no continuity of buildings.

4. Industrial satellite zone dependent upon Paris, composed of industrial communities and their dormitories.

German sociologists, too, in their studies of *Raumforschung*, have tried to define the zones into which the *Stadtregion* should be divided. In particular, O. Boustedt distinguishes three zones: the nucleus (*Kern*) which houses over 1,300 people per square mile and has not more than 10 per cent of its population employed in agriculture; the urbanised zone (over 500 people per square mile, and less than 30 per cent of the active population employed in agriculture); the outskirts (*Umland*) themselves subdivided into inner suburbs, with

---

[1] *Paris et l'agglomération parisienne*, 1952.
[2] I.N.S.E.E. Direction régionale de Paris, Délimitation de l'agglomération parisienne, 1959; cf. J. Bastié, 'Croissance et limites de l'agglomération parisienne', *Inf. Géog.*, 1959. For the comparable zones of London see below, p. 270.

less than 50 per cent in agriculture, and the outer suburbs, with from 50 to 65 per cent.[1]

Norbert J. Lenort[2] has analysed the various German studies of the subject which result in a similar classification: urban nucleus, urbanised zone, marginal zone (*Randzone*).

H. Van der Haegen[3] distinguishes the suburbs by the number of daily commuters over 60 per cent, and subdivides it into immediate suburb or suburban zone (more than 1,000 per square mile and less than 20 per cent of population in agriculture) and middle suburb or peripheral zone (more than 1,800 people per square mile and less than 50 per cent of agricultural population).

Robert E. Dickinson[4] has divided the same urban area into central zone, middle compactly built-up zone (which in Europe dates from the nineteenth century), and outer partly built-up zone. This last, which dates from the twentieth century, is discontinuous and represents the suburban zone.

A threefold division is therefore generally used.

Depending on the case, one is tempted to include or exclude certain of these zones in the definition of an agglomeration. It should be recognised that a subjective element is bound to come into our conception of an agglomeration, as it does into that of a town.

Nevertheless it has been necessary to establish the limits of an agglomeration for administrative reasons. For a long time the need has been felt to coordinate the activity of the police and make it possible for them to work in suburban communities. Likewise, to fight the terrible fires of overpopulated districts, powerful forces are needed, which only an agglomeration can provide. This led, at the beginning of the twentieth century to the definition of Greater London as the Metropolitan Police District.

The conurbation of the Ruhr adopted similar interurban means to coordinate the expansion of towns.

The District of Paris which was first formed in 1961 is now being organised. It includes the departments of the Seine, Seine-et-Oise, and Seine-et-Marne, to promote an overall policy of planning and services under the direction of an administrative council and a delegate-general. In Greater London, one of the difficulties confronting the Abercrombie Plan of 1944 was the fact that there were no less than 143 separate local authorities in the area. In 1965 the Greater London Council was established, incorporating the counties of London and

---

[1] O. Boustedt, *Die Stadtregionen*, 1953.
[2] 'Entwicklungsplanung in Stadtregionen,' in *Die industrielle Entwicklung*, 1961.
[3] H. Van der Haegen, 'De Brusselse banlieue', *Bull. Soc. Belge. d'Etudes Géog.*, 1962.
[4] *The West European City*, 1951.

Middlesex and parts of the adjacent counties of Hertford, Essex, Kent and Surrey.

Almost everywhere people are thinking about integrating the growth of towns into a unit which is the administrative equivalent of an agglomeration.[1] The union of all the elements which were originally autonomous, around a central nucleus presents numerous problems. Ways of life and interests are not the same; too much centralisation may produce paralysis. An agglomeration allows each element to keep its personality; it includes open spaces, recreation grounds, even market-gardens, as well as built-up areas. It is the urban mass, a kind of organic unit whose parts are interdependent, but the expansion of a town into an agglomeration is only a first stage: urban problems are now presented on a regional scale, and the concept of the city-region has been developed.[2]

[1] Cf. for Copenhagen: N. Nielsen, 'Greater Copenhagen'; A. Aagesen, 'The Copenhagen district'; and V. Hansen, 'Some characteristics of a growing suburban region', *Geog. Tidsskrift*, 1960.

[2] J. R. James, 'City Regions', Town and Country Planning Summer School Report, Nottingham, 1962.

PART FOUR

LIFE IN TOWNS

Chapter 25

THE CONCEPT OF URBAN CONCENTRATION

Both to the eyes of a traveller and to those of the map-reader, the town appears above all as a phenomenon of concentration. This impression is received as much on discovering Brasilia spreading out in the middle of its human desert after an hour's trip by aeroplane or ten hours in a car, as it is when, after travelling the roads of Belgium, swarming with traffic and intersected by villages though they are, a more important town is reached. The density of the locality is accentuated, not only the density of population, but also that of the houses, streets and buildings of all sorts. There is an increase in the intensity of land occupation in vertical structures: in a word, of human paraphernalia.

This feeling of urban concentration can be startling when the edge of the town stands out without any transition from the surrounding countryside: this occurs with certain recent creations, such as the petroleum towns of the desert. It can also be the result of historical survival; only a few decades ago Madrid dominated its high, inland plateau without any suburbs, as Rome dominated its underexploited Campagna. The same impression is occasionally given around Paris at the limits of the square development plots which have carried the construction of blocks of villas into the middle of cornfields, and, even more impressive around the city itself, when, from the former military zone now transformed into gardens and sports or exhibition grounds, one looks towards the large, blocks of flats which line most of the Maréchaux boulevards.

Often, however, the encroachment of the town upon the countryside is more insidious: houses multiply, roads turn into narrow streets lined with buildings, crossroads appear one after another: the urban universe is entered almost unconsciously. This is what happens in most towns in England and the United States, with immense suburbs stretching for miles before the urban nucleus itself is reached.

THE EXTENT OF URBAN CONCENTRATION

As all human settlement can be characterised by its intensity, we may attempt an evaluation of urban density; but the problem is a delicate one.

Urban densities are always incomparably higher than rural ones, but they are also much less extensive. This is the very concept of concentration which has already been emphasised. The highest known rural densities are to be found in the countryside of parts of Asia, and may reach from 2,500 to 4,000 inhabitants per square mile on certain small plains, and from 1,200 to 1,500 over much wider areas. The crowded places are in Java, on the Tonkin and Yangste Kiang deltas, and in the basin of Szechwan.

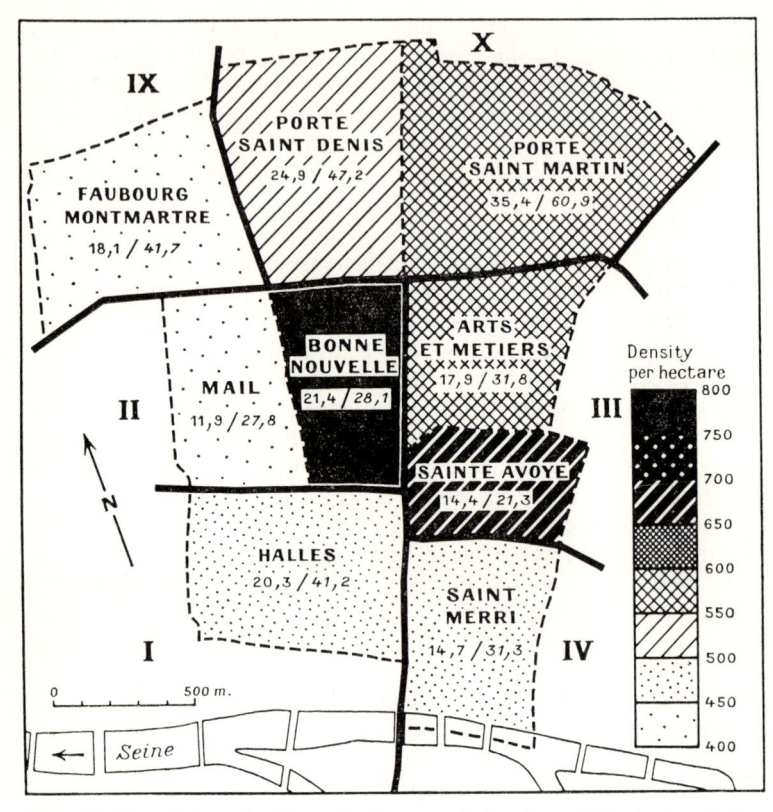

Fig. 17. Population densities of six central districts of Paris in relation to the total surface area

In each subdivision, the left-hand figures are the population in thousands, the right-hand figures the area in hectares. Thus Halles has 20,300 people on 41·2 hectares.

It is difficult to say where the highest urban densities are to be found. Ancient cities seem to hold an unsurpassed record whether one takes the old nucleus without its later additions, or even the town proper as a whole. The Casbah of Algiers, where 91 per cent of the dwellings date from before 1870, crowds into its complicated cubes of stone, barely separated by narrow alleys, over 500,000 inhabitants per square mile, that is 54 square feet per person;[1] this seems to be a sort of paroxysm surpassed only by certain quarters in towns of the Far East; in the town of Victoria in Hong Kong maximum densities

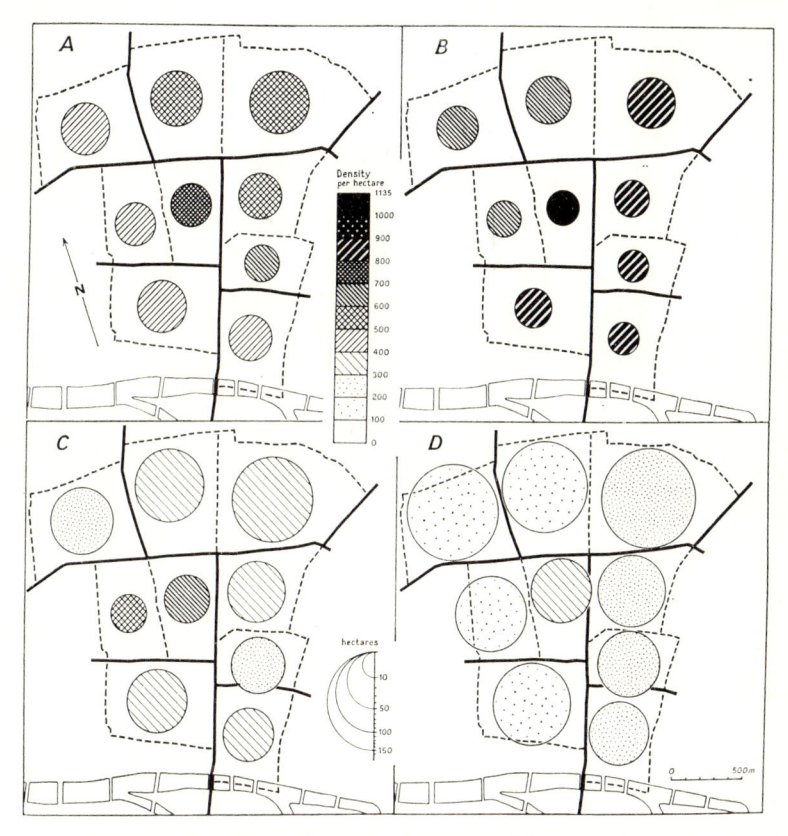

Fig. 18. Population densities of six central districts of Paris

A, in relation to total surface surface area; **B,** in relation to built-up surface area; **C,** in relation to surface area used for habitation; **D,** in relation to floorspace, excluding working areas (the circles are proportional in size to the different surface areas under consideration).

of 650,000 people per square mile can be found.[2] But this is only applicable to parts of towns. On the other hand, as a whole, a really old capital like Paris numbers 85,000 people per square mile

[1] J. Pelletier, 'Alger, 1955', *Cahiers de Géographie de Besançon*, no. 6, 1959, p. 64.
[2] G. Haberland, *Gross-Haiderabad*, Hamburg, 1960, p. 68.

and among the biggest towns in the world there are few which surpass it. Calcutta has a density of 86,000. New York, the heart of the greatest urban agglomeration in the world, which lacks the long history of its elders, only reaches a density of 21,580 per square mile.

These figures are only an indication. They are obtained simply by dividing the total population of a certain administrative district by the total surface area of the same district. Other ways of calculating urban density might perhaps express the phenomenon of concentration more exactly.

In certain cases, in fact, the urban area is widespread and the administrative limits enclose a fairly considerable amount of unbuilt-up land. At Strasbourg it has been pointed out that the gross density was 6,646 inhabitants per square mile, but if forests, fallow land, allotments and wasteland, which cover 53·15 per cent of the total, were subtracted from the surface area of the community, the density in relation to this corrected surface area rose to 14,289 people per square mile.[1] If for some towns in which buildings overflow the actual urban limits this precaution is of no use, in other cases it modifies considerably the results obtained. In this way Cape Town only covers 29 square miles of built-up land, while its municipal territory covers 123: the gross density is 5,957 people per square mile, whereas its corrected density is as high as 25,252.[2] By subtracting from the surface area of New York the large expanses occupied by the waterways which divide the town, the density is augmented from 21,580 to 24,253. For the island of Manhattan alone the subtraction of waterways and the area of Central Park brings the density of human occupation from 53,791 up to over 82,000.[3] In Singapore the heart of the town numbers 194,000 people per square mile, but if the unused stretches of land are removed, the figure of 971,000 is reached.[4]

This precision can be carried even further, to calculate the density in relation to the actual residential surface: in this case, as well as in the case of unbuilt-up areas, spaces occupied by communications and by industrial, administrative and religious buildings, must be subtracted from the communal surface area. This process of correction thus becomes more delicate, and can only be attempted in the case of very familiar towns where sufficiently exact data are available. This method gives a quite different image of urban occupation and constitutes an attempt to show the relationship between the number of inhabitants and the space they occupy for purposes of habitation

[1] H. Nonn, 'Strasbourg: des densités aux structures urbaines', unpublished thesis, Strasbourg, 1962.
[2] Various authors. *Zum Problem der Welt-Stadt*, Berlin, 1959, p. 11.
[3] *The World Almanac and Book of Facts*, 1962.
[4] J. H. Frazer, 'Singapore', *Town Planning Rev.*, 1952, p. 121.

alone. If an even more realistic evaluation is required, a study of 'floorspace inhabited' should be made, that is, considering not only the surface area of a block, but the number of storeys, and among these storeys, including only the rooms actually used as apartments, deducting from the total all space occupied by businesses, offices, storage rooms, etc. The final result obtained is an urban microanalysis, indispensable to everyone in charge of urban affairs and administration or town planning.

In carrying out work of this sort, some people have proposed the calculation of linear densities, proportional to the number of inhabitants and the length of the street and densities of volume, taking into account the height of buildings, not simply their presence on the town map.[1] A similar approach is made by town planners when they calculate the number of people per dwelling, per room, or better still, since the size of dwellings and rooms varies considerably according to the quality and age of the building, per square foot.

Detailed analyses of all these types have been attempted for an average French town, Strasbourg. The conclusion drawn is that any evaluation in figures cannot be dissociated from a description of the environment.[2]

Indeed, very similar urban densities can be obtained from a district of bungalows with small gardens and from another where housing takes the form of large ten or fifteen storey blocks situated in the middle of green spaces; from an African town where one-room huts are crowded closely together and where 'the piling up of blocks hardly leaves room for access roads',[3] and from the central *arrondissements* of Paris where almost all the blocks are over five storeys high; from industrial towns which have no green spaces but are full of factories like Roubaix and Tourcoing, and from residential garden cities.

COMPARISON OF URBAN DENSITIES

In spite of all these difficulties it is interesting and useful to be able to compare the urban densities of different parts of the world, or of different towns.

What is the best method of comparing human concentration where administrative limits are often badly defined? When London is referred to, some people think of the town itself, which corresponds to the county and covers 117 square miles; others think of Greater

---

[1] A. Chatelain, 'Démographie urbaine', in *Livre jubilaire offert à M. Zimmermann*, Lyons, 1944, p. 83–102.
[2] H. Nonn, *op. cit.*
[3] A.-M. Baron, 'Densité de la population des quartiers marocains à Casablanca', *Notes Marocaines*, 1955, no. 6, pp. 3–8.

London which the census confuses with the Metropolitan Police District (M.P.D.), which covers 5,758 square miles. Even this M.P.D. has been described as 'a suitable basis for definition in 1901, but no longer applicable in 1951',[1] so other people go still further and think of London as an even more extensive agglomeration. The County of London houses 3,195,114 people with a density of 26,000 people per square mile; Greater London holds 8,171,902, which is 1440 people per square mile. As to the true 'agglomeration of London', it is estimated that this houses more than 10 million people, and its surface area is so vaguely defined that any idea of its density would be arbitrary. The same thing could probably be repeated in reference to all the big cities in the world. The name 'Paris' designates a town of 40 square miles; as for the district of Paris, it covers over 888 square miles.[2] Almost every study of the United States gives the towns a population which is in reality that of the agglomeration of which the town is only the denser and more active nucleus.[3]

It is therefore important above all to have comparable data and to know precisely to what administrative areas the figures under consideration apply.

There is, however, no homogenity about the extent of the administrative areas. If a certain number of European towns are considered it will be seen that their administrative area varies enormously: Paris, is really of a small size, with only 40 square miles; Milan or Rome, not much bigger with 70 square miles and 80 square miles respectively; London, already much larger with 117 square miles, but what of Berlin which covers 341 square miles? The differences in density calculated in relation to the gross surface area reflect much more clearly the inequalities in surface area than the density of humanity or the type of dwelling. Whereas the density of Paris is over 85,000 people per square mile, that of Milan is only 18,000, that of Rome 23,800, and that of Berlin 9,800.[4]

By examining together the two criteria of communal surface area and of relative density, several types of town can be distinguished.

In countries with a long history of urban civilisation, like those of Europe, or in the rare, ancient cities of China and India, the administrative areas are not usually very extensive, and have grown progressively at the same rate as the population. In the thirteenth century the first walls of Paris enclosed barely one square mile; before the Revolution the new fortifications took in 13 square miles; in the middle of the nineteenth century a law of 16 June 1859 enlarged this

---

[1] T. W. Freeman, *The conurbations of Great Britain*, London, 1959, p. 65.
[2] G. Chevry, 'Paris vu par le statisticien', in *Paris 1960*, Paris, 1961, pp. 19–29.
[3] R. Murphy, *The American City*, New York, 1966, pp. 8–34.
[4] *Zum Problem der Weltstadt*, Berlin, 1959, p. 11, and *passim*.

area to 35 square miles and after the incorporation of the old fortified zone, the town now covers 40 square miles, 6½ of which make up the Bois de Boulogne and the Bois de Vincennes.[1] Such was the progressive pattern of growth of most of the world's towns that go back some way into the past.

In order to follow the increase in the number of inhabitants and the development of the residential area, the administration or public authorities have had to amend the old administrative boundaries: the annexation of peripheral parishes or districts and the inclusion in one set of statistics of several administrative units forming one agglomeration (as was done in Britain in the census of 1951, and in France in that of 1954) bear witness to the care taken to follow the modern trend of urbanisation. The inequality in surface area of large European cities is only an indication of different stages or different ideas.

Comparable examples could be quoted from all over the world: the old town of Pekin covers only 19 square miles and houses 52,000 people per square mile; the agglomeration of Shanghai extends over more than 270 square miles with densities of over 20,000.[2] The city of Tokyo covers 223 square miles with a density of 34,000 and the urban zone has a surface area of 784 square miles with 11,200 people per square mile.[3]

The town of Hyderabad in India increased from a surface area of 20 square miles in the middle of the nineteenth century, with a population density of 10,000, to a surface area of 83 square miles in 1951, with a population density of 13,500.[4] Urban congestion is much higher in large, compact towns like Calcutta and Bombay than in medium-sized towns, often formed from several nuclei linked together progressively over the centuries into a single urban mass like Hyderabad and Madras.[5]

In other continents where urban civilisation has often been imported more recently, either entirely by Europeans who settled overseas or by a combination of contributions of different origins, the urban area is generally much larger and the density less concentrated, but there are exceptions, and homogeneity is no more the rule here than it is in the old countries.

The municipal surface area of São Paulo in Brazil covers 606 square miles and the population density is over 3,600 inhabitants per square mile, but at Recife, the most congested of Brazil's large towns, the

[1] G. Chevry, *loc. cit.*
[2] T. Shabad, *China's Changing Map*, 1956, pp. 105–6.
[3] *Zum Problem...*, *op. cit.*; S. Kiuchi, *Tokio*, pp. 112–26.
[4] G. Haberland, *op. cit.*
[5] J. Dupuis, *Madras*, Paris, 1960.

figures are 54 and over 13,000 respectively. Detailed studies show that the surface area actually occupied by the town of São Paulo is not even half that of the municipality, and the density goes as high as 10,000 or more.[1]

In the United States, a density of 21,580 inhabitants per square mile for New York has already been mentioned. Chicago numbers 17,000 inhabitants per square mile, Detroit 11,964 and Los Angeles only 5,451.[2]

Very similar population densities are found in very different conditions in Africa: Casablanca occupies a surface area of 43·7 square miles with a density of 22,000; at Dakar the communal surface area is 67 square miles, with a density of 27,500, which, if the built-up area alone is counted, rises to 70,000. In Cape Town there are 25,000 inhabitants per square mile of built-up area.

When comparing average densities, it should not be forgotten that internal urban conditions can differ profoundly.

DEGREES OF CONCENTRATION

In the classic type of town, urban densities are arranged in concentric rings. Round a nucleus varying in size and congestion more or less continuous suburbs develop at first, then isolated tentacles distant from one another; continuous concentration gives way to discontinuous concentration. For one and the same town, the larger the area being considered, the more the density decreases.

This analysis can be taken even further. The description of the London agglomeration serves as a typical example: around the small, almost empty central nucleus of the City, there is first a very densely populated urban ring, then two successive rings, the inner one of seventeen towns with populations of over 100,000, more or less interlinked, and an outer ring of towns with about 50,000 inhabitants, more or less distinct from one another but joined to the central urban mass by radiating roads and railways.

The disposition of densities in the São Paulo agglomeration in Brazil presents an equally good example of the same type: in the heart of the city situated around the Praça da Sé where the density is only 22,000 people per square mile and is decreasing, a district of administrative and ecclesiastical buildings or commercial streets represents the less populated but very busy centre of the town; then comes a first ring of highly populated districts with over 46,000 people

[1] *A cidade de São Paulo*, 4 vols. 1950–56; see vol. 2, p. 22.
[2] R. Murphy, *op. cit.*, p. 162, quotes the density of the fifty largest cities in the U.S.A. (1960).

per square mile, a second where the densities range from 25,000 to 39,000 and lastly a much broader zone, stretching mostly to the east and the south-east, and extending tentacles northward and westward, where the densities fall below 25,000, while still remaining higher than 5,000.[1]

Decreasing Densities

(*per square mile*)

Paris		Chicago		Peking	
Town	85,800	Town	17,000	Town	179,000
Seine	29,200	Urbanised Zone	7,600	Town and near suburbs	16,000
Expanded Agglomeration	12,700	Metropolitan District	4,200	Municipal District	1800
District	9,000	Standard Metropolitan Area	1,500		

This seems to be the general pattern for all large towns of a similar type of civilisation but it in no way excludes more localised variations in detail: thus, in Paris we find almost side by side the most congested district in the overpopulated slum of Les Enfants Rouges in the 3rd *arrondissement* (206,000 inhabitants per square mile), and the emptiest district, that of Saint-German-l'Auxerrois, in the 1st *arrondissement* (11,600) which includes almost nothing but historic buildings and gardens. In the surrounding urban *arrondissements*, the population of the residential areas is much denser in the more working-class east, north-east and south-east, than in the more middle-class west. The same applies to the East and West Ends of London.

In heterogeneous towns where several populations of different origins live together, the general arrangement of the pattern alters and the distribution of densities can be much less regular: Casablanca is an excellent example of this fragmented structure. In 1952 the Muslim quarters housed 450,000 people in 2¼ square miles, in the old and new Medina, the shanty towns of the Carrières Centrales and Ben-Msik, and in a few workers' estates.[2] There were 329,000 inhabitants per square mile in old Medina with local maximum figures of 500,000; in new Medina and the shanty towns there were 200,000

[1] *A cidade de São Paulo.*
[2] A.-M. Baron, *loc. cit.* See also J. Martin et alii, *Géographie du Maroc*, Paris, 1964, pp. 174–83.

to 360,000, while the European population of the intermediate districts barely reached densities of 13,000 to 65,000, sometimes less. The arrangement in scattered nuclei in the least favourable positions—old towns with narrow lanes and blind alleys, unhygienic quarters with no amenities, where the numerous new arrivals with no re-

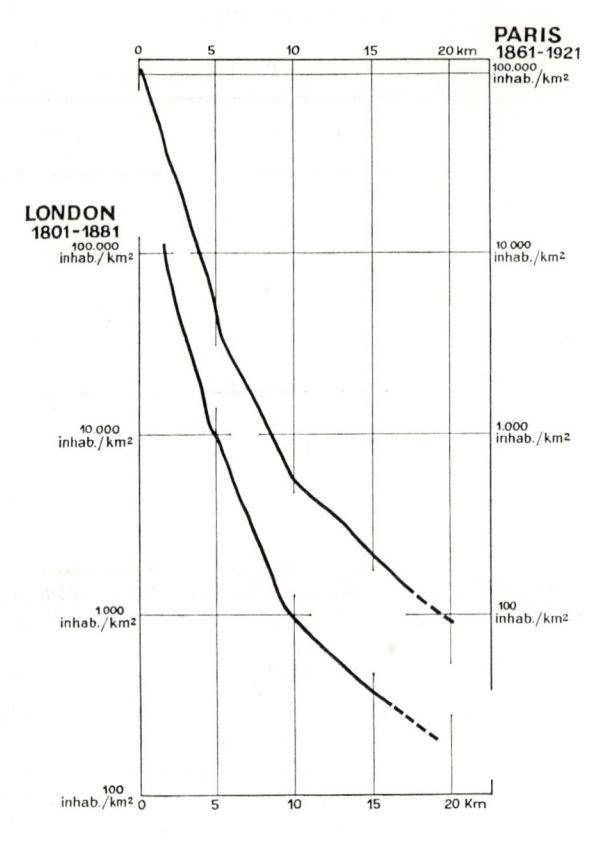

Fig. 19. Variations in density from the centre to the perimeter in London (1801–1881) and in Paris (1861–1921)

sources are herded together—is characteristic of this type of town. The same thing was found in Algiers, where the mass of the Casbah divides the town lying along the bay, with its shanty town fringes one after another on the southern edges of the European quarters on the hills and 'at the bottom of ravines, and in all the highest and most inaccessible areas, often with no view and almost to light, but some-

20 São Paulo: the transformation of the city centre

21 Rio de Janeiro: fragmented site around the bay (Sugar-Loaf in centre); sky-scrapers rising amongst the old villas

22 Brasilia: city of the year 2,000

23 Brazil: a present day shanty-town

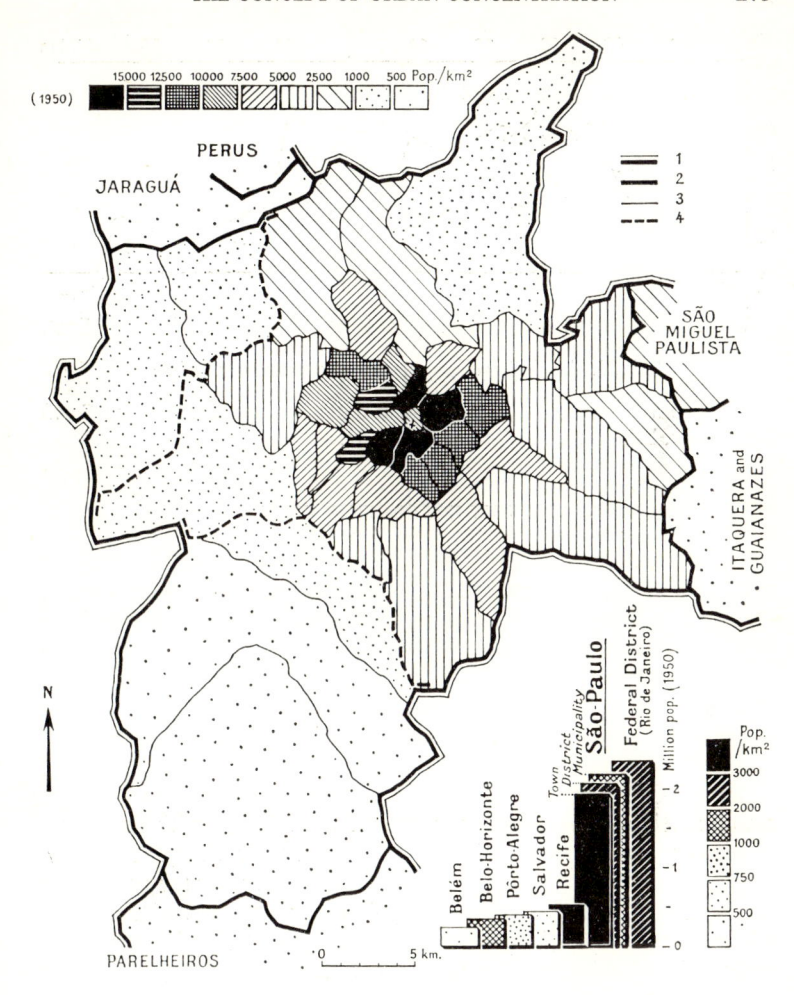

Fig. 20. Densities of São Paulo (different administrative areas) and the large towns in Brazil

I. Limits of municipality of São Paulo. II. Limits of São Paulo district. III. Limits of sub-districts. IV. Limits of town itself.

times also in a magnificent position'.[1] And in Cape Town the districts inhabited by Africans and half-castes literally frame the two great diagonals where the white population lives.

[1] J. Pelletier, 'Alger', *loc. cit.*

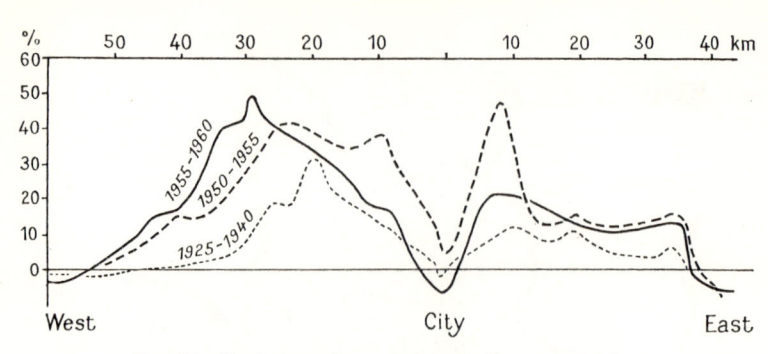

Fig. 21. Evolution of population profile across Tokyo

For a somewhat similar diagram relating to the London area see *South East Study*, London, H.M.S.O., 1964, p. 31.

FACTORS IN CONCENTRATION

The differences recorded between one district and another within the same town, have many and often closely overlapping causes. Differences in density in relation to diversities of population have just been pointed out, but in fact, things are not so simple: to influences arising from racial differences the repercussions of living standards should be added. Pelletier in his study of the social geography of Algiers in 1955 stresses that many very poor Europeans also live in shanty towns while well-to-do Muslims live in the European districts. Overpopulation is more a result of poverty and lack of resources than of a desire for crowded living. Those with the lowest wages, or those who have just arrived from their villages and have not yet found work, find themselves housing wherever they can; they occupy the minimum of space and join other members of the family who are themselves already overcrowded; anyone who has been able to find himself a home can expect soon to see his friends and relations arriving: such is the custom, especially in Black Africa, where one salary often supports almost a whole tribe.

This crowding together of the least fortunate in the most insanitary or inaccessible parts of towns happens all over the world; but in cities with an active economy where remunerative jobs are numerous these immigrations are less massive, and transition is more rapid. The immigrant is not so destitute and can quickly find a paying job. On the other hand, in all developing countries where crowds of people are now rushing towards towns which are already overpopulated in relation to their potential of economic absorption, the congestion of the shanty towns is overwhelming. Many towns in **Latin America** present a scarcely more favourable picture: the *favellas* of Rio, the amphibious villages of Bahia or Recife, and the Barriadas of Lima are not much better than the shanty towns of Africa.[1]

These differences in socio-economic standards of the population and their effects on human concentration are found at every level, whatever the way of life. Residential districts in the suburbs of Paris where 'white-collar' workers predominate are shabbier and more crowded than those where the managerial classes are numerous and where houses and gardens are more spacious. In Strasbourg, buildings of the same age and height (five storeys on average) but differing in quality and comfort, some inhabited by labourers or workers with low incomes (53·1 per cent), some by middle-class or well-to-do families (70 per cent), hold respectively, 0·79 and 0·97 people per room. Serious overpopulation is reached in 5·7 per cent of premises in the first case, and 16 per cent of premises in the second.[2]

Nevertheless, curious inequalities arise, and in the Paris region a community of well-to-do residents like Neuilly can have a higher density than a working-class suburb like Saint-Denis (48,495 people per square mile in the first case, 17,910 in the second), but it is only necessary to walk down the tree-lined streets of the former, and to survey the huge spaces covered in factories and warehouses in the latter, to see that the space at the disposal of the inhabitants is in no way comparable.

Another factor which causes the intensity of occupation to vary considerably is what could be described as 'style of habitat'. This expression includes the nature of the buildings as well as their arrangement: tall houses in narrow streets go with particularly high densities; the opposite applies to low, widely spaced houses.

Certain spots in the centre of Strasbourg (Petite-France, Saint-Etienne) have densities as high as 130,000 to 150,000 people per square miles with 1·25 people per room, while the villa districts near the Orangerie Park have only 18,000, with 0·72 people per room. The style of habitat itself is affected as much by the age of the district as by the ideas of town-planning under which it was erected. In the heart of Strasbourg, four-fifths of the buildings date from before 1850. It is a medieval town which is submitting to modern occupation. But in the low-rented blocks to the east of the town, built after 1923 to house cheaply a population mainly of modest means (56 per cent with low incomes), there are 1·5 and even more than 2 people per room, and densities go as high as 100,000.

Densities are high in ancient historic districts or in old towns because of the narrowness of the roads and the crowding together of

[1] *L'urbanisation en Amerique Latine*, U.N.E.S.C.O., 1962.
[2] H. Nonn, *op. cit.*

houses which is found in all civilisations—in the medinas of the Arab countries as well as the crowded districts in the ancient nuclei of Asian towns: Old Delhi has 350,000 people per square mile while in the new city the density is only 21,800; in the heart of Peking the density reaches 179,000 while in the more recent area immediately surrounding it, there are only 30,000 people per square mile.

Moreover, in districts like this, overcrowding creates overcrowding: lack of comfort, old buildings unadapted to modern ways of life, are accompanied by particularly low rents which favour the proliferation of the poorest people and create areas of overpopulated slums which are present in other towns besides the very ancient ones. The dilapidation of districts near the centre of Chicago, overrun by the Negro population which now forms one-seventh of the total population of the city itself, is another aspect of the same phenomenon.

Lastly, the function of districts is felt equally in the use of space, the occupation of buildings and the density of the inhabitants. The heart of a large city, even if it numbers many buildings, is almost empty of permanent inhabitants: the City of London now houses less than 5,000 people against 128,000 in 1801. Commerce, offices and administration occupy the central districts; industrial buildings and communications routes make holes in the urban mass. In fact, the districts which are purely residential should have the highest densities. The commercial centre of Berlin numbers 6,200 people per square mile; the nineteenth-century residential districts, over 100,000; the suburban villa belt, between 10,000 and 33,000.

CHAPTER 26

PROBLEMS OF SPACE

The spontaneous development of towns and their continuous attraction of masses of people have often led to quite monstrous growths. Some of these deserve the adjective monstrous by reason of their very size, which creates insoluble problems; others because of their extreme specialisation like the mining towns whose life, based only on the mine, is lived partly underground; almost all of them, whether large or small, suffer from the lack of organisation which results from their disordered growth.

DIFFICULTIES OF EXPANSION AND ADMINISTRATION

The prime difficulty encountered by a growing town is that of adjusting its administrative boundaries to conform with its demographic and spatial growth. Most large cities come up against implacable opposition from the bordering authorities who refuse to allow themselves to be absorbed. New York has not extended its territory for forty years; Philadelphia has only expanded by one-tenth of a square mile since 1854; San Francisco still covers the same area as it did in 1856, and Chicago has not expanded since 1889; London, since the creation of the London County Council in 1888, has retained the area that was delimited in 1855.

Some cities have had greater success and have succeeded in altering their original boundaries: Sydney absorbed eight suburban industrial districts in 1948, Zurich staged a similar expansion in 1934, and Manchester in 1930. In India, the city of Bombay, which had increased its area but slowly between 1872 and 1949, suddenly expanded more than fourfold and now, with the realisation of 'Greater Bombay', covers 93 square miles; in the case of Calcutta the growth was quicker, with suburban annexations in 1888 and 1923, but the trend of opinion in favour of a 'Greater Calcutta' has not yet achieved satisfaction.[1] In Moscow a law passed in 1960 increased the size of the city, by engulfing five neighbouring townships, from 137 to 338 square miles.[2]

We have already dealt with the stages in the growth of Paris, but this city, apart from some tiny annexations of military land, has had virtually no extension since 1860. Thus unity of the urban agglomeration is far from being attained. If one takes only the urbanised area in the strict sense, it included 117 *communes* in the *départements* of

[1] W. A. Robson, *Great Cities of the World*, 1954, pp. 61, 141, 577.
[2] Th. Shabad, *Soviet Geography*, 1960, no. 12, p. 90.

Seine and Seine-et-Oise; but the Greater Paris region covers 331 *communes* spreading into the four départements of Seine, Seine-et-Oise, Seine-et-Maine and Oise. Until recent years there was no authority capable of taking efficient and coordinated action amongst this two-tiered multitude of juxtaposed administrative divisions; and there was no common boundary for the provision of such public services as the railway, water supply, gas or electricity. The creation of a 'Paris District', corresponding to the départements of Seine, Seine-et-Oise and Seine-et-Maine is an attempt to remedy this fragmentation.[1]

A similar situation prevails in most of the great conurbations: the problems of London have already been outlined, but the same thing happens in the Manchester conurbation, which includes seven County Boroughs and twelve Municipal Boroughs, twenty-nine Urban Districts and two Rural Districts. In the six English conurbations there are no less than 232 separate housing authorities.[2] In the United States things are no better: the five greatest urban agglomerations in 1952 were divided up amongst 3,386 local government authorities, of which 748 were municipalities, 108 townships, six counties, 419 school districts (which are automonous bodies capable of levying taxes) and 235 'special districts'.[3] In Los Angeles, whose monstrous growth has taken it from 26 square miles in 1850 to 420 square miles in 1950, some two million people are still living outside the city limits whilst within these limits there are thirty independent enclaves.[4]

This profusion of small units of administration, and the overlapping of their boundaries and functions, makes the control of urban areas difficult and leads to inextricable financial complications. In France, it only needs the arrival or the disappearance of a factory for the budget of a small industrial *commune* on the edge of a large city to be completely upset; and common enough are the complaints of poor residential *communes* which must provide housing, schools and administrative services for a population that earns its living elsewhere and often also does its shopping outside the district.

THE NEED FOR SPACE

Another problem presented by this never-ending expansion is the taking over of huge areas for building. In the United Kingdom it has been estimated that 3,500 square miles of new building land will be

[1] J. Bastie and M. Brichler, 'Délimitation de l'agglomération parisienne, *Population*, 1960, pp. 433–56.
[2] J. Beaujeu-Garnier, *Les Iles Britanniques*, pp. 211 ff.
[3] G. A. Wissink, *American Cities in Perspective*, 1962, p. 68.
[4] W. W. Crouch, '*Los Angeles*', in *Great Cities of the World*, ed. W. A. Robson, 1954, p. 306.

required during the next few years to cope with planned urban extensions; and as the towns lie in the lowlands and in the most accessible areas, this means the sacrifice of 10 per cent of the cultivated area, or the equivalent of three average counties. It is further stated that, if all the urban population of the country were rehoused under modern conditions, one-sixth of the lowlands would be covered with houses. Working on rather less expansive lines and confining himself to the next twenty years only, Stamp calculated that an area equivalent to the Grand Duchy of Luxemburg would be necessary to cater for the needs of urban expansion, but even that would be not much more than half the area that was taken for building in England between 1927 and 1939.[1] In France, it is estimated that the housing needs of the capital city for the next ten years are equal to all the houses now in existence in Paris. Where can they be put? Must the last green spaces, the last islands of agricultural land in between the spreading tentacles of the conurbation, be sacrificed? Though the problem of the amount of agricultural production is not acute in France, the cost of losing these valuable market-garden lands is very serious. In Japan, the area occupied by towns and cities is equal to one-fifth of the cultivable area.[2]

These considerations apply to small countries; in the United States in 1950 the 157 major towns of over 50,000 inhabitants only covered 1 per cent of the total area of the country. Nevertheless, the question of *land* is one of the major elements in the problem of urbanisation.

The geological nature of the site plays a highly important rôle. In London the excavation of docks was easy in the alluvial and gravel terraces of the Thames, but in Rio de Janeiro the superficial deposits overlie extremely hard granites, which render the construction of an underground railway very difficult. A city such as Lille, whose very name ('the island') indicates its origin in marshy surroundings, has great difficulty in constructing tall buildings; concrete piles are necessary, sunk through the spongy alluvium of the former marsh, and excavations that uncover a whole system of malodorous sewers. Similar difficulties are encountered on the coastal plain on which the modern city of Nice has developed.[3]

Industry often demands large sites that cannot be found near the old urban centres; the picturesque hillsides of Marseilles are unsuitable, and all the great industries have found flat sites on the edge of the Etang de Berre; the narrowness of the South Wales valleys, with their strung-out mining towns, prohibits the construction of

[1] L. D. Stamp, *Man and the Land*, London, 1955.
[2] C. D. Harris, 'The pressure of residential and industrial land use', in W. L. Thomas, *Man's Role in Changing the Face of the Earth*, Chicago, 1960.
[3] R. Blanchard, *Le Comté de Nice*, Paris, 1960, pp. 156–7.

large factories except at confluence points; Grenoble and Lyons, whose original sites were perched up on the valley sides, have developed huge industrial and residential suburbs on the alluvial river plains that were strictly avoided by the old towns.

This matter is so important that certain towns have become industrial in modern times simply because they could offer convenient and easily serviceable sites: the industrial zone of Amiens, below the edge of the Picardy plateau, is a case in point, whilst the industrial estates in Great Britain are an example of the systematic application of this principle.

In the state of Connecticut, a considerable conflict developed between the town of Newington, which wanted to allocate a large area for industrial use, and a small landowner who wished to build a house—and this example highlights both the practical and legal aspects of the problem. Some industrial firms have acquired from the outset areas far larger than they require for immediate use, intending to reserve them for future expansions: at Morrisville, in Pennsylvania, the U.S. Steel Corporation bought an area, for the Fairless steelworks, twice the size that it needed, and this one example out of thousands that could be quoted underlines the importance for towns of having space available for both housing and industry.[1]

Small or medium-sized towns often offer better facilities for expansion that can be provided by the large cities, and the principle of industrial decentralisation is practised both in France and in the United Kingdom, so that, for example, overspill population and industry from a city like Glasgow can help to give new life to several small Scottish towns. The Chinese have taken similar action with regard to Shanghai, and the Soviet authorities have filled many new Asiatic towns with large factories transferred from European Russia.

Around the great metropolises, the availability of land increases with distance from the central nucleus, as the example on the following page shows.

Finally, we must refer to the type of loose urbanisation that is frequently encountered in new countries, where there is a very prodigal use of urban land. The best examples are in California, where curiously enough it is not Los Angeles that holds the record with 65 per cent of its land unoccupied, but the bay side of the San Francisco peninsula with 75 per cent, and the San José–Santa Clara area with 86 per cent.[2]

[1] D. A. Muncy, 'Land for industry', in H. M. Mayer and C. F. Kohn, *Readings in Urban Geography*, Chicago, 1959, pp. 464–77; R. Murphy, *The American City*, New York, 1966, Ch. 17, pp. 357–64.

[2] H. F. Gregor, 'Spatial disharmonies in California population growth', *Geog. Rev.*, 1963, pp. 100–22.

Residential land use around New York, 1954–55

Time-distance zones from Manhattan[1]	Total area (acres)	Rate of built-up land
1 [2]	14,390	79
2	70,021	73·7
3	136,942	66·8
4	140,047	36
5	51,734	15·7

[1] Each zone is fifteen minutes journey-time greater than its predecessor, from Manhattan centre.
[2] Including Newark.
Source: E. M. Hoover and R. Vernon, *Anatomy of a Metropolis*, 1959, p. 130.

THE PRICE OF LAND

There seems to be no limit to the prices fetched by these precious pieces of urban land—$40,000 a square yard in New York, £7,500 a square yard in the City of London and £750 in slum-clearance areas. In Paris, land in the Champs-Elysées will fetch 10,000 fr.[1] a metre, and the price will be at least 5,000 fr. in the 16th and 8th *arrondissements*, 1,300 fr. in the peripheral 13th and 20th *arrondissements*; in the suburbs the price depends on the quality of the housing—60 fr. a metre, on average at a distance of 4 to 6 miles from the city gates, but 100 fr. or more in the best residential areas. Further out, 12 to 20 miles from the city, and within a mile or two of a railway station, the price falls to 30 or 40 fr. Some land, that is well serviced for industry, may actually have a higher value than housing land: thus in the vicinity of the Porte d'Ivry, the price of a square metre is 2½ times as much as in nearby residential areas.

Prices may rise at a breakneck speed, for they depend not only on demand but also on speculation. Along the Champs-Elysées, a square metre of land that was worth 0·15 fr. in 1830 and 5 to 6 fr. in 1914, fetched 10,000 fr. in 1963, and this represents a rate five times as fast as the rise in the cost of living. Perhaps one could hardly regard this as normal, for the Champs Elysées has been a very select quarter for a long time. But the story of land values around the Rond-Point de la Défense is even more striking: in 1914 the price was less than 5 centimes a square metre, and it rose to 10 fr. in 1950 and 1,000 fr. in 1959, whereas between these last two dates the cost of living barely doubled. As the city's tentacles extend, so land values

[1] These figures relate to the franc of 1963.

rise in suburbs further and further afield. The result is a disturbing distortion of land values as between purely rural areas and areas that may be destined before very long for urban development. A hectare of land along the Avenue Montaigne has the same value as 8,000 hectares of excellent wheatland in the Soissonais, and it represents the purchase price of one-third of an entire, albeit somewhat poor, *département* such as Basses-Alpes.[1]

The same phenomenon is encountered in most other countries. In the United States there is a complete reversal of land values in some small Middle West towns, produced by the transport revolution, which is creating an entirely new urban network. Formerly in the days when the railways were worked by wood-burning locomotives, towns grew up round stations at intervals of six to ten miles; but now, with diesel locomotives, and the competition of the motor car, the station mesh is at intervals of sixty to a hundred miles. Furthermore, the emphasis is now on road rather than rail junctions—in North Dakota for example, 90 per cent of all petrol transport was by rail in 1946, but by 1958 86 per cent was by road tankers. The result is an inflation of land values in the new road centres, with a normal rise in the price of good agricultural land, but a decline in land values in and around the small and now abandoned towns, where many houses simply cannot be sold and where, because of the decline in income, the normal public services can no longer be maintained without State subventions.[2]

In Chicago, in 1910, the value of land in the Loop district, little more than three-eighths of a square mile, represented 40 per cent of the total value of the 212 square miles of the entire city. Along the suburban railway lines, there were peak land values around all the stations and lower values in between, but in certain cases where the nuclei around the stations were less well equipped with shopping facilities (as at West-Mont, Clarendon Hills, Highland and Hollywood) the values were noticeably lower than in other places.[3]

Speculation in land values, by raising prices too high, may result in a paralysing effect on development. In Paris the price of a square metre of land represents about a third of the value of the square metre of floor space constructed on it. In Bombay, between 1939 and 1952, the price of land quadrupled, and while for a single-storey house in the suburbs the price of the land represented about 10 per cent of the total cost, in a good class residential quarter it represented 10 per cent

---

[1] J. Bastie, 'Capital immobilier et marché immobilier parisiens', *Ann. de Géog.*, 1960, pp. 225–50.

[2] *Labor mobility and population in agriculture*, Iowa, 1961, pp. 96 ff.

[3] C. R. Hayes, 'Suburban residential land values', in H. M. Mayer and C. F. Kohn, *Readings in Urban Geography*, Chicago, 1959, pp. 556–61.

of the cost of *each flat*. In Singapore, in the zone of small houses about 4 miles from the city centre, without even a good main road, the price of land represents one-quarter of the total cost.[1] One can well understand what happens on the edge of expanding cities: around Rio de Janeiro and São Paulo, for example, the land is sterilised for years, waiting for purchasers; the timber is cut, road access is provided, and then the price rise is awaited.[2] On the peninsula north of Bahia, there are large areas that are neither cultivated nor inhabited; the owners are content to sell them off in plots as the city extends with a consequent increase in demand and rise in price. It is the same on the western side of Paris, where big development companies bought part of the Montesson plain in 1927 and let it out as farmland in anticipation of its future value.[3]

These questions of land ownership may even play a part in orientating urban development: the great landowners around Nancy at the end of the nineteenth century refused to sell out to the steel companies, so the capital of Lorraine remained aloof from the regional industrialisation movement of that time. At Armentières, a small northern French town much in need of new industry, several families bought up in small lots the only large area on which industry could have been established, and this with the object of frustrating the development of any new manufactures that might compete with the traditional textiles.

The growth of industry is controlled both by the location of available sites and by the price of such land. In the middle of the urban area the constraints are severe and cleared sites few, but in the suburbs things are much easier, as the following table, from an American source, shows:

Comparative costs of factory building

	In the town	In the suburbs
Price of an equal area of land	$2,600,000	$3,000,000
Cost of construction[1]	3,403,000	2,861,000
Total cost per unit area	$25·25	$13·34

[1] The suburban factory also has a private branch railway 150 yards long.
Source: E. Hoover and R. Vernon, *Anatomy of a Metropolis*, p. 271.

[1] *Problèmes fonciers urbains et politique d'urbanisme*, U.N.O., 1953, pp. 87, 121.
[2] Around São Paulo, between 1916 and 1943, the price of land multiplied 20-fold, whilst it only rose 8 times in the central area. R. M. Morse, *From Community to Metropolis*, 1958, p. 279.
[3] J. Bastie, *loc. cit.*

All good urban planning must start with legislation to prevent abuses due to land ownership and to limit the profits that can be made by land speculation that often works to the detriment of the public. This financial contingency of land prices, combined with the difficulty of finding the necessary space and with the variety of physical limitations that beset towns, is something that weighs heavily on urban administration.

UPWARDS AND OUTWARDS

When land is scarce, there is a tendency to build upwards in order to accommodate the maximum number of dwellings or offices. In fortified towns it was necessary to remain within the walls, hence the narrowness of the streets and the crowding of the houses in these historic centres. In a city such as Toledo, and in some of the older streets in Naples, Rome and Florence, modern means of transport are impossible. English towns were fortunate in escaping this constriction,[1] and naturally the newer quarters that lie outside the old walled cities of the continent have spread to a greater extent.

The relief of the land may exercise a similar influence, for in narrow mountain valleys where space is at a premium, tall narrow houses are the general rule; the same thing is true of hill-top towns and of towns that cling to steep coastal or valley slopes. The old quarters of Genoa rise in terraces above the port, and new blocks are higher still; the tall dark houses of Le Puy are packed around the cathedral, whilst Tulle is strung out along its narrow valley.

High prices have the same effect, and this is why, even in the middle of vast plains one sees the town centres of America or Australia rising skywards. Though one might conceivably invoke lack of space to explain the skyscrapers of Manhattan, one can hardly use the same argument to justify the architectural characteristics of the cities of the central plains.

Rio de Janeiro and its satellites, Copacabana, Ipanema and Leblon, are undergoing this kind of transformation. In 1940 this conurbation housed 1,519,010 people, and in 1960, 3,223,407. The response to this doubling in size, on a site fragmented by granite mountains that come right down to the sea, has been to pull down the three- or four-storey houses in the centre of Rio and build in their place narrow skyscrapers; the pleasant villas of the southern beach towns are being bought up to suffer a similar fate. The results are not particularly happy, for the streets have not been widened: they are now narrow canyons and the ocean breezes no longer blow between the tall blocks that were never envisaged when the city was originally

[1] A. E. Smailes, *The Geography of Towns*, 1958, pp. 102 ff.

laid out. Moreover, this tall façade now constitutes a barrier between the sea and the valleys behind.

Something of the same kind, in a very different environment, is happening during the reconstruction of certain areas in Paris: small detached town houses with gardens, in the 16th *arrondissement*, and small single-storey cottages in the 15th, have been replaced by eight- or ten-storey blocks of flats. The Seine frontage, south-west of the Passy bridge, has been completely transformed in recent years, and the 15th *arrondissement* is changing completely in appearance, as the following population statistics show:

Quarter	Block No.	Population 1954	Population 1962
57th	2,829	396	702
60th	3,092	37	254
60th	3,096	111	307
61st	3,235	495	647
61st	3,237	1,076	2,592

Thus the whole Paris region, in response to the demand for housing, is now bristling with tall blocks of flats, which have all the more effect on the general appearance of the city because they have generally been built in localities where it was possible to buy land, such as small unbuilt patches in the suburbs, and the sites of detached houses or cottages in the peripheral districts. Such building is generally at the expense of private open spaces. The built-up area of Paris has risen from 36 per cent in 1913 to 54 per cent in 1957, and this growth has taken place largely on former private gardens which have shrunk from 1,574 acres in 1913 to only 370 at the present time, a reduction of more than 76 per cent in fifty years. We have already called attention to the abnormal crowding of buildings in the city of Paris, and it seems as if, in the absence of strong countermeasures, the situation will get worse rather than better.[1] At the present time only 31 per cent of the total area is represented by roads and open spaces, as against 43 per cent in a residential city like Washington.[2]

In flat country, where land is abundant and cheap, towns normally spread out, with wide roads and ample squares, and low-built houses. Many examples are to be found in western France, such as Le Mans, Nantes and Bordeaux; they spread almost indefinitely in uniform

[1] M. M. Fratoni, 'Les espaces verts à Paris', unpublished thesis, University of Paris, 1963.
[2] G. Schwarz, *Allgemeine Siedlungsgeographie*, Berlin, 1961, p. 394.

suburbs of single-storey, semidetached houses, each with its garden or yard at the back, and all built to a regular plan with equal frontages (like the famous 'Mancelle' house with a frontage of 20 feet), tiled roofs and but a moderate degree of comfort. In 1930 Le Mans, with 78,000 inhabitants, possessed a road mileage not much less than Lyons, which had a population of 580,000.

This brings us very near to the typical Anglo-Saxon form of town, with its endless suburbs. Undoubtedly this is a reflexion, within the limits imposed by surface relief and financial considerations, of the wishes of the inhabitants and of their conception of civilisation. The Anglo-Saxon peoples prefer individual houses and on the whole view with horror the idea of large blocks of flats; an investigation made shortly after the last war across the whole range of social classes showed that between 70 and 98 per cent, with some variation between the different social groups, preferred an individual house with garden. From another continent, Pretoria, which is above all a city of villas, provides a further example; here, when several large blocks of flats were built in the centre, only one-third of those who took up residence (and they were mostly young couples) stayed more than a year. The Latin races, on the other hand, seem to be able to live quite happily in flats.

The Asiatic peoples also have a preference for low houses and spread-out cities. True, one must also take account of the building materials employed and constructional techniques, as well as the quite recent rural origin of much of the urban population in most of the newer towns of both Asia and Africa. One can hardly imagine the wooden houses of Bangkok or Pnom-Penh, or the mud-brick hovels in some parts of Syria or Egypt, or the swarming houses of the Japanese, attaining any great height. But when stone or other more solid materials are used it is easier to build upwards; in the Middle East, some people, like the Armenians, build their traditional houses with three or four storeys.

A distinction must be drawn between historical evolution and the development of building techniques, which play a general part, and the customs and tastes of the people, which may be more important than any other factor. Cutting across a national urban pattern which reflects developments based on materials, techniques and tradition, one can trace a general tendency, operating between the late nineteenth century and about 1930, that may be called the 'Europeanisation of the townscape'. It was expressed in the construction of European quarters, or of whole towns, laid out in geometrical pattern with roads of moderate width, with large squares ornamented with statuary, with houses of two to six storeys, lavishly decorated in the well-to-do streets. This sort of thing is found not only in the historic cities of Europe but also in the old cities of other continents; and whole towns were built in this fashion in colonial territories. It was brought about in the first instance by the rebuilding of the city centres, the widening of the major roads or the construction of peripheral boulevards. It has given character to large parts of the modern towns of Nice, Strasbourg, Lille, Le Mans, and to the central areas of Roubaix, Tourcoing, Manchester and many others. Rio de Janeiro was virtually gutted and reconstructed during this period, while in São Paulo the movement proceeded more gradually. Rabat and Dakar were created, and Algiers, Casablanca and Tunis were doubled in size. The cities of the Middle East and Asia developed their European quarters. Alongside the Indian city of Bombay, with its tall houses and carved wooden balconies, there grew up streets lined with five- or six-storey buildings of reinforced concrete, and green squares dominated by public and other buildings of red brick and white stone, with gables and spires in typical English style. Hong Kong built its banks and big shops on the edge of the teeming Chinese city.

A second phase, which began slowly but spread rapidly after the Second World War, may be called 'Americanisation'. Buildings began to rise higher and higher, emulating the skyscrapers of Manhattan; façades began to lose their stony or concrete opacity and to glisten with glass and sheet metal. Some cities have adopted the new style in its entirety and have built skyscrapers in the midst of earlier forms of development; others have taken a less drastic course and have built structures of moderate height, with cleared spaces around them.

In South America the first of these courses has been generally adopted, though with a latin flavour to the skyscrapers, as at São Paulo. This city, which had only just blossomed out in European style, has rushed into high building, and it is odd to see this displacing not only the colonial-style houses with gardens but also the central blocks which were hardly half a century old.

Few modern cities are able to resist the temptation to build upwards. It is symptomatic of the trend that even a provincial city like Amiens, characteristically built low, with its cathedral towers the only vertical element, could be persuaded to build the twenty-storey Perret tower. Even in Paris, where the regulations governing building heights were for long extremely severe, blocks of ten, twelve or even twenty storeys are being built—fortunately outside the historic central area. And in a country so wedded to low buildings as the United Kingdom, the City of London has added several more storeys during its reconstruction. At the present time, however widely one travels, it is rare to find a city that has not at least a few skyscrapers.

Often they are hotels, as at Conakry, Bangkok or Istanbul; sometimes entire districts as in Belo Horizonte or Sydney. Almost all towns are building higher, not merely because it is the fashion but also because of the increasing pressure of population.

Even the communist world has not escaped this twofold revolution in building styles. In Moscow in 1926 only 0·7 per cent of the buildings were of more than eight storeys, but after the postwar reconstruction this figure had risen to 40 per cent. The University of Moscow projects its fifty storeys far above the onion-spires of the old churches and the surrounding lawns. However, in areas where space is plentiful relatively low building is still general. All the large towns that have been created during the last three decades—for example, Karaganda, Komsomolsk, Kirovsk, Balkash, Kramatorsk and Igarka—cover large areas with low buildings. Except in the large cities, which have blocks of flats, there is much encouragement of private house-building through the medium of State loans.[1]

Given the evolution that has been described above, does it follow that the city of the future will be a synthesis of all the different aspects, a city both spread out and built upwards? Brasilia, which prides itself, according to its founders, on looking forward to the year 2000 and presenting a model for future urbanisation, may perhaps be regarded as typical. In it there are glass-walled skyscrapers and gracefully rounded palaces, with tall blocks of flats alternating with bungalows set amidst vast open spaces, wide roads and broad stretches of lawn. The vertical features are lost in the horizontal expanse. But it must not be forgotten that Brasilia is growing in a country which has vast unoccupied areas, and that we are dealing here with a site specially chosen to accommodate a new capital city.

LAND USE WITHIN TOWNS

A town is not a homogeneous area. A journey from the congested streets of the central business district to the airier suburbs will reveal a whole series of different aspects; but the progress from centre to periphery is not uniform, and the intermediate stages vary appreciably from one town to another.[2]

The main causes of this diversity are the three fundamental aspects of the urban function—the tertiary activities (commerce, transport and administration), the industrial activities (mines, factories, workshops, craftsmen) and residences. The secondary and tertiary occupations may be more or less equal, or one may be more important than

[1] J. Saushkin, *Moskva*, Moscow, 1955; *L'Union Soviétique*, Moscow, 1957, pp. 216–23.
[2] For all aspects of Part Four of this book, cf. R. Murphy, *The American City*, chapters 12–19, giving examples and details of American cities.

the other; the secondary may be almost non-existent or the tertiary but feebly developed; the residential function is always present but it may offer a great variety of forms.

If one divides a town into sufficiently small sectors, a mosaic is produced in which each fragment is a more or less homogeneous entity. Certain specialised areas are thus picked out, such as the quarters in which there is a predominance of either secondary or tertiary occupations, in which tertiary occupations are combined with residence or the secondary group with residence, or in which there is nothing but one or other type of residential accommodation. The question then arises, does the localisation of these 'quarters' arise fortuitously or does it reveal a schematic functional pattern? Three theories have been put forward:

1. The *theory of concentric zones*, formulated by E. W. Burgess for the city of Chicago,[1] envisages a series of zones, each with a different specialised function; the central nucleus which includes the major commercial, social and administrative functions and also the transport termini; the transition zone surrounding the centre and forming a zone of deterioration in which poor residential quarters (mainly occupied by Negroes and recent immigrants) are mingled with small factories; the working-class residential zone, peopled by those who wish to live near their work; the middle-class residential zone of individual houses and apartment blocks; and finally, on the periphery, the outer residential ring or commuter belt, consisting of a series of scattered nuclei, located particularly along the radial transport lines that converge on the city centre.

Even though Chicago spread over a wide and more or less uniform plain, and in a more or less concentric fashion, this theory is perhaps too rigid, for the industrial zones are found lining the railways or on the lakeshore, cutting across all the various belts; the Negro quarters which were originally confined to the second zone have spread out southwards, where, because of segregation, a wide empty quarter has developed in front of their advance, apartments and even detached houses being sold at a low price as the black wave moves forward; and lastly, the elegant residential district known as Gold Coast has grown up on the lakeshore, cheek by jowl with the business and industrial sectors.

2. The criticisms levelled at this first theory lead to a consideration of a second, the *theory of sectors*. In this it is suggested that development following lines of transport is superimposed on the concentric

[1] E. W. Burgess, 'The growth of the city', in R. E. Park and E. W. Burgess, *The City*, Chicago, 1925, pp. 47–62; and E. W. Burgess, 'Urban areas' in T. V. Smith and L. D. White, *Chicago*, Chicago, 1929, pp. 113–38.

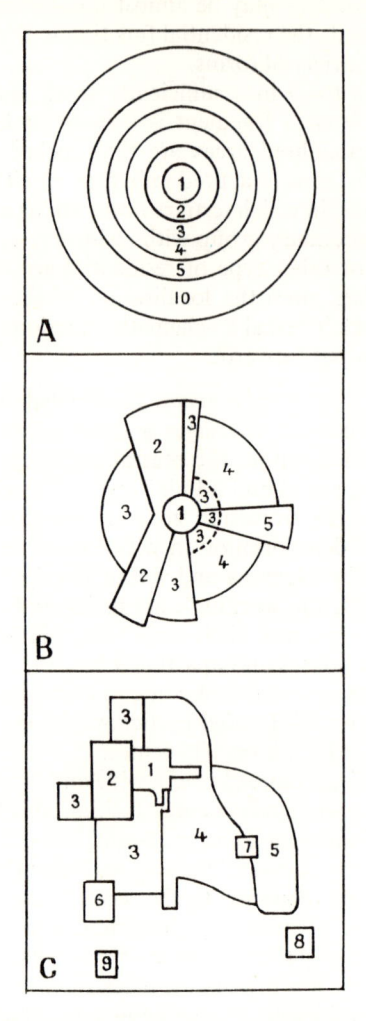

Fig. 22. Three types of urban 'zoning' (from *Readings* in *urban geography*, p. 281)

A. By concentric zones. **B.** By sectors. **C.** By multiple nodes.

1. C.B.D. **2.** Light industry. **3.** Working-class housing. **4.** Middle-class housing. **5.** Upper-class housing. **6.** Heavy industry. **7.** Secondary business district. **8.** Suburban residences. **9.** Suburban industry. **10.** Immigrant zone.

belts as sectors radiating from the city centre. This theory is more supple than the first, and, at least in a more or less homogeneous physical and human environment, seems closer to reality.[1]

3. Towns, however, may develop in an area of accidented relief, and they may house populations of different racial origins, or they

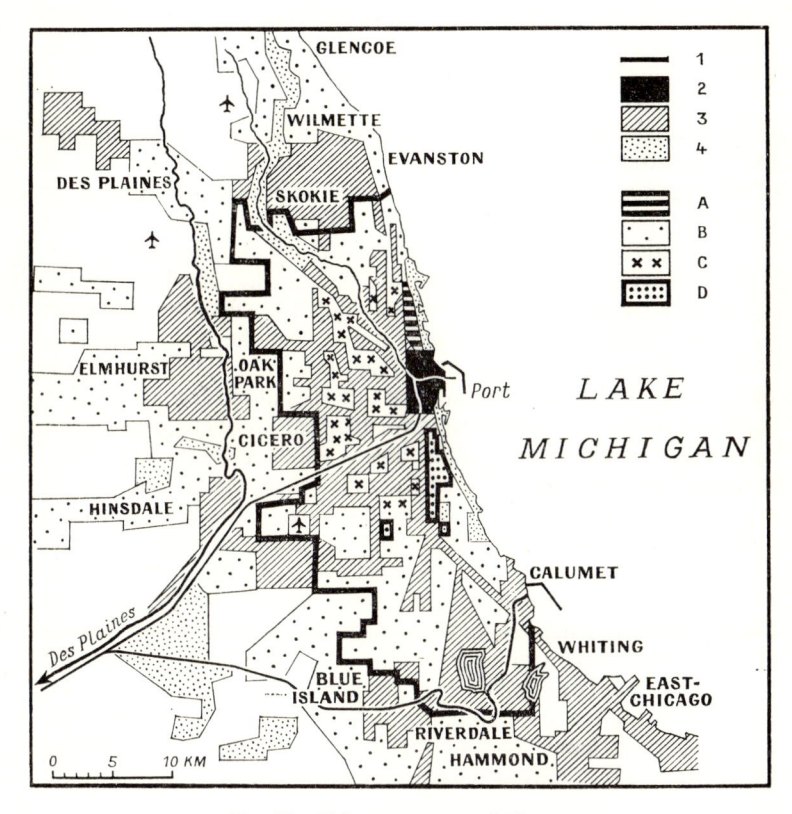

Fig. 23. Urban structure of Chicago

1. City boundary. **2.** C.B.D. **3.** Industrial zone. **4.** Parks—Residential zones: **A,** upper class; **B,** middle class; **C,** working class and immigrants; **D,** Negro quarters.

may include two quite different civilisations, as in some colonial towns, or two different periods of growth, as in the case of almost all ancient cities that had their origins under quite different economic and

[1] H. Hoyt, *The Structure and Growth of Residential Neighbourhoods in American Cities*, Washington, 1939.

political circumstances from those that pertain today. In this case we have a structure consisting of *multiple nodes*.[1]

It seems, however, that in the light of actual city studies rather than theoretical conceptions, it is very difficult to define urban land use in terms of this or that geometrical figure. We may simply emphasise certain conclusions that are readily verifiable.

Every town has a centre of greater or less complexity, that has well defined characteristics and plays an essential part in the life of the citizens. We shall return to this central nucleus later on.

The tertiary activities are partly concentrated in the centre and along the main roads radiating therefrom, and partly scattered through the residential quarters, generally as secondary nuclei, the size of which depends on the size of the town and its population. Very large towns are characterised by the presence of many such nuclei; most of the residential districts have their own shopping and business centres.

As distinct from shops, which are widespread, markets, wholesalers and warehouses are of less frequent occurrence and more narrowly localised. They are only found near the main arteries of communication. Quaysides have huge buildings externally simple but carefully laid out inside and often with specialised accommodation such as cold stores, warehousing for tropical products, grain, wine, or bananas. In the vicinity of the railway goods stations and produce markets there are rows of sheds, garages, and vegetable stores. A whole system of rail tracks, cranes and various forms of transport clusters around these buildings, and a shuttle service of lorries distributes the goods. A visit by night to Les Halles in Paris, or to Covent Garden in London, gives some idea of the activity that characterises such markets in the heart of a great city. More familiar to most people is the atmosphere of the waterfront at a great port, and a typical image is presented by Liverpool, with its gigantic square blocks housing the great shipping and marine insurance companies and the corn exchange, backed by the massive and dusty warehouses, in a state of noisy uproar during the day and silent by night.

INDUSTRIAL ZONES

The localisation of industries within towns is a much more complex matter. In towns that are primarily industrial, the factories may be scattered over the entire urban area, intermingled with working-class housing. Thus in Roubaix-Tourcoing, which has 67·4 per cent of its active population engaged in industry (46·4 in textiles alone), the red-brick mills with their tall chimneys dominate the town centre and

[1] C. D. Harris and E. L. Ullman, 'The nature of cities', in H. M. Mayer and C. F. Kohn, *Readings in Urban Geography*, pp. 277–86.

PROBLEMS OF SPACE

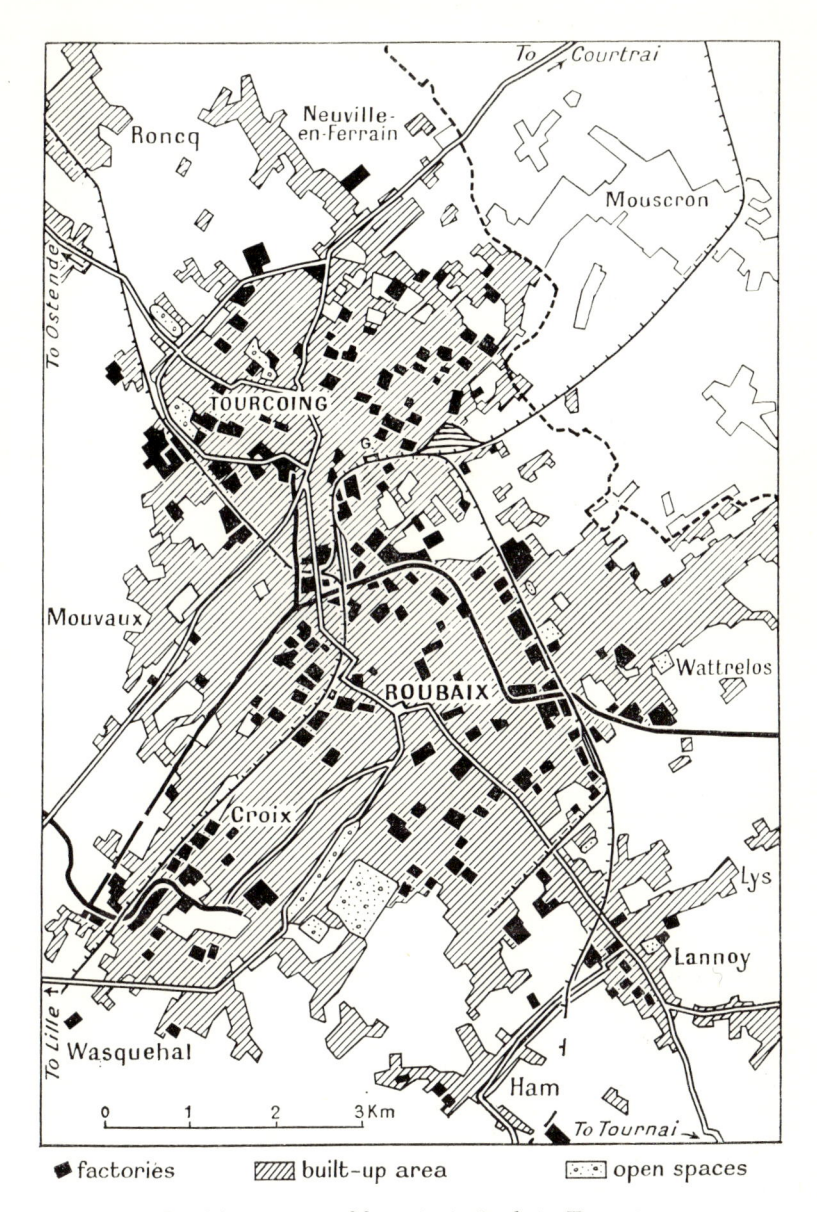

Fig. 24. Location of factories in Roubaix–Tourcoing
The pecked line is the Belgian frontier

294   LIFE IN TOWNS

are found along almost every street, a situation that is repeated over and over again, and with similar red brick, in the towns of South Lancashire.

In more evenly balanced towns, the location of the industrial quarter depends on the nature of the industries. Dirty and noisy industries that need a lot of space are forced into the outskirts by urban bye-laws or simply by the price of land, and they generally lie

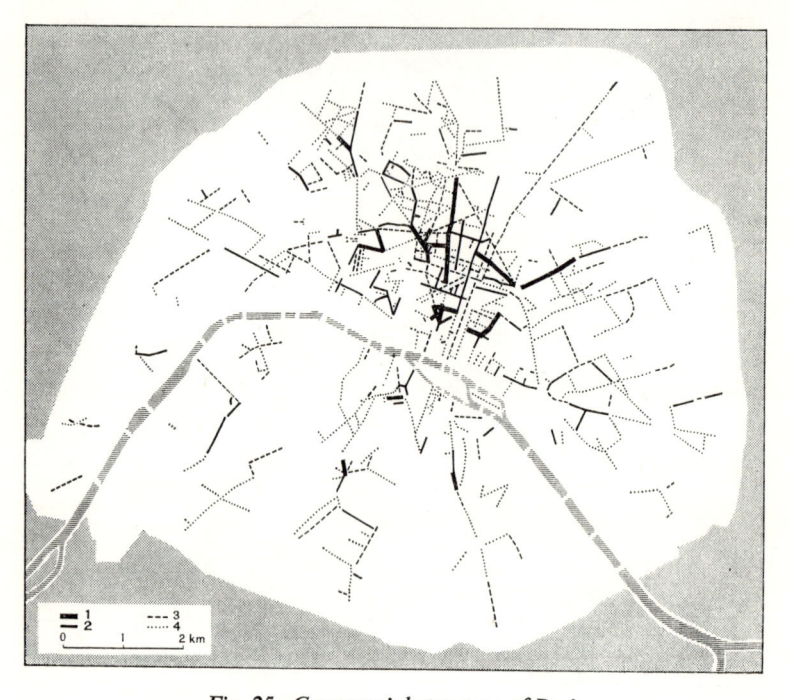

Fig. 25. Commercial structure of Paris

1. More than 30 shops to 100 metres of road. **2.** From 26 to 30 shops. **3.** From 21 to 25 shops. **4.** From 16 to 20 shops.

along transport routes; if they need heavy and bulky raw materials they may attach themselves to waterways, but if their requirements are more valuable or easier to transport, they will use railways and roads. In general, in west European towns, the prevailing westerly winds result in most noxious industries being located on the eastern side. On the other hand, clean and quiet industries, based on the availability of skilled local labour and using power supplies that produce no dust and no smoke, are found right inside the towns. Such industries as dressmaking, tailoring or jewellery are indeed often found in quite small workshops, housed on upper floors or at the back of commercial or residential premises and leaving no mark at all on the urban landscape.

In Paris there is a very sharp industrial contrast between the city proper and the inner suburbs; manufacturing industries are much less in evidence in the city, and only the clothing industry has a larger proportion of workers in the city than in the suburbs.

Active population of Paris in 1954

(per cent)

Occupation	Paris city	Seine département excluding Paris
Primary group	0·1	0·6
Industry	37·8	50·8
Building	4·9	6·2
Heavy metallurgy	0·5	0·9
Mechanical engineering	8·0	15·5
Electrical engineering	2·6	4·2
Chemicals	2·3	4·3
Clothing	5·2	3·2
Tertiary group	62·1	48·6
Commerce, banking, insurance	24·6	18·9
Public services, military, administration	12·4	11·6

In London, the banks of the lower Thames are lined with heavy industry, including seventeen power stations, six gas works, several cement works, paper mills, oil refineries, sugar refineries, sawmills, and the great Ford factory at Dagenham with its blast furnaces and coke ovens; along the railways and roads which run parallel to the river are smaller and more specialised industries such as engineering, synthetic rubber, and chemicals, while other industrial concentrations occur in the valleys of the Lea, Cray and Wandle, where there are engineering works, metal refining, paper and furniture factories; and along the main line railways there are food industries, electrical engineering, gramophones and radio, and so on.

The localisation of the typically industrial parts of the urban landscape thus appears to be very strongly influenced by the arteries of communication, and since these in their turn depend largely on the physical environment (relief and drainage) it is almost always possible to discern a close correlation between the major factory locations and the detailed relief of the town site.[1]

[1] V. G. Davidovitch in *Voproci Geografii*, vol. 38, pp. 27–77.

Another localising factor is the availability of large areas of low-priced land. It is this that often confines industries to certain parts of valleys or to particular sites alongside railways or roads, or pushes

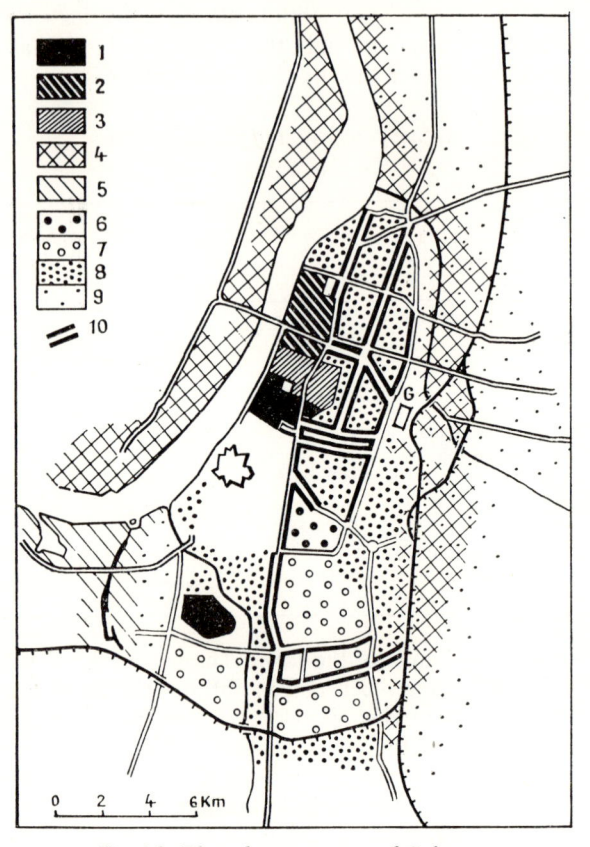

Fig. 26. The urban structure of Calcutta

1. Administrative sector. **2.** Wholesaling. **3.** Business sector. **4.** Industrial zone. **5.** Port installations. **6.** European quarter. **7.** High-class residential area. **8.** Middle-class residential area. **9.** Low-class residential area. **10.** Major roads.

them to the city limits. The location of the industrial zones of Calcutta is very characteristic, for they are found alongside the railway and canal on the eastern side of the city and on the banks of the Hugli (except on the quays that fringe the city centre and the old fortress).[1]

[1] N. R. Kar, 'Calcutta als Weltstadt' in *zum Problem der Weltstadt*, pp. 127–58.

The Seine is lined with factories and warehouses except during its passage through the heart of Paris, between the Austerlitz and Passy bridges. In Tokyo, as in many other cities, the reclamation of alluvial areas has yielded sites for large industries.[1]

Small diversified industries, confined to small premises and employing only small numbers of workers, have much less effect on the urban landscape, even though they may be quite widely disseminated. But they often have a tendency to congregate in certain areas; in Paris, furniture is made in the Saint Antoine district, whilst the clothing industry is particularly concentrated in the 2nd and 3rd *arrondissements*, the furriers in the 9th and 10th. In Birmingham the jewellery and gun-making industries were formerly highly concentrated in districts north and north-west of the city centre.[2]

Finally, within and around all the other quarters, and making up the whole fabric of the town, are the residences of the population, with a great variety of forms and types that we shall study in detail later on.

An urban area tends naturally to become organised as a collection of diverse elements, and this is by no means merely the result of the complexity of modern urban life. We may take a trip into the past, as it were, by visiting a Moroccan *medina*, for example in Fez or Marrakech. Here we shall find shopping streets, with laden donkeys and their drivers jostling the passers-by, with their multicoloured meat stalls, the warm odour of spices and the chattering crowds; we shall find groups of weavers, with their freshly dyed hanks of yarn, brilliantly coloured, hanging out to dry in the sun, we shall hear the noisy tapping of the copper-beaters, the dull grinding and hammering of the carpenters, deadened by the sawdust; we shall follow narrow lanes running between the blank outer walls of palaces which open inwards to cool and scented courtyards, or alleys in the poorer quarters which are swarming with small children; and we shall pass the great carved doorways that lead into the contemplative spiritual calm of the mosques. All the characteristics of the modern town are here, in embryo, in this glimpse of life as it was in the Middle Ages.

THE CENTRAL AREAS

In every town, old or new, large or small, the centre has its own particular appearance and problems. It is the core or nucleus, particularly if, as often happens, it lies at the focus of the main roads; this is emphasised in many French towns by the road signs at the city limits—*Centre-ville et toutes directions*.

[1] S. Kiuchi, 'Tokio als Weltstadt', *op. cit.*, pp. 112–26.
[2] M. J. Wise, 'The evolution of the jewellery and gun quarters in Birmingham', *Trans. Inst. Brit. Geog.*, 1949, pp. 58–72.

The centre is also the focus of the tertiary activities of the town—local government, banking, commerce, containing also buildings concerned with public religious or cultural life. It is here that shop rents are highest[1] and land sells for the highest price,[2] so that depending on the local circumstances, one may find, often side by side, the greatest congestion and the largest buildings. Urban centres thus offer a variety of appearances, depending on the type of town.[3]

First, the great capitals and metropolitan cities, in which the central areas are grossly swollen, and of which the heart is the 'city'. The City of London is typical. The narrow streets are lined with tall, soot-blackened stone buildings of which one in two houses a bank; they teem with life in the daytime and are empty at night: the surging crowds move in at 9 a.m. and depart at 5 p.m. Here are the strong-rooms and offices of the world's greatest money market, and around St Paul's cathedral and the Bank underground station, which serves the Bank of England, is the most famous 'square mile' in the world. Outside this central core lies a wider and more varied zone, but still dominated by tertiary occupations; it forms the business and administrative area stretching from Paddington in the north-west to the loop of the Thames at Westminster on the south-east, covering about 15 square miles, or 15 per cent of the County of London and 0·2 per cent of the whole conurbation.

In other cities, the fundamental distinction is between the old ones, especially those whose site was chosen for defensive purposes and was therefore fortified, and those of more recent foundation. The former are liable to what one might call historical congestion; behind the facade of the crowded old quarters that are the admiration of the tourists, the poorer people are still living in the environment of a bygone age—in medieval squalor. Whenever practicable, and if the town has not had much recent growth, some of the business activity has been concentrated here, usually by some reconstruction of the buildings. More often, however, local government offices and the bigger shops have migrated to more spacious quarters, and the town has developed a second nucleus. Thus we find the old Le Mans perched up on its hill, with its town hall and cathedral, overlooking the modern prefecture and business streets built on the slope that runs down to the confluence of the rivers Huisne and Sarthe; in Strasbourg the vast squares that date from less than a century ago lie adjacent

---

[1] W. William-Olsson, 'Stockholm: its structure and development', *Geog. Rev.*, 1940, pp. 420–38.

[2] R. E. Murphy and J. E. Vance, 'Delimiting the C.B.D.', in H. M. Mayer and C. F. Kohn, *op. cit.*, pp. 418–46.

[3] J. Beaujeu-Garnier, 'Méthodes d'études pour le centre des villes', *Annales de Géographie*, 1965, no. 406, pp. 694–707.

to the crowded streets of the 'petite France' quarter. The same phenomenon of the duplication of town centres is found in colonial territories where European townships have been created alongside the older native centres, as in North Africa.[1] Modern reconstructions, sometimes the result of wartime devastation, have brought about extensive changes in city centres. In the City of London, some of the bombed sites are still car parks, whilst other now bristle with tall, glass-fronted buildings. In Lille, the old slum quarters have been systematically cleared—incidentally enhancing the value of certain historical monuments—and in the reconstruction housing and commerce have been mingled.

In modern cities, the business life hinges on a central district that may have achieved this function spontaneously. Amongst the great cities of the United States, the famous Loop district of Chicago provides the finest example, but every small town has its Main Street, made famous in literature. Elsewhere, as in the colonial towns of Soviet Asia, an administrative and social centre has been deliberately created.

In the United States, the town centre has been called Central Business District (C.B.D.);[2] here are grouped businesses and shops of various kinds and sizes, and the services; no industry, except perhaps printing, no residences, no public or religious buildings, no warehouses or wholesalers. In appearance the district often consists of a series of skyscrapers in the centre of an otherwise low-built and spread-out town, even though the flat relief may offer no obstacle to outward expansion. It has a function very different from that of the European town centre, for the social and recreational aspects are completely missing—no coffee terraces, no places of amusement.

It is not easy to gauge exactly the rôle of the centre in the life of a town. It was estimated in 1956 that 37 per cent of all the employees in the metropolitan region of New York worked in the C.B.D. of Manhattan. In Paris, a 1950 estimate gave 525,000 as the number of people going to work each day in the central districts; the 1st, 2nd, 8th and 9th *arrondissements* received about nine-tenths of their daily working population from outside.[3] In London, at the census of 1951, the population of the central boroughs by night was 502,590, but the number working in the same boroughs by day amounted to 1,508,908. Statistics relating to shopping would vary enormously from one town

---

[1] On this subject see R. Danger, *Rapport d'enquête sur la ville d'Alep*, Aleppo, 1931.

[2] E. M. Horwood and R. R. Boyce, *Studies of the Central Business District and Urban Freeway development*, Seattle, 1959.

[3] M. Brickler, 'Les migrations alternantes dans la région parisienne', in *Paris, 1960*, pp. 151–70.

to another. In United States towns, the proportion of the total sales within the urban area effected by the shops of the C.B.D. amounted in 1954 to 46.2 per cent at Evansville, 43·8 per cent at Newark, 35·6 per cent at San Francisco, but only 24·8 per cent in New York, 26·8 per cent in Philadelphia and just over 16 per cent in Chicago. The degree of concentration thus varies with the structure and function of the city, and probably also with the area that it covers, for in the vast spread of Los Angeles the relative intensity of the C.B.D. functions is quite slight, and only 13·6 per cent of the retail shopping is concentrated in the centre.[1]

Though it is difficult to put it into figures, the activity of the C.B.D. is also characterised, as is obvious to any observer or even to every pedestrian, by an extraordinary intensity of traffic movement, by a daily inward and outward flow of employees that helps to cheapen the provision of services within the district, and by an equally astonishing daily influx of shoppers and other visitors. The movement towards the city centre is greatly enhanced if its crossroads function necessitates a passage through the centre in order to get from one part of the city to another.

Statistical studies made in the United States have enabled certain general conclusions to be drawn. It appears that the most intense inward movements occur in medium-sized towns, and that they decrease in proportion to the increased size of the town. Thus for towns of 100,000 to 250,000 inhabitants, the proportion of incomers between 7 a.m. and 7 p.m. on a normal working day reaches 665 for every 1,000 of the population, a figure which drops to 201 for cities of over 3 million inhabitants. In the town of Charlotte, N.C. the number of entries was actually greater than the number of inhabitants, at 1,047 per thousand whereas in New York the proportion was only 147 per thousand.

A second conclusion is that the relation between persons who pass through the C.B.D., those who work there and those who are there at the peak period (about 3 p.m.) can be expressed as 4:2:1. The number of persons actually working in the C.B.D. reaches a minimum of 107 per thousand in Los Angeles. The numbers present in the C.B.D. during the peak period in mid-afternoon is about 115 per thousand in medium-sized towns of 100,000 to 250,000 inhabitants and falls to 70 per thousand in very large cities.[2]

These movements necessitate an extraordinary provision of transport and help to explain the unavoidable congestion in town centres; they are responsible for many special regulations such as prohibition

[1] G. A. Wissink, *American cities in perspective*, pp. 124 ff.
[2] D. L. Foley, 'The daily movement of population into C.B.D.', in H. M. Mayer and C. F. Kohn, *op. cit.*, pp. 447–53.

or severe limitation of parking, loading and unloading (which in New York must be done before 5 a.m.). In the C.B.D. of Manhattan,

① Old business centre
② Modern business centre
--- Limit of the non-residential zone

1 Av Anhangabaú
2 Viaduto do Chá
3 Praça da Sé
4 Parque D. Pedro II
5 Railway station

ACTIVITIES
Commerce
  1
  100
  200 Shops, depots etc.

Industry, handicrafts
  1
  25
  50 Workshops, factories etc.

Liberal professions
  1 50 150 Chambers, consulting rooms etc.

Finance
  1 5 10 Banks, loan offices etc.

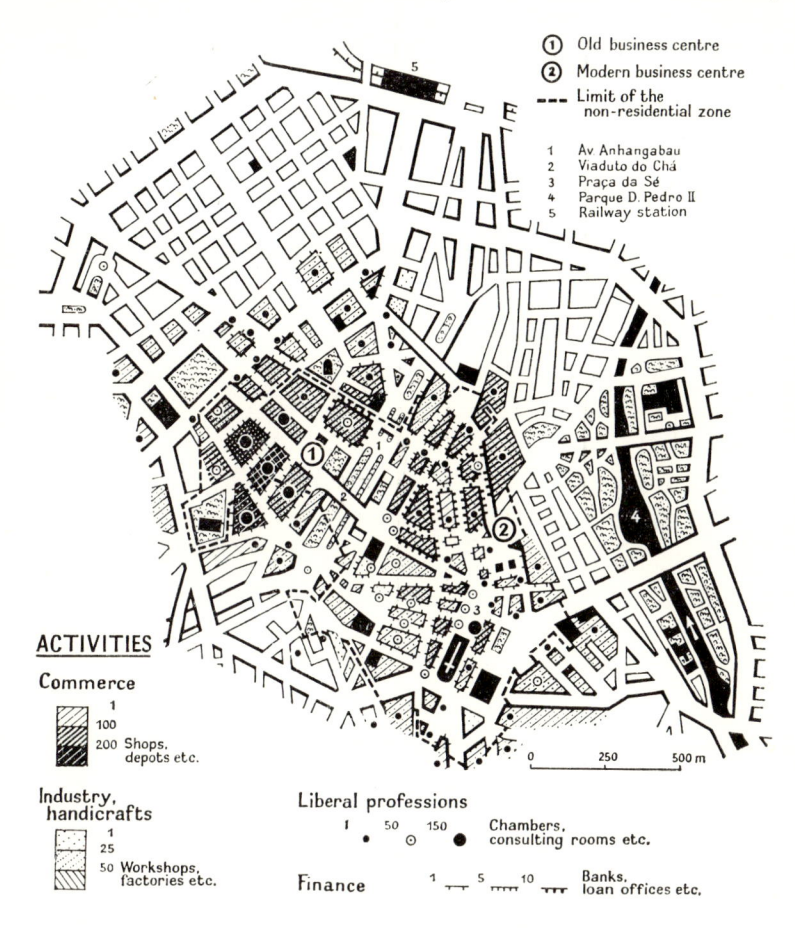

Fig. 27. The business centre of São Paulo

Commercial activity: retail and wholesale trade, offices and business headquarters.
Industrial activity: workshops and factories.
Liberal professions: lawyers, doctors and dentists, engineers.
Financial activity: banks, loan offices, money changers.

in 1956, of a total entry of 3,316,000 persons (of whom 1,970,000 were simply passing through) 736,000 came by car or taxi, 246,000 by omnibus, 233,000 by railway and 92,000 in lorries (Figs 28, 37).

This congestion in the centre, combined with the continuing extension of the urban area, has provoked a reduction in movement towards the centre and an increasing dispersion of business activities into the peripheral zones. At Atlanta, in 1948, 68,100 people entered the C.B.D. every day to work, and 132,900 went in for shopping or

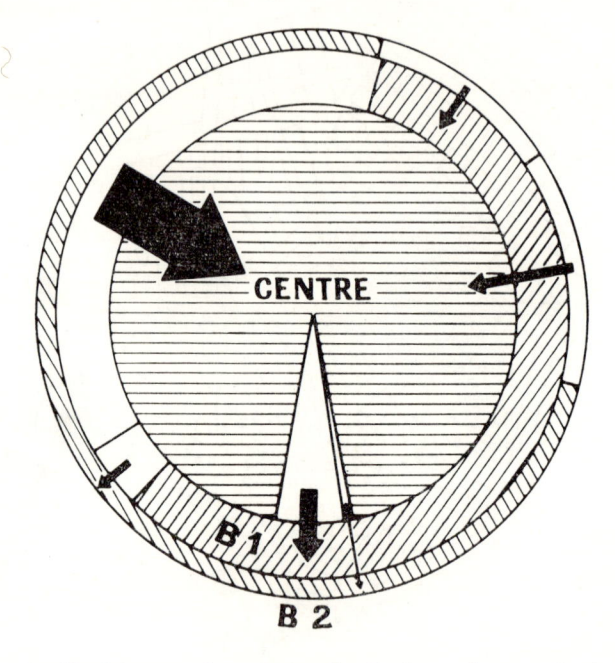

Fig. 28. Diagram of population distribution and commuting in the New York region

The ruled areas represent the proportion of the working population engaged in the zone in which it lives; the blank areas the proportion that works in another zone. The thickness of the arrows is proportional to the number of daily migrants. The outer circle represents the total working population living and working within the region as a whole. (From figures given in *Anatomy of a Metropolis*.) For the zones see Fig. 29.

other business, making 201,000 out of a total population of 695,000; by 1953 the population had risen to 800,000 but the concentration on the C.B.D. had diminished to 57,000 workers a day and 110,900 shoppers or visitors.

This is not an isolated example: all the available statistics for United States' towns confirm it. Between 1954 and 1958, in ninety-four towns totalling 91,937,000 inhabitants in the latter year, sales in the shops of the C.B.D. only rose by 3·4 per cent, whereas in the zones outside the C.B.D. they rose by 53·8 per cent. Los Angeles is once again somewhat exceptional, since for twenty-five years the

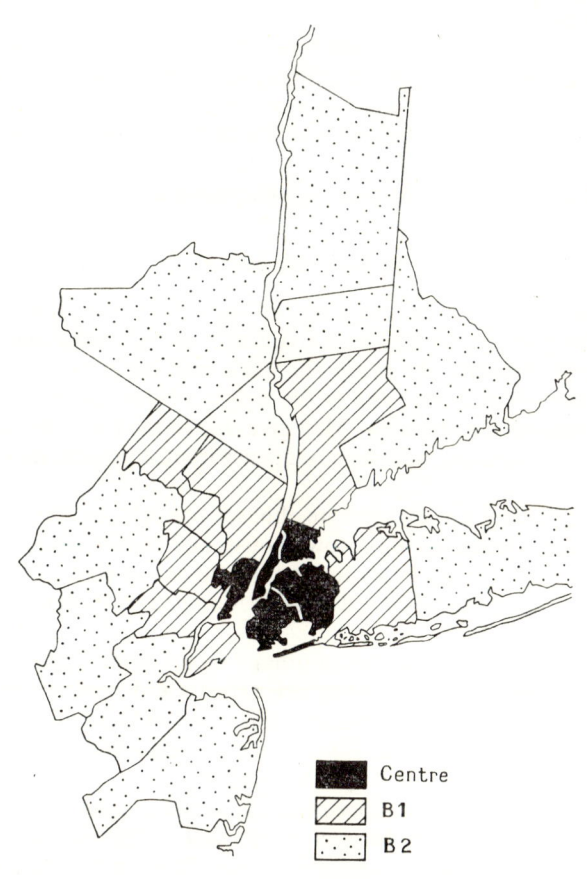

Fig. 29. Zones of the New York district
(cf. Fig. 28)

Centre: the core. B1, the inner ring. B2, the outer ring.

shopping in the centre has been declining and between 1946 and 1957 sales dropped by 30 per cent. Between 1948 and 1954, the sales of retail stores in the C.B.D.s of Chicago, Detroit, Philadelphia and Los Angeles fell by between 5 and 12 per cent, and in other large cities it rose between two and fifteen times less fast than in the cities as a whole.[1] In Paris between 1954 and 1960 the number of shops fell by 2,975, though it rose by 2,200 in the *département* of Seine-et-Oise; the trade in meat grew ten times more rapidly in Seine-et-Oise than in Seine, and the latter lost 109 bakeries, while in Seine-et-Oise 163 new ones opened.[2]

Thus the congestion of city centres eventually reaches a limit. Even though new office blocks are still rapidly rising in the centre of London, in New York, and, despite the very severe regulations, in Paris, inconveniences of congestion, and the difficulty of movement, of parking and of life in general are such as to provoke a spontaneous outward movement on the part of both people and business.

RESIDENTIAL AREAS

Dwellings occupy the major part of the urban area, and they are found everywhere, though least frequently in the centre and in the warehouse quarters. In many shopping and business streets they occupy the upper floors; and in the industrial quarters they occupy spaces between the factories. Man gives life to the town, for without him it could not exist, and in consequence it must provide him with a home.

Although they are everywhere, houses are seldom alike. There have been many studies of the town house, and though many interesting monographs could be written on the subject in detail, little more than a summary synthesis and classification can be presented in this book.

The towns of Asia and Africa often contain many quite primitive huts and hovels, usually built by new arrivals and conforming to the pattern with which the occupiers were familiar in their villages. Even in European towns the rural house is still found in areas that have only recently been engulfed by urbanisation.

Amongst the houses that one generally thinks of as urban, there is a fundamental distinction between those that house one family and generally have only one or two storeys, and those that contain several families—in which case the variety is almost limitless. Single-storey or two-storey houses almost always dominate spread-out towns. They are characteristic of certain types of human grouping for which they form the majority or indeed almost the whole of the residential accommodation. They vary from the sordid little brick or cob houses in the courts of the textile towns of northern France, through the miners' rows of the older English coalfields and the monotonous working-class streets of Lancashire and Yorkshire towns, to the

---

[1] G. A. Wissink, *op. cit.*, pp. 130 ff.
[2] P. Benaerts, 'Le Commerce', in *Paris 1960, op. cit.*, pp. 103–10.

luxury villas of more elegant residential quarters. They are especially characteristic of Anglo-Saxon countries and of European colonies in the tropics: Noumea, Conakry, Brazzaville, and Washington are made up of villas, and large parts of Dakar, Léopoldville (now called Kinshasa), the British sector of Singapore and the French quarters in the towns of former Indo-China are composed of an almost endless succession of detached or semi-detached houses with gardens.

This predominance is more striking the further one goes from the city centre. The New York conurbation shows this to perfection, but it is the same story over the whole of the United States—detached one-family houses occupy 56·6 per cent of the whole urbanised area, but only 38·5 per cent in the inner zones and 75·9 per cent in the suburbs.[1]

Types of housing in New York metropolitan region, 1950

(*Percentage of all houses*)

	Single family	Two family	Several families
The region	32·2	15·0	52·8
Central zone	14·6	13·1	77·2
Manhattan	1·8	0·8	97·4
Rest of Central zone	18·9	17·2	63·9
Inner ring	51·7	19·2	29·0
Outer ring	69·8	15·2	14·7

Source: J. Bogue, *op. cit.*, p. 715.

The impression is inescapable that the more recent the urban growth and the greater the available space, the more does this type of dwelling spread. A comparison of Californian towns with their forerunners of similar population totals in other parts of the United States, is very striking, and the high proportion of houses built between 1940 and 1950 shows that there is no weakening of this trend, rather the reverse (Table on p. 306).

As for dwellings housing more than one family, they may range from those of two or three storeys, housing two or three families, very common in French provincial towns, to the great apartment blocks, cubes of glass with prefabricated partition walls, laid out like a chess board, that give a basic minimum of comfort and the elementary resources for human intercourse to many thousands of families who are lost in anonymity, herded together in an ultra-modern

[1] E. W. Hoover and R. Vernon, *Anatomy of a Metropolis*, p. 138.

Percentage of dwellings

Towns	One-family houses	Built 1940–50
Fresno	81·8	38·0
Knoxville, Tenn.	68·2	20·7
San Jose	78·2	34·4
Wilmington, Del.	30·2	20·2
Sacramento	70·3	39·9
Salt Lake City, Utah	59·8	25·1
Los Angeles	63·8	33·2
Chicago, Ill.	28·4	10·0
San Francisco, Oakland	47·6	28·0
Boston, Mass.	29·9	7·2

Source: H. F. Gregor, 'Spatial disharmonies in California population growth', *Geog. Rev.*, 1963, p. 117.

medium that has been described as almost inhuman. Many French towns now have these monstrosities: Sarcelles and Massy-Antony near Paris, Le Haut-du-Lièvre at Nancy are but a few examples. All kinds of intermediate gradations exist—blocks built of brick, of rough or cut stone, or of concrete, more ornate and only five or six storeys high if they are fifty years old, rather more daring, with eight or ten storeys, if built more recently, giving shelter to several families who can get to know each other; more or less comfortable, and well ventilated, for they date from the time when the front was more important than the back and the middle class was more important than the working class.

It would be possible to classify residential zones by the type of housing, though it must be emphasised that housing areas are rarely homogeneous; and that there is frequently an almost inextricable mixture of different types of building and different classes of people. This is the case especially in towns that are undergoing a transformation. In Paris little slum islands still inhabited by the poorer classes stand adjacent to rebuilt streets where blocks of luxury flats have just been completed. The contrast is found, in a more crystallised form, in some parts of London, where, behind the handsome red-brick or plaster-faced blocks that line the main roads and house the well-to-do, there are small two-storey houses of cheaper yellow brick along the back streets, the homes of much less wealthy folk. The same sort of thing occurs in Paris, where the contrast is between the tall stone houses of the mid-nineteenth century boulevards and the smaller and older houses that remain from the period before the cutting of these straight and wide roads. In south-east Asia the towns sometimes have properly made-up roads lit by electricity and lined with a narrow fringe of respectable buildings—but just behind, attracted by this mirage, are hordes of poor families. In Bombay, the last luxurious buildings on Malabar Hill directly overlook a collection of hovels at the entrance to the Indian town.

Almost all towns have their slums, except perhaps the quite modern planned towns. But poverty has many faces, and we can distinguish the slums that result from the decay of formerly more prosperous quarters and those that are being created every day. The former are generally in the central areas of towns; they represent the housing areas of a former period, uncomfortable and often insanitary, or perhaps former middle-class housing that has decayed. In them swarms a population of immigrants, of people regarded as racially inferior, of new arrivals from the country districts, of elderly people existing on reduced incomes, or of poorer wage-earners with large families. In Paris North Africans occupy the southern slopes of the Saint-Geneviève hill and the rue Mouffetard area, which is ripe for demolition, and the sordid flats and decrepit houses of the 18th *arrondissement*. In one district in the 11th *arrondissement* which was regarded as a slum as long ago as 1917, and in which 36 per cent of the houses have no convenience, there are 1·2 people to a room and a density of 560,000 to the square mile, four times greater than in the neighbouring district; this is one of the worst areas in the whole of Paris. Between 1954 and 1957, the population increased by 53 per cent with the arrival of whole families, often numbering five, six, or seven persons; 40 per cent of the population is French, 40 per cent North African including many Jews, 20 per cent foreigners (Russians, Poles, Italians, etc.), there are large numbers of small children, and many more adults than young people; 63·8 per cent of the men are just labourers.[1]

In United States towns where there has been an influx of Negroes, rents have dropped, gardens have become neglected, houses are no longer repainted, and the white people have cleared out. In New York the greatest concentration of Puerto-Ricans is found in north-eastern Manhattan, adjoining the Negro quarter, and in some sections, they make up between 41 and 76 per cent of the population.[2] In England coloured immigrants have almost 'colonised' some older housing areas in certain cities—as in the Brixton, Notting Hill and Willesden districts of London, or Sparkbrook in Birmingham. But

[1] J. F. Thery, 'Les habitants de la cité Lesage-Bullourde (XIe)', *La vie urbaine*, 1959, pp. 187–240.
[2] R. T. Novak, 'Distribution of Puerto-Ricans on Manhattan Island', *Geog. Rev.*, 1956, pp. 182–6.

on the whole the congested slums of English towns have nothing to do with the influx of immigrants, and they result simply from the decay of nineteenth century housing. The problem is a serious one, for a study made in 1951 revealed that a million and a half people were living at a density of more than 1·5 persons per room; in Birmingham 21·3 per cent of the houses were classed as unfit and ripe for demolition, in Manchester 20 per cent and in Hull 30 per cent.[1] In some cases the war had ended the scourge of the slums in no uncertain fashion, as in Stepney for example, where almost the entire borough was destroyed, in which before the war one-third of the houses were due for demolition.

The old quarters of the port area of Naples, or those of the trans-Tiber district of Rome, or of Palermo, show inexhaustible swarms of children running in and out of dark cellars and grim alleys that the realistic Italian film industry has not hesitated to expose. In Bahia, Brazil, part of the great influx of miserable creatures from the countryside has found shelter in the Pelurinho hill district, where prostitution and tuberculosis are rife.

Many of those who are drawn towards the town cannot get in. They seek out the most derelict spot, where their occupation of the land will not be disputed, and here they set up a few planks, a roof of corrugated steel sheets, a few bits of sacking, the odd brick or two, a reed or paling fence—and thus begins a shanty-town or *bidonville*. In Rio de Janeiro these *favellas* occupy the steeply-sloping granite hills, and in 1957 no less than one-third of the entire population lived in them. In Lima, Peru, some 47 per cent of the inhabitants of the *barriadas* were born in the countryside, and mostly up in the mountains, whilst 27 per cent of them were without regular employment.[2] In Bahia, the shanty-towns have occupied all the marshy and malarial valleys and have even extended over the shallow waters of the bay, where, as in the canals and lagoons of Hong Kong, Bangkok, Pnom-Penh, and Saigon, they are built on piles driven into the mud. In Algiers, they form a ring round the town, on the back slopes of the hills, often close to the cemeteries. When there just is no land available, not even a flooded mud-flat, the wretched creatures live in the junks that clutter up the rivers and ports of Far Eastern towns. When neither roof nor floor-space is available, there are people even more unfortunate than the inhabitants of the shanty-towns, for they have nowhere to lie but on the roadside, and here they curl up, with their small bundles under their heads, and covered by a blanket. In the

[1] For a more recent survey, with maps, see F. T. Burnett and S. F. Scott, 'A survey of housing conditions in the urban areas of England and Wales: 1960', *Sociological Review*, 1962, pp. 35–79.
[2] *L'urbanisation en Amérique Latine, op. cit.*

towns of India, or in Recife when drought strikes the *sertão*, there is indeed a crisis worse than that of the slums.

Working-class housing is commonly grouped around the business and shopping centres and near to the industrial establishments. Frequently it forms a belt around the central business district, with offshoots around groups of factories or along factory-lined roads. It is usually somewhat confined in space and occupies the naturally less favourable sites (e.g. flat areas with no view and no breeze as opposed to hillsides or sea-front, or on badly exposed slopes), or sites that are rendered less desirable because they are subject to smoke pollution from the city (as on the eastern sides of most European towns), or sites along suburban railway lines that are very easily accessible and so densely peopled.

These zones are usually monotonous and often grim. Cheaply built with bricks or concrete, they have austere frontages, and everything is skimped, whether it be the width of the roads, the height of the ceilings, the number of rooms per house, or the yards (and gardens if in the suburbs). The lack of imagination in the layout and appearance results from the tradition of economical building, both in Great Britain ('bye-law housing') and in Germany, where streets, neighbourhoods, indeed whole sections of towns are almost identical.

Such working-class quarters are often of a uniform age and plan. They include the *cités ouvrières* round coal mines or around some great factory like a Lorraine steelworks or the Michelin works at Clermont-Ferrand. The changes that have occurred in France since 1920 in the mining districts bear witness to the social revolution that has taken place. For the depressing 'miners' rows' of the pre-1914 period, there were substituted streets of neat little houses, each with front and back garden; these were still somewhat deceptively monotonous, but a further advance has brought in groups of medium-sized houses, generally arranged in pairs and set amongst newly-planted trees, that are really very pleasant.

Though some of these housing estates may contain only the workers in the factory or mine that built them, most working-class districts have a mixed population. In a typical working-class quarter of Strasbourg, for example, south-west of the railway station, with three-roomed flats in four- or five-storeyed blocks, 12 per cent of the accommodation is 'tied' to business or industrial firms; wage-earning office and factory workers form 53·1 per cent of the total, families with a moderate income 25·8 per cent and well-to-do people only 6·5 per cent.[1]

In the Far East, a Chinese is never away from his business; neither is the Greek or Lebanese trader in the towns of the Mediterranean or

[1] H. Nonn, *op. cit.*

Black Africa; this results in a type of house, quite small, in which business and petty trading overlaps with the living quarters; the house may have an upper floor, or if of bungalow type it may have a little courtyard and an outhouse. Such houses rapidly multiply along a single street, and may even form whole quarters as in Singapore or in Cholon, the Chinese quarter of Saigon.

What we might call the well-to-do residential districts vary from middle-class housing to the most luxurious, occupying all the best situations. In Paris the area lying between the Place de l'Etoile and the Bois de Boulogne is one of the best residential districts in the capital, but the Avenue du Bois reaches the peak, both in the social hierarchy and in the rents.

Within these zones there is a complete absence of industrial buildings, and commerce, if it exists at all, is of a superior type. Shops, other than those purveying the ordinary daily necessities, are of a luxury goods type. But in many streets, and even in whole sectors, no shop front mars the ground floor of the beautifully laid out houses or the appearance of the trellises and gardens. Traffic movement, especially of pedestrians, is light, and these peaceful oases lie between the main roads where the city's lifeblood flows.

These zones are sometimes reached through a series of transitional areas in a way that would satisfy the urban theorist, but at other times their distribution is quite capricious. We may quote examples from Paris: one passes logically from the working-class districts of the 18th *arrondissement* to the middle-class neighbourhood in the east of the 17th, then to the upper-class area in the west of the 17th, which borders the aristocratic 16th *arrondissement*; but sometimes tradition and snobbery disobey logic. Many of the old houses of the île Bourbon, and along the Seine quays facing the Louvre, have been internally reconstructed and are now luxury residences set in the midst of the literary, artistic, financial and political quarter; but behind this brilliant facade, away from the river, other buildings of the same age have been allowed to deteriorate.

In general it is true that the well-to-do quarters lie away from the most active areas of the town, whether these be industrial or commercial; they develop in the most favoured areas and they often comprise whole sectors of a town, with a varied and pleasing appearance. Thus in Rome the Prati quarter is like a cool, verdant garden, with fine houses, away from the noise and bustle of the city centre. In Bombay, Malabar Hill is the elegant residential area, as is the extremity of the Dakar peninsula, delightfully cooled by the sea breeze. In the suburbs, space is less precious, and the villas are more individual; the roads are wide and the gardens large; large blocks, where they exist, are stone-faced and have elegant escape-staircases.

Though the middle-class residential suburbs still make much use of public transport to get to town, the wealthier citizens have one or two cars per family and are not subject to such restraint.

The localisation of these residential areas varies with the functional nature of each town; in industrial towns the wealthy quarters are very small. In Roubaix, for example, the bulk of the town's population is working-class, with relatively low incomes, and the small area of luxury houses lies on the south side, taking advantage of the relief provided by a small escarpment. On the other hand, in mainly residential towns such as Washington, the residential areas are much more uniform in character.

Shifts of location take place in a more or less regular sequence: the oldest residential areas, which are gradually usurped by other forms of urban land use, give way to areas further out, whilst the development of private transport, partly due to technical progress and partly to the increased purchasing power of the people, has the same effect. In Edinburgh, the famous Princes Street has almost become a boulevard of multiple chain stores, and already the movement that made George Street the principal residential area has passed further on, to Queen Street. In the suburbs of Rio de Janeiro, Copacabana, formerly queen of the residential areas, has become outmoded, and one must now go further along the coast to Ipanema and Leblon to find the millionaires' villas extending almost into the coastal forests.

In such complex urban areas as we find in Europe, it is difficult to find any degree of uniformity within the well-to-do districts. The 16th *arrondissement* in Paris is a good example, for here there are individual town houses, villas buried in the gardens of the former large estates, and fine blocks of luxury flats of different dates, all intermingled and having nothing in common except a particular socio-economic structure in relation to the city as a whole.

Socio-professional structure of the 16th arrondissement

(*percentage of the active population*)

	Auteuil	Chaillot	Muette	Porte Dauphine	Paris
Liberal professions and administrators	19	15	19	17	8
Industrialists and big businessmen	4	5	5	6	2
Middle classes	15	10	12	10	11
Personal servants	16	30	28	32	13
Salaried personnel	20	16	15	14	22
Working classes	14	13	11	11	31

Source: G. Belleville, *Morphologie de la population active à Paris*, Paris, 1962.

Though one can remark many curiosities in the location of these residences when studied in detail, there are certain recognisable generalisations that can be made, or even certain rules that can be laid down, if one takes a broad view. The arrangement of the residential quarters of Paris may be cited as an example. In the east a working-class population predominates, occupying the urban fringe in the 17th, 19th, 20th and 13th *arrondissements*; this is the famous 'Paris crescent', with 23 per cent of the population and many industrial establishments, with poor tenements in a state of decay or even abandoned. The 'pont de Flandre' quarter which contains two canals, a gasworks and the cattle market and abattoirs of La Villette has a particularly miserable environment; of its active population, 54 per cent are workpeople (as against an average of 31 per cent for the whole of Paris), and the proportion of labourers in this total is twice as high as in the whole city; indeed the lowest working classes form the most important section of the population. It has been noticed by Germaine Belleville, in an excellent study,[1] that in 1872 the working classes were much more widely disseminated within the city, and at that time they formed 36 per cent of the population of Paris as against 17 per cent at the present time; they have been re-grouped outside the city, and the situation resembles that remarked on by J. Bastie with reference to the rising price of land, 'which creates an incessant urge to seek cheaper land further out for the least profitable uses such as the housing of the poor'.

The centre of the city comprises mainly the artisan classes, shopkeepers and manufacturers, whilst in the west there are two principal groups, quite different but complementary to one another—the professional and administrative classes on the one hand and personal servants on the other.

Out in the suburbs, there is a collection of industrial and working-class *communes* in the north and north-west (except for Neuilly which by its socio-professional composition resembles the western *arrondissements*); to the east and south, on the other hand, the mainly industrial *communes* are rapidly left behind and one enters an outer ring of mainly residential *communes*.

In New York one can also distinguish three zones, on the basis of occupations and *per capita* income. The main urban nucleus consists of 'black-coated workers' and shopkeepers; the average income is moderate, with the boroughs of Queens and Bronx being rather more favoured than Brooklyn and Hudson, where there are quite a few working-class people and the average income is lower. Outside the five central boroughs, the inner ring corresponds to a zone of very high incomes: in Nassau, Westchester and Bergen, which are the

[1] *Op. cit.* p. 321 ff.

wealthiest, the active population belongs mainly to the professional, administrative and business classes, whilst in Richmond salaried workers and wage-earners predominate and the average income is much lower. The outer ring consists mainly of working-class people, including factory workers and personal servants, and the average income is sensibly lower[1] (cf. Fig. 41 on p. 419).

In towns of moderate size in which industry is expanding, there are sometimes very marked segregations within the residential areas. Thus in the case of Le Mans, the industries are mainly on the southern side, in the valley of the Sarthe, and here there are large working-class suburbs like Les Pins and La Gautrie. At Oxford, the establishment of the automobile industry has been responsible for the creation of what is in effect a second town, Cowley, quite separate from the ancient university city.

The new urbanism, whether in the socialist world or in the west, has been forced to find new solutions to the problem of the location of the different types of district within the town. To this problem we shall return later.

MOVEMENT AND TRANSPORT

The disposition of the various quarters within the urban area, and the unceasing growth of towns of all sizes, necessitate the provision of a transport system that should be continually becoming more efficient, more coherent and, above all, less expensive. But the conditions of movement within a town vary greatly with the natural or physical conditions, with the layout of the system, and with the technical and financial circumstances of the enterprise that controls it.

The site of certain towns presents immediate difficulties. To enter Manhattan from either west or east means crossing arms of the sea, and the system of suspension bridges (amongst the largest in the world) and tunnels is at times inadequate to cope with the traffic which converges from the wide motorways. At Sydney the existence of only one bridge (even though, spanning the bay with its enormous arch, it carries a six-lane roadway and four lines of railway) creates considerable traffic blocks.[2] At Liverpool, in succession to the nineteenth-century ferries which carried traffic across the Mersey estuary (and still do carry many passengers) a long tunnel under the river, opened in 1934, assures a road link with the twin town of Birkenhead; but it carries so much traffic that ventilation is a problem, and there are plans for a second and even a third tunnel. At Bangkok and at Conakry a single raised causeway crosses the marshes or the many canals, giving a bottleneck access to the hinterland. River

[1] E. M. Hoover and R. Vernon, *Anatomy of a Metropolis*, pp. 165, 226.
[2] O. H. K. Spate, *Australia, New Zealand and the Pacific*, Oxford, 1956, p. 36.

11*

crossings in large cities are not always easy, and bridges, even in such considerable numbers as in London or Paris, tend to slow up traffic.

Water is not the only natural obstacle; mountains or even little hills can add to the difficulties of movement. At Rio de Janeiro it has been necessary, in order to get from the southern beaches to the city centre and beyond, to undertake enormous engineering operations; tunnels have pierced the granite hills, and the narrow coastal plain has been widened, to the detriment of the natural beauty of the site, by filling up part of the bay in order to accommodate a dual carriageway. The hills of Fourvières and Montmartre are reached by zigzag roads—or by funicular railway.

Even if there are no site complications movement can still be very difficult in old and congested towns. Italian guides perform quite remarkable feats of virtuosity in taking vehicles through the narrow streets that characterise the centres of most of their towns. Paris would be quite impossible today were it not for the nineteenth century gutting and creation of wide radial and transverse boulevards. In order to facilitate movement in Brussels, it was necessary to carve up the city and make a system of circular boulevards. Stockholm had to sacrifice part of its older quarters in the creation of a new road system cutting across the city and crossing the sea arms and lakes. The enormous urban sprawl of Los Angeles has a system of motorways as its arteries; indeed, roads occupy no less than one-third of the entire conurbation and two-thirds of the central area.[1]

The increasing difficulty and indeed danger involved in crossing the centres of towns has led to the construction of ring roads to carry traffic that does not need to enter the town; such urban by-passes are the first stage, the second being the building of special motorways that avoid the towns altogether. The two stages have been passed through in recent years in France, Great Britain and Spain, but Germany and Italy much earlier adopted the motorway principle (*autobahn, autostrada*) which is also very well developed in the United States. The necessity for avoiding town centres is particularly evident in Europe, with its dense population, high degree of urbanisation, and, especially in the west, its intense motor traffic.

In the countries of eastern Europe and the U.S.S.R., conditions are different. The authorities have been much concerned to create a road system and to reshape the main urban arteries without there being, at any rate for the time being, a volume of private motor traffic sufficient to demand such facilities.

In new countries, road construction in and around the town does not keep pace with the building of houses. Even in Asiatic Russia,

[1] H. F. Gregor, 'Spatial disharmonies in California population growth', *Geog. Rev.*, 1963, pp. 118–22.

with all the technical resources there available, the intensity of new town construction and the rapidity of urban expansion have meant that in many cases the roads are nothing but muddy ditches in wet periods and dust bowls in times of drought. The same thing can be said of the bulk of the recent town extensions in Latin America and Africa. In the latter, the European quarters all have made-up roads, but these are conspicuously absent from the African quarters.[1]

The convergence of men and commodities on town centres is represented on the map by a radial road system. All towns are crossroads to a greater or less extent, and the difficulties of access are sometimes enormous; the centre is like a gigantic heart, pulsating morning and evening as it sucks in and disgorges its flow of people and vehicles. In Paris two motorways only, inadequate in their traffic capacity, and a number of main roads, provide entry and egress; movement to and fro at the peak hours is difficult enough, but at road junctions within the city the traffic is almost inextricable—and this when in any case the vast majority of journeys between the suburbs and the city are made by railway or by métro.

The movement of people within towns and between the periphery and the centre, is often made possible by diverse means of transport, road, surface, underground and overhead railways, and waterways. Such transport may be public—in which case it is usually possible to get precise information about its organisation, timetables and traffic—or private; in the case of the latter we have generally nothing but guesswork and rough estimates to go on. The larger the city, the greater the proportion of the transport effected by the public systems; in 1934 in Rotterdam, Hamburg and Berlin, which had population totals in the ratio of 1:3:7, the number of passengers on the public transport systems was in the ratio of 1:5:25.[2]

It is in the communist countries that public transport has the greatest degree of dominance. Mass movement is considerable, for the economy is well developed, especially in the U.S.S.R. and the countries of eastern Europe, where private transport plays but a small part. Conversely, in the western world, the higher the standard of living, the higher the proportion of private transport, with the U.S.A. breaking all records. The number of passenger miles travelled on the Moscow underground is three times as high as on the New York subway.

The development of urban transport has revolutionised town life. In Paris the first horse bus ran in 1828, and there were a thousand of them in 1860, each carrying seventeen to twenty-four passengers and

---

[1] J. Denis, *Le Phénomène urbain en Afrique Centrale*, Paris, 1958, pp. 326 ff.; M. Banton, *West African City*, London, 1957, passim.
[2] G. Schwarz, *op. cit.*, p. 491.

running on more than fifty routes prescribed by the public authorities; 74 million passengers used them. In 1853 the first tramways were laid down, and these were greatly extended after 1874; 1900 saw the inauguration of the Métro, and 1905 the first motor omnibus.[1]

Road transport is the most widespread and varied form, for in addition to the innumerable two-wheeled and four-wheeled vehicles there are the tramcars and omnibuses, and in Far Eastern towns the famous rickshaws, formerly propelled by men on foot but now largely replaced by bicycle-drawn carriages or 'pedicabs'. Together, all these vehicles contribute to the intensity of urban traffic. The larger cities have sought to relieve congestion on the surface by the construction of underground railways, and indeed this seems to be the only means of rapid movement and of avoiding the road bottlenecks.

In New York the 'subway' carries three-quarters of all the passengers who use any form of public transport; its lines are underground only in the central part of the city and are elevated above ground level outside the central area; they radiate from the twin business quarters of Wall Street and Fifth Avenue/42nd Street, and provide express trains that only stop at every third or fourth station and average 25 m.p.h. As in Paris, there is one fare only, whatever the distance travelled, but the system differs from that of the French capital by not providing much in the way of interchange facilities between the various lines. Several other United States cities have their 'metropolitan' railways; the most important is that of Chicago, but it carries only 20 per cent of the total traffic of the public transport system. There is however in Chicago an underground goods railway also. The London underground system was the first in the world; its fare system is related to the length of the journey; there are many lines, with many branches extending into the outer suburbs, but it only carries 15 per cent of all the public passenger traffic. The Moscow underground is noted for the magnificence of its stations; it has only three radial lines and one circular, but it carries a very heavy traffic. The part played by the Paris Métro is very important indeed; in 1938, surface and underground transport were almost equal, each carrying about 750 million passengers a year. During the war, the Métro broke all records, reaching 1,508 million passengers in 1945 when road transport, suffering from petrol shortages, carried only 201 million. In 1957 the Métro carried 1,124 million, and road transport 897 million. With the increasing congestion on the surface, the pressure on the Métro is increasing.[2]

[1] Guignot, 'Naissance et développement des transports en commun, 1660–1960', in *Paris, 1960*, pp. 171–80.
[2] J. Benedetti, 'Le Métropolitain et les transports parisiens', *La conjoncture économique dans le départment de la Seine*, 1958.

Metropolitan railways in the World's Great Cities[1]

Date of opening	City	Number of lines	Length in miles Tunnel	Length in miles Total	Number of stations	Passengers carried (millions)	Passengers per mile of line (Annual: millions)
1904	New York	38	132	236	507	1362	6·0
1900	Paris	15	99	117	363	1153	9·9
1890	London	6	89	225	230	676	3·0
1935	Moscow	4	31	37	41	730	19·5
1919	Madrid	5	17	17	45	392	22·7
1914	Buenos Aires	5	19	19	53	340	18·3
1902	Berlin	10	37	47	102	267	5·6
1908	Philadelphia	4	12	28	47	146	6·5
1912	Hamburg	5	7	71	60	158	2·2
1943	Chicago	4	17	28	65	116	
1942	Osaka	1	7	7	11	130	19·0
1927	Tokyo	2	12	14	28	167	9·8
1923	Barcelona	3	8	9	28	96	10·5
1902	Boston	8	9	22	33	90	4·0
1933	Stockholm	3	4	19	33	52	2·4
1954	Cleveland	1	0	13	12		
1952	Toronto	1	3	4	12	68	14·7
1950	Rome	1	4	7	11	11	1·5
1928	Oslo	3	1	16	52	20	1·2
1932	Sydney	1	2	5	6	102	2·0
1897	Glasgow	1	6	6	15	33	5·1
1930	Athens	1	1	15	16	52	4·9

[1] The figures relate to the year 1957. Since then an underground line has been opened in Lisbon, and one is planned for Rio de Janeiro.

Other towns, that are crossed or served by waterways, may adopt somewhat unusual forms of urban transport. The Venetian gondolas are famous, but they are supplemented today by motor-driven boats or water-buses. At Istanbul the number of boats arriving at the maritime station makes it look like Saint-Lazare or Waterloo, for the commuters come from all the little settlements on both European and Asiatic shores of the Bosphorus, and return in the evening after work.

For suburban transport there is much more competition between public and private means, but the railway is a considerable rival to the road. Road transport depends largely on the number of vehicles available and the distances involved; and the competition between road and rail creates many problems, with important financial repercussions.

In the United States, the private car seems to be winning the race. At Boston in 1956, of more than half a million daily passengers, 57 per cent travelled by private car and 34 per cent in private or public omnibuses, making 91 per cent by road and only 9 per cent by rail. Around New York, 350,000 commuters used the railway in 1930 but only 208,000 in 1960. This craze for motor transport is the cause of an enormous squandering of resources; it requires 140 buses to provide the same carrying capacity as a normal train. The cities of New York and Philadelphia look ruefully at their budgets, for it would cost seven times as much at Philadelphia and sixteen times as much at New York to develop a system of motorways adequate to carry the ever-increasing traffic, as to increase the capacity of the railway lines. And the railways actually have to be subsidised to keep them running at all, whilst a considerable part of the annual budget is devoted to the constructions and maintenance of the motorway system (33 per cent in the Connecticut budget of 1957).[1] Moreover, the rail network around New York is much denser and of more practical value than around London or Paris.[2]

In Europe conditions are rather different. Within Paris, 55·9 per cent of the daily commuters use public transport, 7·6 per cent private cars, 7·6 per cent bicycles, 1·7 per cent motor bicycles or scooters, and 27·2 per cent walk to work. In the suburbs the proportions are slightly different, with a smaller proportion of public transport (50·7 per cent) and a much higher proportion of bicycles (21·5 per cent). For the whole of France the figures are more than 10 million daily commuters, of whom 22·5 per cent used public transport, 32·1 per cent bicycles, and 37·5 per cent their own feet; obviously it all depends on the distances involved.[3]

A study of commuting in Amsterdam showed how important is the question of distance,[4] though of course one must bear in mind the extraordinary importance of the bicycle in the Netherlands. Taking a series of concentric zones around the city, at distances of 7, 11, 15 and 19 miles from the centre, it was estimated that within a radius of 7 miles, half the commuters used bicycles and 16·5 per cent the railway; in the second zone, from 7 to 15 miles out, tramways took pride of place, whilst beyond 19 miles all commuting was by train. In the whole daily movement, private cars represented no more than 7 per cent of the traffic; this figure is in line with that for Paris and

---

[1] J. Gottman, *Megalopolis*, New York, 1961, pp. 631 ff.
[2] D. Neft, 'Some aspects of rail commuting New York, London and Paris', *Geog. Rev.*, 1959, pp. 151–63.
[3] M. Brichler, 'Les migrations alternantes dans la région parisienne', in *Paris 1960*, pp. 151–70.
[4] R. E. Dickinson, 'The geography of commuting', *Geog. Rev.*, 1957, pp. 521–38.

somewhat higher than the average for the whole of France, which is 4·5 per cent.

Considerable changes have recently been in evidence, however. Around Milan, Italy's major commuting city, 51·6 per cent of the passengers travelled by train in 1958, 18·3 per cent by tramways, 23·3 per cent by buses and 6·6 per cent by individual means. By 1960 these figures had become 40·2 per cent by railway, 12·4 per cent by tram, and 19·6 per cent by bus, showing an overall reduction in the use of public transport, whilst individual transport had risen to 27·6 per cent (11·8 per cent cars and 15·8 per cent two-wheeled vehicles).[1]

The Russians are very concerned about the problem of urban transport. It is considered that two hours is the maximum that should be spent on the daily journey to and from work; and much effort has been put into the construction of mass transport media, both within the great cities and between them and the smaller towns with which they are linked. The zone of satellite towns around Kiev, for example, is closely connected by a system of radiating railway lines. Suburban electric railways exist round all the large conurbations. Sometimes preference has been given to omnibus routes as between Chakhty and the neighbouring mining towns, or as around Novochakhtinsk. The Caucasian spas are served by an electric railway, the bathing beaches of Sotchi-Khosta by local railways, buses and even speedboats.[2]

The transport of larger and larger numbers of people from further and further out presents towns with grave problems, but chiefly those that arise from the rush hours. These peak periods of traffic flow oblige the transport companies to put into service a large number of vehicles that are only in use for four or five hours a day, and cause congestion on the roads and railways, particularly in the centre of towns, as we have already seen. At Milan, between 8 a.m. and 8 p.m. 330,000 vehicles cross the interior of the Cerchia dei Navigli; tramways which with their fixed track add to the difficulties, buses and private vehicles jostle and mingle in endless bottlenecks. Entry into the city in the morning, and the return when the factories and offices close, can only be made at the expense of long delays. Indeed, for millions of men and women the world over, the twice-daily queue has become just another fact of life. In Rio de Janeiro one may sometimes wait an hour or two for a bus, and in Bahia, where economic activity is on a much smaller scale, the bus station alongside the cathedral is an incredible sight between 6 p.m. and 7 p.m. And Paris and London are in some danger of facing a complete standstill.

[1] E. Dalmasso, 'Problèmes de transport dans la région milanaise', *Méditerranée*, 1963, pp. 39–51.
[2] *L'Union soviétique*, pp. 122–6; V. G. Davidovitch, *Voproci Geografii*, vol. 38.

In these circumstances, the fatigue of long journeys, with all the waiting about, the bustle, the crowds, the 'standing room only', can be very exhausting, especially for women. In Paris, the average journey time for each individual is one hour and twenty minutes, but it may sometimes reach two or two and a half hours, that is five hours a day. In Milan, many workers leave their homes at 5 a.m. and do not return until 10 or 11 p.m.; and those who commute between the coalfield towns or places in Belgium and the textile towns of Lille, Roubaix and Tourcoing keep similar hours. There are even some miners in Germany who do a journey of sixty miles on crowded roads, and some of the workers in the Liège region travel from the Belgian coastal area.[1] Such movements are affecting more and more people; around Milan, commuting traffic amounts to 140 million passengers a year by railway and tramway, in Naples, to 103 million, in Rome, to 48·4 million, in Turin to 14·8 million.[2] In the Paris region more than 2·2 million people must leave the *commune* where they live in order to get work. In the U.S.S.R. journeys between the centres of large towns and their suburbs number 3,000 million a year, and the average distance travelled is six to seven miles by bus, twelve to fifteen miles by train; the loss of time is evaluated at 1,000 to 1,500 million hours for the actual travel or 2,000 to 3,000 million hours if one counts waiting time as well.[3]

Finally, the expense of commuting is considerable. In France, an investigation showed that the method of transport used was related not merely to distance but also to social class; only the upper classes used motor cars, whilst bicycles were most used by wage-earners and to a smaller degree by salaried employees and the middle classes. These last two groups were the chief users of public transport; the working classes rather less so. In all some 36 per cent of Parisians used 'free' transport (walking or bicycle), compared with an average for the whole of France of 72 per cent. In Paris, nearly two-thirds of the working population must budget for travelling expenses. This can be burdensome, and every means is sought of reducing its effect, by granting travel allowances, and by subsidising public transport so that fares do not rise in proportion to other increases in the cost of living; and by issuing weekly tickets or books of tickets at reduced prices. In 1900 the journey to work cost a labourer the amount that he could earn in twenty-one minutes, whilst the ordinary fare would have cost the equivalent of thirty-one minutes' work. In 1960 a workman's ticket cost him the equivalent of only six minutes of his

[1] J. Schieffer, *Marché du travail européen*, Paris, 1961, pp. 40 ff.
[2] E. Dalmasso, *loc. cit.*
[3] O. K. Kudryavtsev, 'Passarjirskie cviazi gorodov-Sputnikov', in *Goroda Sputniki*, Moscow, 1961, pp. 150–66.

labour, and an ordinary ticket nine minutes. If one compares the cost of fares with the wages paid to the employees of the public transport companies, one finds that during this same period the fares have been reduced in the proportion of $3\frac{1}{2}$ to 1. It is thus the public at large who defray a part of the cost of urban travel.[1]

SUPPLIES

With their large populations, towns are concentrated centres of consumption; they need a constant flow of supplies of all kinds—foodstuffs, raw materials for industry, consumer goods, and fuels or other means to produce heat, light, power and the means of locomotion by road and rail. It is wellnigh impossible to evaluate precisely the quantity and value of these various necessities or the means employed to obtain them and distribute them within the urban area. However, some general remarks may be of interest, and some provisional conclusions can be drawn.

Every town needs light and heat; the supply of energy is thus one of the major requirements. The needs are out of all proportion to those of the surrounding countryside, and this is especially so in less well developed countries, in which the rural areas are devoid of electricity. A few examples, from recent years, are given below:

Percentage of houses with Gas and Electricity

Country	Gas Urban	Gas Rural	Electricity Urban	Electricity Rural
West Germany (1956)	69·7	22·6	88·0	86·6
Austria (1951)	55·4	0·5	97·2	82·7
Spain (1950)	—	—	86·4	73·8
France (1954)	77·6	49·8	95·4	89·5
Greece (1951)	—	—	53·2	2·9
Hungary (1960)	—	—	92·6	62·9
Sweden (1945)	—	—	99·6	86·2
Canada (1951)	39·4	—	99·3	65·9
U.S.A. (1960)	31·4	3·8	98·0	95·0
Brazil (1950)	65·7	14·3	60·0	3·6
Colombia (1951)	—	—	63·5	4·2

Source: *Statistical Yearbook*, United Nations, 1962, pp. 611 ff.

[1] *Paris 1960*, pp. 151–69; M. Brichler, *loc. cit.*

A French geographer has suggested delimiting the Paris region on a basis of the consumption of electricity.[1] The conurbation, which contains 17 per cent of the French population, consumes 18 per cent of the electricity, 19 per cent of the coal, 21 per cent of the petroleum products and 43 per cent of the domestic and industrial gas used in France, or 20 per cent of the total French consumption of fuel and power.[2] In Uruguay, Montevideo, with 29 per cent of the population, consumes more than one-third of the electricity.[3]

A large proportion of the fuel that produces the energy is heavy and bulky, and often moves by water, especially by sea, as in the case of coal and oil. But coal may also provide railway traffic, whilst oil moves more and more by pipeline (called 'oléoduc' in France). Nearly one third of Paris's coal supply comes by water and the rest by rail; of the oil supply, one half comes by pipeline from the lower Seine and the rest by barge or by road tanker. Chicago gets most of its coal by rail, especially from the Illinois coalfield, but Milwaukee is served mostly by Appalachian coal brought in the great 'lakers'.

Gas has generally been produced at the place of consumption, using coal as its raw material. The first installations go back to the early nineteenth century in Europe (London, 1807, Berlin, 1825), and they were set up alongside waterways. Later there was concentration on a smaller number of larger production units having a much greater efficiency. Furthermore, gas obtained from the coke ovens and blast furnaces of the great metallurgical districts is now much used, as also is natural gas; in both these cases pipelines are used to carry the gas to the consuming centres which may be far away from the mining and industrial areas.

The gas supply of the Paris region comes from Lorraine, from the northern coalfield, and from the natural gas field of Lacq, in south-western France. Most of the great cities of the United States in the central plains and in the north-east are supplied from the natural gas fields in the centre of the continent and in Texas, by long underground pipelines which have a total length of 250,000 miles; one pipeline links Texas with Los Angeles.[4] In the U.S.S.R. the situation is similar; Moscow has received natural gas from Saratov, 524 miles away, since 1947, and Kiev gets its gas from Dachava, over a distance of 319 miles. A pipeline of 800 miles is being constructed from Stavropol to Moscow, which will be the most important line in Eurasia and will give Moscow a supply of natural gas twenty times

---

[1] Kirche, 'Les limites de l'agglomération parisienne d'après la consommation d'électricité, *Bull. Ass. Geog. Fr.*, 1949, pp. 138–42.
[2] J. Bastie, *Paris en l'an 2000*, Paris, 1963.
[3] G. Schwarz, *Allgemeine Siedlungsgeographie*, Berlin, 1962, p. 483.
[4] R. M. Highsmith, *Conservation in the United States*, Chicago, 1962.

as great as it gets now. The Stavropol gas already serves Rostov-on-Don and it will later be piped to many other towns. In England natural gas from the Sahara, brought by tanker to a new installation on Thames-side below London, is now used in the London area and is also piped to the Midlands; and the recent discovery of gas under the North Sea may well lead to a whole new network of pipelines.

Electricity is also, like gas, often produced near the great urban centres of consumption, in thermal power stations. Four-fifths of the Paris consumption is supplied by the series of large power stations located on the banks of the Seine and the Oise, which use waterborne coal. But electric current is also transmitted to the towns by cable from coalfields or other thermal sources or from hydroelectric installations. In the U.S.S.R. three great systems are interconnected—the Urals, the south and the central region—and these will soon be extended by another great system incorporating the new hydro-stations at Kuibishev and Volgagrad. Further, it is intended to link all these production centres to the towns of central Siberia. In California the main thermal stations using natural gas are located in the large cities, but in the north-east of the U.S.A. the thermal stations are at the ports where the coal is brought in. It was as long ago as 1925 that a 'super-power zone' was created in the New England–New York region, in which thermal and hydro sources were linked by a grid serving the towns and cities of Megalopolis.

In tropical regions, which are generally much less developed, electricity consumption is quite small. In the mainly non-industrial cities of north-eastern Brazil, the development of hydroelectricity at the Paulo Affonso falls on the São-Francisco river has considerably improved the amenities. In the case of a great industrial city like São Paulo, the electricity demand multiplied eightfold between 1927 and 1952, and from 1946 onwards there was something of a crisis, with severe restrictions on the consumption of current for several years. Until 1954 the current was almost all generated by water power, but them the thermal station at Piratinga came into use; since that time the resources of the São Paulo basin have been developed, and the electricity grid has been connected to that of Rio.[1] In many small towns in the interior of Brazil, electricity is only available from small diesel generators which almost shut down during the night.

When there are rivers that can easily be harnessed, as in central Africa, the towns can have the benefit of electricity supplies; the station at Edéa on the Sangha feeds Duala, one on the Djoué serves Brazzaville, and others on the Sangha and the Zongo serve Léopoldville.

[1] M. S. Radesca, 'O problema da energia eletrita', in *A cidade do São Paulo*, pp. 99–120.

The last of these towns consumed 20 million KWh in 1945, and 100 million in 1956, of which 78 per cent was used in industry, 20 per cent in European homes and 0·5 per cent by the Africans.[1]

Fuel and power supply thus represents a particularly important aspect of the difficulties created by urban concentration. This becomes very evident if a small aberration in the natural conditions should upset what is often a very delicate equilibrium. Thus during the very severe winter of 1962-3, many French towns ran out of coal, and the railways were unable to cope with a sudden exaggerated demand; England was similarly affected. The situation is much more serious if a deficiency simultaneously affects industry, the public services and domestic consumption. In Paris, one-third of the coal supply goes to domestic consumers, for whom it represents one-third of their fuel supplies. In São Paulo, one quarter of the electricity is used by private consumers.

The problem of fuel storage is one of increasing gravity. In the case of Paris, underground gas storage has been developed near Beynes, using a domed sand formation that is capped by clay; this reservoir can store more than a month's supply. Coal is stock-piled along miles of waterside quays and in other large yards, whilst petroleum needs huge tank-farms.

Another crucial point is the price of the various sources of energy, the raw materials of which have often been transported over long distances. Competitive prices are an important consideration for large industrial consumers, and may influence their choice of power supply. In the United States electric current in the region of the Tennessee Valley Authority costs two and a half times less than current sold in New York. Natural gas reaching the Texan coast at Galveston is roughly the same price as that which is piped to Chicago, but it is only two-thirds the price of gas in New York and three and a half times less expensive than gas at Boston. The price of coal from the Pittsburgh field is 30 per cent lower at Cleveland than at Chicago, and naturally enough the price of petrol is lower at towns along the Gulf coast than it is in the industrial cities of the north-east and the Great Lakes region. New York comes out quite well in this price comparison, by reason of its excellent water transport facilities and its good system of pipelines.[2]

It is much more difficult to gauge the importance of other kinds of supplies for urban areas. So much depends on the size of the agglomeration, as well as on the nature of its functions and the purchasing power of its inhabitants. A large manufacturing centre will have need

[1] J. Denis, *op. cit.*, p. 330.
[2] J. B. Kenyon, *Industrial Localization and Metropolitan Growth*, Chicago, 1960, pp. 99 ff.

of large quantities of raw material which may either be locally produced or imported from afar. Briey, Longwy and Scunthorpe make steel from local iron ore; Dunkerque, Port Talbot and Sparrows Point get their iron ore from many sources abroad, some of them thousands of miles away. Saint-Jean-de-Maurienne makes aluminium with bauxite from the French Midi, Portland and Spokane use bauxite from South America and the West Indies, reduced to alumina at places along the Gulf of Mexico and then transported by rail to the Pacific North-West; the smelted metal is then sent to the industrial belt of the north-east. Ports on major trade routes are exceptionally well favoured for raw material imports, and this is why so much heavy industry is concentrated in and around ports and along the great navigable rivers.

Food industries also absorb a large quantity of produce which must arrive in a perfectly fresh condition; such industries are therefore widely scattered through agricultural regions. In the United States, food industries give employment to one and a half million people, as many as are employed in the vehicle-building industry—but they are dispersed in 42,000 factories, whereas there are only 5,000 factories in their rival industry. A city such as Baltimore has an important canning industry that receives the produce of quite a large region, particularly for tomatoes and peas; the producers have regular contracts to supply the city factories.

The concentration of agricultural produce on food preparation factories is a process analogous to that by which a town assembles the food supplies for its population. The nature of the urban food supply problem will vary with the size of the town and the general economic development of the country in which it lies. Many small towns are essentially rural markets, and a large part of their daily needs may be covered by local production. This is evidenced in the existence of weekly or twice-weekly markets, to which the local cultivators and horticulturists bring their produce; the farmers have their daily milk round, and the local forest yields wood; at the end of the summer apples and other fruits are bought in large quantities and put into store in the cellars and store rooms of the large country houses; the housewives make jam and other preserves, and the whole town is self-supporting save for exotic commodities. These traditions are still firmly entrenched in the European countryside, and even many large towns still retain the close and direct contact between local producers and consumers through the market, the occasional fairs that attract the housewives, and the daily milk supplies. In large cities like Lille, Le Mans, Nice and many others, local producers always set up their stalls in the markets, where the professional salesmen will sell both the local produce that he has collected himself and other foodstuffs that have come from the four corners of the country or even of the world. At Vichy, in the gigantic covered market, the retailers occupy the ground floor and the local peasants the galleries on the first floor. At Lille the local producers' milk rounds still contribute an appreciable proportion of the total consumption.

Even in the case of very large towns, local production from the urban periphery is still important, as the profusion of market gardens, orchards, and glasshouses well shows, together with the intensification of dairying, even sometimes in apparent defiance of the natural environment. Thus the demand for milk on the Côte d'Azur has caused a revival of cattle-raising in the immediate hinterland, which is obviously more suitable for sheep; around Paris and Lille, stall-fed cows give high milk yields, though the unnatural conditions wear out the beasts very quickly; one-third of the milk supply of Leeds was until fairly recently produced within the city limits under similar conditions. Around London and Brussels, the glint of the glasshouses, seen from the air, is an indication of the intensity of the demand for tomatoes and other fresh and early crops.

The great cities of the American Megalopolis are surrounded by counties in which the proportion of land cultivated is quite small but the value of its produce is extremely high. Five per cent of the total value of agricultural production in the United States is here obtained from only 1·8 per cent of the total cultivated area. Each large city has its controlled area reserved for milk production; market gardening ('truck-farming') takes place right up to the urban fringe, and several crops a year are obtained from small but intensively cultivated areas. Since 1920 the progress of vegetable cultivation has been continuous, fostered by a change in dietary habits due in part to propaganda regarding vitamins; during the main season, three-quarters of the vegetables consumed in New York come from farms within a radius of 100 miles that are within easy reach of water, rail or road transport; 35 per cent of the vegetables are transported by lorry (truck). Lastly, speculation in egg and poultry production has reached such dimensions that one finds in this region the chief county in the United States for the production of ducks (Suffolk, Mass.) and for chickens (Sussex, N.J.).

Around some of the large Brazilian cities the Government has encouraged the settlement of Japanese colonists, with the object of producing fresh vegetables for the urban population. In the Far East, every town has its girdle of market gardens; 80 per cons of the fresh vegetables consumed in Tokyo, amounting to 900,000 tent a year, are derived from a zone of gardens that begins at the urban limit and is concentrated within a radius of thirty miles from the city centre;

the belt forms the most important market gardening belt in the whole of Japan.[1] In African towns, the European quarters demand their fresh vegetable supplies; sometimes these are imported at great cost, as happened at Dakar before the establishment of market gardens run by natives in the alluvial depressions not far from the town.

A large part of the local production around the towns of Latin America, Asia, Africa, and even Europe, is taken directly to the markets by the producers; the colourful costumes, the strong scent of the piles of fruit, the odour of spices, of dried fish and smoked meat leave an indelible impression on the mind of anyone who has visited these bustling markets in the tropics.

Local food production will vary with the fertility of the soil, with the nature of the climate and with the capabilities of the people; beyond such production, towns may be obliged to seek their supplies from far afield. The development of means of transport and the rising standard of living have encouraged a broadening of taste which has given rise to many cross-currents of trade in produce of all kinds.

Take the Icelandic capital, Reykjavik, for example. Apart from a limited fruit and vegetable production from glasshouses heated by natural hot springs, and local resources of animal products and fish, the city must import its supplies from abroad. And it is interesting to note that in western Europe tropical fruits and the citrus fruits of the Mediterranean are much in demand, whilst in the tropics, the temperate apple fetches high prices. There is a kind of annual rhythm in the supply of tomatoes or pears, for example, which can be found on the tables of Paris homes at all times of the year; in the case of the pear the circle of producers extends from the orchards on the hills surrounding Paris to those of the Argentine, crossing all latitudes at the appropriate moment of ripening. And such eclecticism is not only to be found in the most advanced countries: the cycle of orange supplies in Singapore offers a similar example, for the fruit comes in succession from California, Western Australia, South Africa, Cyprus, Israel and Spain.[2]

Urban populations require both quality and quantity. We may well speak of 'Paris's stomach'. The population of 8 million consumes annually nearly 250 million gallons of wine, a million tons of potatoes, 800,000 tons of green vegetables, 700,000 tons of bread, 400,000 tons of meat and 400,000 tons of milk. And though 350,000 tons of vegetables and 50,000 tons of fruit may be produced locally, a large part of the food supply must come from afar. Vegetables and

[1] J. D. Eyre, 'Sources of Tokyo's fresh food supply', *Geog. Rev.*, 1959, pp. 455–74.
[2] J. E. Spencer, 'Seasonality in the tropics: the supply of fruit to Singapore', *Geog. Rev.*, 1959, pp. 475–84.

fruit come mainly from Brittany, the lower Rhône, the Garonne valley, Rousillon, the Mediterranean coast, and northern France (mainly potatoes); milk is drawn from a vast area, and the average distance travelled is about sixty miles. Butter and cheese are sent from Normandy, Poitou and the Charente area, and from an eastern zone which stretches from Alsace to Burgundy; beef comes mainly from western and central France, mutton from Loiret, Burgundy and Champagne; fish arrives from Boulogne, La Rochelle and the fishing ports of Brittany. Many foreign countries, too, contribute supplies, such as Italy and Spain for fruit, Netherlands and Denmark for butter and cheese, Great Britain for whisky and certain other specialities; and further afield parts of north and tropical Africa, the West Indies and the rice fields of the Far East.[1] For the transport of all these commodities, there is much competition between the railway, which takes about 45 per cent of all the foodstuffs other than meat, and road services which take 65 per cent.

In towns of lesser importance the food situation depends largely on natural environment; a town like Nice, situated on the edge of an area that is not favourable for agriculture, and having a wealthy population with expensive tastes, must import from afar a large part of its supplies, and though some vegetables, fruit and milk may be produced locally, flour comes from Marseilles and Basses-Alpes, beef from Charolais, poultry from Bresse, and eggs, butter and cheese from a wide variety of sources; extra milk comes in from lower Dauphiny or the Garonne basin. In these circumstances the cost of transport is high, but at least the local climate, particularly in summer, allows a reduction in the consumption of certain items, and local vegetables are abundant and cheap.[2]

Conversely, a town such as Lille is situated in one of the most fertile countrysides in France, and though it may be necessary to supplement the local meat supply with imports from other regions, there is a local abundance of cereals, milk and dairy products and fresh vegetables. It is necessary however, to import many things that come from more favoured climates, such as early vegetables, many kinds of fruit, and exotic products which are in any case something of a luxury.

One could match these three French examples in many parts of the world. In Tokyo a typical Japanese town, rice, the staple food, is obtained from northern Hondo and the coastal plains of the Sea of Japan, four-fifths of the fresh vegetables—of which 125 varieties are grown—come from just outside the city and the remainder from

---

[1] J. Antoine and J. Morice, 'La consommation parisienne', in *Paris 1960*, pp. 233–42.

[2] R. Blanchard, *Le comté de Nice*, Paris, 1962, pp. 211 ff.

northern Hondo and Hokkaido. The consumption of fruit is increasing, and though the quantity is only one-third of the vegetables the value per unit of weight is three times as great, and this is encouraging the planting of orchards in almost all parts of the country. Within the local area, the produce is transported by every conceivable means —on human shoulders, in light carts drawn by the producer himself, by a bicycle or by a docile buffalo, and increasingly by lorry. For produce from further afield the railway is the most important means, and good rail transport is one of the factors—an assured market is another—that has encouraged the growing of vegetables and fruit in areas at some distance from the capital city. Water transport is only used for a few of the less fragile products.[1]

In the United States, the enormous market provided by the New York conurbation not only encourages local market gardening in the few scraps of agriculturally utilisable land that remain within easy reach of the city, but also draws on the resources of half a continent; Florida and California wage a desperate battle to supply fresh, preserved and frozen fruit and vegetables to the cities of Megalopolis. And New York imports from abroad not merely for its own use; it also acts as a redistribution centre for such particularly expensive commodities as caviar and French champagne.

In India the large cities must look far afield for the wherewithal to feed and employ their people; Madras draws its food supplies from all over India, and the raw materials of its industry from abroad as well. The railway continues to play a dominant rôle. Local distribution, whether to Hindus, Muslims or Europeans, is effected through the 'bazaars'.[2]

Once the huge quantities of foodstuffs have arrived in the city, they must be redistributed. It is estimated that in Paris some 1·4 million tons of produce a year pass through Les Halles (the central markets) and 300,000 tons is marketed by other means. Sales to the actual consumers are effected through a host of small shops, in permanent covered markets (St Germain and St Honoré for example), by itinerant salesmen in the periodic markets such as that in the boulevard Port-Royal, and by deliveries from the large shops and chain stores. In Tokyo an organisation for the distribution of foodstuff was brought into being after the great earthquake of 1923, in order to ensure the freshness of the supplies and to hinder price speculation. There is a large market at Tsukiji on the quayside and six subsidiary markets distributed through the conurbation on which the trading is centralised. These municipal markets handle two-thirds of the fresh produce; there are also forty-three small private markets, authorised

[1] J. D. Eyre, *loc. cit.*
[2] J. Dupuis, *Madras*, pp. 487 ff.

since 1958, in the new outer districts of the city, and these handle 30 per cent of the distribution.[1] In New York, the wholesale markets for certain products have recently been decentralised from Manhattan and located in the peripheral boroughs; this has been done partly to avoid the high costs of transport in the city centre (which themselves arise from the traffic congestion), and partly to get more space (as in 1959 when the fruit market was moved out to Bronx).[2] In Paris there has been agitation for some time to overcome the anachronistic and irrational congestion at Les Halles by the construction of rail-served markets on the outskirts, and this is now being done.

The convergence of the food, raw materials and other supplies necessary for the life of a city requires the presence of adequate means of transport. All towns that can do so make use of water transport; this is natural enough for seaports and towns on navigable rivers, but inland ports have sometimes been created at the expense of considerable and costly engineering works. Thus Lille created a port in 1948, linked with the canal system of northern France (which has its maritime outlets at Calais and Dunkerque) and dealing annually with 1·6 million tons of traffic, a total that gives it third rank in France as an inland port, after Paris and Strasbourg.

To return to the example of Paris, its port handles 13 million tons of traffic, of which 12 million are inwards; this is eloquent testimony to the enormous appetite of the population for supplies of all kinds. In addition the railway freight stations handle 30 million tons. There is a great difference in value between the water and rail traffics, for of the port traffic 50 per cent is building materials, 20 per cent oil and 12 per cent coal. Road transport, too, plays a notable part, but its total tonnage is small; of the traffic that travels more than thirty miles, 13 per cent moves by road, 18 per cent by water and 70 per cent by rail.[3]

Chicago provides an excellent example of a city whose prosperity has been based on the excellence of its transport facilities. Second city of the United States, it produces 7·4 per cent of the national industrial output, and though dating only from 1830, it now has 10,000 factories. The raw materials of its industry comprise 28 per cent of local agricultural, forestry and mining products, and 63 per cent imported materials such as coal, ores, rubber, oil and natural gas. Its port is of great significance, handling 70 million tons of traffic a year, of which two-thirds uses the Great Lakes and 20 per cent the canal system. It is linked by pipelines to the great oil and gas fields of the central plains and the Gulf states, and it is the world's largest

[1] J. D. Eyre, *loc. cit.*
[2] J. Gottmann, *Megalopolis.*
[3] *Paris, 1960.*

railway centre, with twenty-eight radial lines converging on the terminal stations and a girdle line that carries 30 per cent of all the rail traffic. Chicago is also the most important road focus in the United States, and a large part of its deliveries is effected by lorries, 30 per cent of which enter the city empty in order to pick up loads for delivery elsewhere. More than 4,000 factories have private railway branches. The city contains 9,100 wholesale establishments and 50,000 shops scattered through seventy-five shopping districts. The roads carry an unceasing flow of lorries laden with pigs, cattle, raw materials, machinery, and at night, milk. Few of the world's cities can show a greater intensity of traffic.[1]

THE PROBLEM OF WATER

This is one of the major problems of urbanisation, and deserves separate treatment from the remainder of the supply problem. No town can exist without water, and quality is as important as quantity. We may replace coal by gas or oil, and oranges by apples, but there is no substitute for water. A town may smelt iron or spin wool, but water is necessary both for human consumption and for industrial use. It is but rarely that the water situation is exactly right; some climates are cruel, and water must be sought from afar; elsewhere there is too much water and the town must be defended against it, often without being able to use it because of its quality.

The possession of water is so important that it influences administration and even politics. Los Angeles has had to go very far for its supplies: an aqueduct constructed in 1913 brings water 146 miles from the Sierra Nevada, and the availability of this supply led neighbouring areas (like San Fernando and the suburban areas between the city and the western beaches) to request annexation by the city so that they could benefit from its water supply. The same reason—lack of water supply—led Brooklyn to ask for incorporation in the city of New York in 1898. In some areas there has been a positive battle for water, like the struggle between New Jersey and New York for the use of the Delaware river, a dispute that went to the Supreme Court in 1929; a compromise was reached, but in 1952 the State of Pennsylvania took up the cudgels and New York was forced to make further concessions.[2]

In France, Paris has encountered difficulties in its scheme for getting water from the upper Loire basin (see below, p. 336). The

---

[1] J. H. Garland, *The North American Mid-west*, New York, 1955; H. M. Mayer, *The Port of Chicago*, Chicago, 1957.
[2] A. V. Burkalow, 'The geography of New York city's water supply', *Geog. Rev.*, 1959, pp. 369–86.

Dunkirk industrial zone, in the Nord *département*, has had to get water from the *département* of Pas-de-Calais, on the other side of the Artois hills; this has created friction between the two *départements* and still the quantity is insufficient. In Great Britain, Liverpool has met much opposition in Wales to its construction of reservoirs, and Manchester's battle with the Lake District is still in progress.

Towns require the concentration of considerable quantities of water in small areas, and the demand is constantly increasing. In Paris the consumption per head of population was 5 gallons a day at the beginning of the eighteenth century, 77 gallons in 1935 and 100 gallons at the present time; the demand grows at about 4 to 6 per cent per annum, by reason of the increase of population and industry and the progress of sanitation. Consumption varies widely as between cities and suburbs; the citizen of Paris may consume his 100 gallons a day[1] but his suburban counterpart only uses 46 gallons, and a similar discrepancy exists at Lyons, where in the city the consumption is 86 gallons and in the suburbs only 44. Climate also plays a part and southern towns consume more than northern towns: thus Nice consumes 170 gallons a day per head of population, Antibes 209 gallons; the Londoner is content with 44 gallons, the Roman needs 220. In the United States, the citizen of Chicago uses 233 gallons, and of Los Angeles 264 gallons a day. But local habits also count, and the people of Rennes only use 30 gallons a day.[2] So also does the technical equipment of the water supply organisation: Bombay, one of the best provided cities in all India, has an average consumption of 40 gallons, but many towns of the Middle East are in serious difficulty and consumption is much lower, at only 9 gallons in Aleppo (in 1955), 22 at Damascus and Bagdad (in 1947), and 46 at Alexandria.[3]

The requirements vary tremendously. A normal figure per person for drinking, cooking, washing and sanitation is 10–12 gallons a day, but this quantity can easily be doubled if the domestic sanitary arrangements are good or if the house has a garden. Furthermore, the public services of a town use between 25 and 50 per cent of the total consumption, whilst losses from one cause or another may be assessed at 20 per cent; and finally, perhaps 10 to 30 per cent will go to the town's factories, the quantity depending on the nature of the industries. Industry demands more and more water. The production of a ton of cotton cloth requires the use of 7,800 cubic feet of water; if the cloth is wool it will need 20,000 cubic feet, and a ton of rayon will need 28,000. A steelworks uses 6,000 cubic feet of water for

[1] J. Beaujeu-Garnier, 'La consommation de l'eau à Paris', in *Mélanges offerts au Professeur Pardé*, Paris, 1966.
[2] R. Doucé, 'L'alimentation en eau potable', *Urbanisme*, 60, pp. 44 ff.
[3] A. R. Hamide, *La ville d'Alep*, Paris, 1959, p. 115 ff.

PROBLEMS OF SPACE 333

every ton of steel made, an aluminium producer 46,000 cubic feet per ton. Electric power stations use 100 cubic feet per second to produce 3 million KWh a day; but this is either re-used after being cooled in huge cooling towers, or if the station is alongside a large river, the water is returned to the river, slightly warmed after use—an essential procedure in the case of a river like the Seine, from which power stations take one-third of its normal daily flow.[1]

Where is all the water to come from? The ideal, in principle, would be to have two supply systems, one collecting and delivering drinking water and the other concerned with water for industrial use or road cleansing, that need not be so pure. In Paris the consumption of non-potable water is almost 60 per cent of the quantity of potable water, and a normal figure is between 50 and 60 per cent. But the installation of two systems is costly, and may be technically difficult.

The best quality water comes from springs. Paris tapped the Dhuys springs in Champagne in 1865, those of the Vanne in 1874, and those of the Avre in 1893; further sources were tapped between 1900 and 1925; but all these only provide one-third of the drinking water, and the other two-thirds are abstracted from the Seine and the Marne and filtered and chlorinated before entering the public supply system. In a similar way Vienna gets its water partly from Alpine sources and partly by abstraction from the Danube. At Hamburg, the Elbe provides only 14 per cent of the city's water, the remainder coming from underground sources; underground supplies also feed most of the towns of lower Saxony, and give Berlin 88 per cent of its water.[2] Four-fifths of London's water supply is in effect purified sewage, for it is taken from the Thames and the Lea, into which the sewage of all the towns upstream is discharged; but of course it is purified in huge open-air reservoirs before it gets to the public.

In Mediterranean countries, water supply is often difficult; the traditional means of supply are tanks and wells, as in the old town of Nice, where such methods began to prove inadequate as long ago as 1860; since then the springs of Saint-Thècle were tapped in 1864 and the water conveyed by a seven-mile aqueduct, those of Vésubie in 1885, with an eighteen-mile aqueduct which also serves other coastal towns as far as Menton. The growth of Nice, however, has necessitated the provision of still more copious supplies, and after having bored successfully in the Paillon alluvium, the first of three underground water-bearing horizons was tapped in 1929, in the old alluvia of the lower Var; here are valuable reserves that may allow for the future

[1] R. Doucé, *loc. cit.*; J. B. Keynon, *Industrial localization*, pp. 104 ff.; E. M. Rawstron, 'Power production and the river Trent', *East Midland Geog.*, 1954, 23–30.
[2] G. Schwarz, *op. cit.*, p. 486.

development not only of Nice but of other places along the Corniche coast.[1] Other Mediterranean towns are much less well-off. Aleppo, at first supplied only by wells and tanks, increased its intake by tapping, in a not very hygienic fashion, some springs about seven miles north of the town, and the problem was not really solved until it was decided to take water from the Euphrates, over fifty miles away.[2]

In the United States, surface waters supply some 2,000 towns that include about one-half of the nation's population. Of the thirty-nine towns that had more than 250,000 inhabitants in 1950, only three—Houston, Memphis and San Antonio—were supplied solely from underground sources. The bulk of these towns are to be found in New England, in or on the edge of the northern Appalachians, especially along the Fall Line, and in the basins of the Ohio, the middle Mississippi and the lower Missouri, as well as in the northern part of the great plains in Montana and the Dakotas. In most cases the use of surface water is inevitable for there are no other local sources; the rocks may be impermeable, without underground aquifers, or the underground sources may be saline or polluted.[3]

Philadelphia, Washington, Pittsburgh and St Louis use river water; Chicago and Buffalo are on lakes. As for New York, it has had to construct a whole series of dammed reservoirs in an area extending between 25 and 125 miles north-east and north-west of the city; the works were begun in 1842 and are still in progress. Water from twenty-seven reservoirs is transported through more than 340 miles of aqueducts and tunnels, and the amount now available is one-third greater than the current consumption; it is estimated that it will suffice until the end of the century. Despite this, water from underground sources on Long Island, after being used for industrial refrigeration and air conditioning, is returned to the ground so as to keep up the level which was getting dangerously low.[4]

In the countries of Africa and Asia the problem of water supply is often serious; wherever Europeans have settled, they have installed water systems that serve the European houses and either terminate at the boundary of the African quarter or end up therein as fountains. But large masses of people have no such service. In Léopoldville the purified waters of the Lukunga provide the bulk of the potable water, together with abstractions from the Congo, but many African homes have only the polluted waters of the Funa or the Ngombe.[5] In the

[1] Blanchard, *op. cit.*, p. 210 ff.
[2] A. Hamide, *op. cit.*, p. 115.
[3] J. F. Borchet, 'The surface water supply of American municipalities', in H. M. Mayer and C. F. Kohn, *op. cit.*, pp. 569–84.
[4] A. V. Burkalow, *loc. cit.*
[5] J. Denis, *op. cit.*, p. 328.

Far East, the foreigner looks on with amazement at two individuals side by side in the thick and muddy waters of a river, one relieving himself and the other cleaning his teeth. In the major Brazilian cities, the lack of water during the dry season is a dreadful curse, and even a city like Rio, which is fortunate in having a series of reservoirs behind the scarp of the Serra do Mar, has its supplies cut off for several hours a day. As for the working-class suburbs of Bahia or Recife, they have no water for weeks or months on end; whilst in the interior the people drink river water—the São Francisco for example—almost untreated except by the more careful families, and even stagnant pools are used.

Bad water is responsible for numerous ailments; Shanghai's cholera epidemics resulting from impure water supplies are well known, and many intestinal diseases are spread by polluted water.

The provision of running water varies very greatly from one country to another, but it is obviously always greater in the towns than in the countryside. In India, only 6 per cent of town dwellers have potable water, and even those who drink purified water often wash themselves and their clothes with polluted water. It has been estimated that this situation results in 2 million deaths a year, and 50 million infections.[1]

Percentage of Houses with Running Water

Country	Towns	Rural areas
Austria (1951)	44·9	21·2
Spain (1950)	58·9	13·2
France (1954)	75·4	34·3
Greece (1951)	23·0	0·6
Hungary (1960)	52·0	4·2
Sweden (1945)	88·5	39·3
United Kingdom (1951)	97·9	79·9
Canada (1951)	94·1	39·5
U.S.A. (1960)	98·9	79·0
Brazil (1950)	39·5	1·4
Colombia (1951)	62·3	5·1

Source: United Nations Yearbook of Statistics, 1962, p. 611.

An insufficiency of water supplies may be due to one of two causes: in underdeveloped countries, simply to lack of adequate works and equipment—a situation that is capable of amelioration; in more advanced countries, to a growth of population disproportionate to

[1] J. Beaujeu-Garnier, *Géographie de la population*, vol. 2, p. 382.

the available resources. In the United States, it was estimated in 1955 that 42 per cent of the large waterworks companies were confronted with a consumption equal to or greater than the average intake, and that no reserve existed in case of exceptional demand or to cater for the needs of the future. Though the situation is least grave in northern and south-eastern U.S.A. it is quite dramatic in California, where already the great cities are getting their supplies from the Sierra Nevada, well over 100 miles away, and the forecast has been made that in fact there are almost no tappable supplies left; indeed California's prodigious population growth may well be halted simply by lack of water.

The situation in Paris is also serious; consumption is already so high that the slightest summer drought can have an almost catastrophic effect. Already the higher parts of the city and some of the suburbs are deprived of water for several hours of the day during July and August. Before the last war there was a scheme for abstracting water from the Loire, but it ran into opposition from the riparian authorities and would have been very expensive. A more recent plan, which also arises from the need to control the Seine floods, envisages the construction of a series of dams on the upper Seine, the upper Marne and some of their tributaries from the Langres plateau; this will be even more costly than the previous scheme, but it has the additional advantage of regularising the flow of the Paris Basin rivers. The works have been begun, and the additional water supplies will enable Paris to cope with the existing demand. But what happens when the city's population rises to 12 or 15 million, as is thought probable? The water engineers are very pessimistic.[1]

The great industrial region of northern France is also almost strangled, not only by limitations on the considerable supplies of pure water but also by the problem of effluent evacuation. Marseilles gets water for its factories from the river Durance, sixty miles away, and the establishment of an industrial zone at Rennes caused the city to reorganise its water supplies. In England, Birmingham gets its water from central Wales, Liverpool from North Wales, and Manchester from the Lake District.[2] In London there have been many appeals to the public spirit of the citizens, encouraging them to reduce their consumption of the precious liquid—and whenever there is the slightest suggestion of a dry spell, so that people want to water their gardens, they are forbidden to do so. In the case of the oil-towns in the deserts, they are not all as lucky as the Saharan centres that have discovered 'fossil' water in their oil bores (though how long will these

[1] J. Bastie, *Paris en l'an 2000*, Paris, 1964.
[2] L. D. Stamp and S. H. Beaver, *The British Isles*, 5th edn., London, 1963, pp. 96–101.

24 Ghardaia: market place

25 Moulay Idris: ancient Moroccan town

26 Istanbul: old houses and the Blue Mosque

27 Cairo: citadel, Mehemet mosque and south-eastern quarter seen from the minaret of Ibn Tulun mosque

supplies last?), and the towns of the Persian Gulf oilfields have to get water from afar, so that indeed water costs more than petrol.

There is one final aspect of the water problem. The water, after use, does not just disappear into the atmosphere; the polluted water must be evacuated, and this may be just as difficult a problem as the getting of it.

DISPOSAL OF WASTE

Towns absorb a great deal, but they also transform and reject a great deal, and getting rid of waste, whether it be sewage, organic matter or any other kind of debris, creates many problems.

Much of the water that is used, particularly by industry for cooling purposes, is not polluted and does not enter the public sewage system. Some of it is recirculated several times, and some is pumped back underground; much simply returns to the river from which it was taken, as in the case of electric power stations.

Effluent that contains chemical impurities is more difficult to deal with. Despite stringent regulations, it is often disposed of untreated, and bitter controversies often result from such action. People who live along the banks of Belgian rivers complain that they are polluted by the effluent from the industrial region of northern France; industries situated on the shores of the Great Lakes put their effluent into the nearest available water, and as a result fish die, the shores are covered with slimy black mud, and Longfellow's 'shining big-sea water' is but the poetic image of a bygone age. Disposal of effluent is every bit as great a problem in industrial areas as obtaining a fresh supply.

The greater part of the polluted water, however, enters the public sewage system. In theory, in well-equipped towns, it is led into sewers of large dimensions and so to sedimentation tanks or purification works. But a complete system of sewers is far from being the general rule. In Paris about 97 per cent of the buildings are linked to the sewage system, which has a length of over 1,200 miles; in Berlin there were 4,100 miles of sewers in 1938, but their discharge was little more than half that of the Paris system. When one moves to smaller towns the figures are less impressive; of the towns in Lower Saxony only 15 per cent have complete sewage systems and only 30 per cent of all the houses are sewered. Sewers do not necessarily follow as suburbs extend. Many French provincial towns are only sewered in their central sectors, and both old and new suburbs have to rely on cesspits of greater or less antiquity.

The size, technical difficulty and cost of the necessary works are such that many developing towns do not undertake them, and even in rebuilt old towns the situation is far from satisfactory. Take Aleppo as an example: in the old city, underground galleries formerly used by the defenders of the citadel as a means of egress to the countryside, have long been used as sewers, but they became clogged up and were sources of infection; another system was constructed in 1901, in the form of ditches, only just beneath ground level and covered by loose flagstones, that spilled their contents over the fields and gardens around the city. It was not until 1921 that a modern system began to be constructed, which now has a length of fifty miles and has its outfall in the river below the city. In comparison with many others, this is an encouraging example.

Around many tropical towns, great black vultures wheel in the sky, waiting to swoop down on the rubbish tips; they are just a part of the urban scene. There are the urubu vultures of north-eastern Brazil, the griffon vultures of Africa and India; in Bombay Parsee corpses are thrown to the vultures in the mysterious towers of silence.

The sewage system of great cities generally leads to special treatment works. The Paris system, which also collects storm-waters (about 10 per cent of the total) converges on Clichy; here, after preliminary treatment, about three-fifths of the effluent is discharged into the Seine; of the remainder, 60 per cent goes through the sewage farms in the valley below Paris and is used to irrigate market gardens, and 40 per cent goes to the purification works at Achères and is then discharged into the Seine. The purification process is so efficient that the effluent from Achères has been passed as drinkable. Nevertheless, what with treated sewage and industrial effluent, the riverside dwellers below Paris are always complaining that on hot summer days the river literally stinks.

In Japan industrial pollution of irrigation water on the outskirts of some towns has begun to harm the rice fields.

Solid refuse has quite another destination. In some modern blocks of flats, pulverisers reduce its bulk and enable it to be flushed away by water, but this is a novelty that is not as yet widespread, at least in Europe. The bulk of it finds its way into the dustbins or garbage cans that are emptied daily or at less frequent intervals, their contents being carted away by more or less specialised vehicles. Only thirty years ago, in many French provincial towns, concessions were granted to certain individuals to collect household refuse; they went round with a horse and cart, emptied the bins and made what they could out of the rubbish collected, selling it for fertiliser or to the rag-and-bone merchants. In many towns of the Middle East and Asia, this kind of system still prevails. At Aleppo, a large part of the rubbish is sold as organic manure to the cotton growers and market gardeners beyond the urban fringe. Too often, however, the dust carts simply discharge their contents on to crude and naked tips at some distance from the town, on the edge of a wood, or in a hollow, or down a slope, generally in places that would be quite pleasant but for these disgusting heaps.

Large cities, however, have adopted other methods: some of the rubbish is burnt, and the heat may in some cases be used in a district heating scheme, as happened in some German towns before the war. In Paris such a system has been in operation since 1927, and thirty-three miles of pipes now convey hot water to certain central and eastern

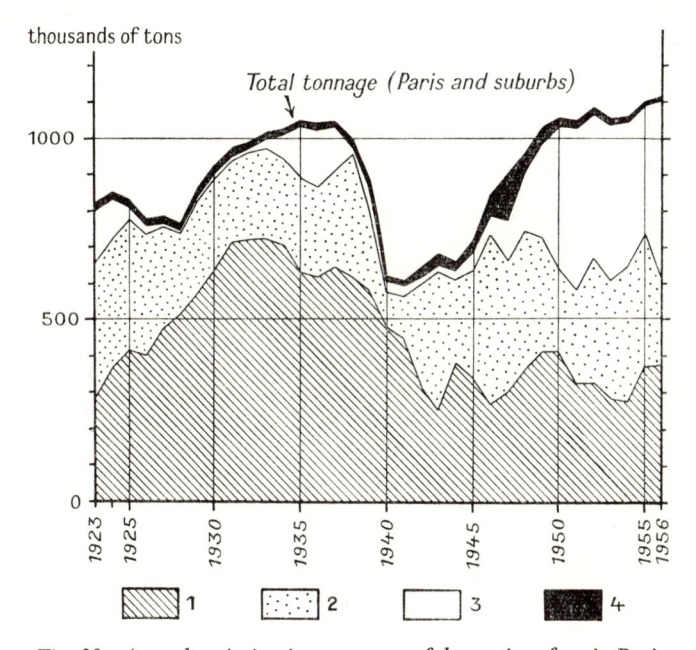

Fig. 30. Annual variation in treatment of domestic refuse in Paris

1. Incineration. **2.** Controlled tipping. **3.** Spread on agricultural land. **4.** Salvage and loss.

districts of the city. In the U.S.S.R. and eastern Europe the system is in regular use; Warsaw has two large central-heating plants where coal and domestic refuse are burnt to provide hot water for a large part of the city.

Another method is to pulverise the refuse and use it to spread over cultivated ground; this too is done in Paris—but the local farmers are not keen on this form of 'manna', and most of it goes in covered lorries by night to the field of Champagne, and even here the recipients recoil from the smell and uninviting appearance of their land after the stuff has been spread. However, an ever greater problem is provided by glass bottles, that are thrown away in countless numbers, unburnable and impossible to pulverise. They just have to be dumped, and around Paris many old limestone quarries are being thus filled in.

All told, the Paris conurbation produces 1·2 million tons of domestic refuse a year; about two-fifths of this is dumped in heaps, just over a third is pulverised and used as fertiliser, and just over a quarter is burnt to provide energy and central heating.

Lastly, we must not forget the debris resulting from the demolition that is constantly taking place within towns. The town that is really alive is the scene of continual destruction and rebuilding. Part of the material may be used again, but the bulk of it must be taken away and dumped. The honourable company of rag-and-bone men contribute in their own way to the clearance of urban rubbish, but builders' rubble is another matter. Much of the 'export' from the port of Paris is material from demolitions being taken away in barges. Around London, worked-out gravel pits, which themselves have provided concrete aggregate for new buildings, became the receptacles for the millions of tons of rubble created by the *blitzkrieg*. Seaboard towns have the double possibility of dumping their rubbish in the sea and at the same time extending the land surface; this is what has happened at Rio de Janeiro, where the coastal boulevard is on reclaimed land, and the very heart of Noumea, in New Caledonia, is built on a filled-in marsh; the waste from the nickel smelter at Doniambo has created a terrace on which factory extensions have been built and which may in future be used to extend the local airfield runway. For inland towns, however, where water supply is limited and where space is rare and expensive, it is difficult to keep cleanliness, health and amenity in step all the time.

URBAN SATURATION

One would be tempted to think that the term 'urban saturation' could only apply to great metropolitan cities. It is true that small towns are less prone to think of themselves in this light, for smaller populations pose fewer problems, and moreover since they are often situated in the heart of open country, their citizens can find the freedom and verdure of the countryside only a few paces or a few streets from their homes. Their problems largely stem from their limited financial resources; it may be difficult to construct a complete sewage system for a town of half a million inhabitants, but it is no less difficult to provide a water supply for a small town of 10,000 people who have but meagre financial and technical means at their disposal.

The growth of several small mining towns in Lorraine is actually paralysed by the impossibility of increasing the water supply. Thionville, for example, with 45,000 inhabitants, needs 700,000 cubic feet of water a day, but its supply consists of 70,000 cubic feet from springs, about 200,000 cubic feet from bores and headings in the Moselle alluvium and 140,000 cubic feet pumped from the neighbouring iron mines, making a total of only 410,000 cubic feet; the deficit must be made up by direct pumping from the river Moselle. The towns of Metz and Nancy are in a similar situation, and are reduced to abstracting and purifying Moselle water to satisfy domestic consumers. To an increasing extent towns are finding that they have no alternative but to use river water that is always more or less polluted and of which the flow is limited.

We may conclude, therefore, that at all levels of size there is a saturation limit for towns, and that this limit depends on natural conditions and financial circumstances. The latter may be ameliorated, but it is difficult to cope with the deficiencies of nature.

What forms does this saturation take? First, obviously, is overcrowding; once all the space within the city limits is built up, the city cannot expand save by pulling down and rebuilding or seeking more land beyond its boundaries. Both these solutions have been envisaged by the planners of Moscow, for example.[1]

Overcrowding is detrimental to the physical and mental health of individuals. It comes about through the sacrifice of even the smallest open space, and through crowding families into veritable rabbit-warrens; by reducing the size of rooms and the height of ceilings, the width of passages and of staircases. It is not the density by unit of area that matters, or even the density per unit of floor space or per room, but the density per volume of living space. Congestion has long been complained of in Paris, but the deficiency of open space is likely to grow rather than to diminish.

Open space (square feet) per inhabitant

Paris	15·0
Lyons	38·7
Lille	70·0
London	96·9
Berlin	140·0
Vienna	269·1
Washington	538·1

[1] G. C. Mischenko, 'Goroda i pocelki sputniki Moskvi', in *Goroda Sputniki*, pp. 40–90.

The lowly position of Paris in this list is emphasised further by noting that there is one square metre (10·7 square feet) of recreation ground for every fifteen children, and the deficiency in public and private open space and sports grounds is already 75 per cent.[1]

The second solution, which entails seeking building land outside the city limits, is liable to provoke an extravagant sprawl. The Californian towns offer the classic example: land for new houses is sought beyond all previous urban settlement in order to get it cheaply; but from the moment the new development is launched, it must be linked with the existing agglomeration; prices rise and the next comers must go a little further out still. This continual leap-frogging produces chaotic, 'atomised' towns, in which the problems of transport and public services are insoluble save at very considerable expense.[2]

Transport difficulties weigh heavily on the future of towns. In order to permit traffic to flow within the centre as well as to converge on it from all directions, one is reduced to the absurdity of destroying the town in order to allow it to grow. Streets must be widened, motorways constructed—and houses destroyed. Pavements must be narrowed and their trees removed, as is constantly happening in Paris; it will soon be necessary to make tunnels of the basements and construct overhead walks in order to allow the pedestrian to go about his business and his pleasure. And this increased intensity of traffic is accompanied by an increase in noise and in atmospheric pollution; the health and equilibrium of the citizens are gravely menaced. The city has become the focus of an abnormal life, from which all who can escape for weekends in the country, for Sunday walks, and for frequent holidays.

Urban microclimate has been the subject of many studies, but the pollution of the air offers a wide field for investigation.[3] The temperature within the built-up area is higher by two or three degrees (F), and air movement is strongly canalised and often paralysed by the lines of buildings, which act as a screen. The exhaust from motor vehicles, the smoke from domestic chimneys, and the fumes and dust from factories, make the air we breathe as noxious as a sewer.

[1] A. Millon, 'Les espaces verts à Lille', *Bull. Soc. Geog. Lille*, 1962, pp.5–41; M. M. Fratoni, 'Les espaces verts à Paris', unpublished thesis, Paris, 1963.
[2] H. F. Gregor, *loc. cit.*
[3] P. A. Kratzer, *Das Stadtklima*, Brunswick, 1956; see also T. J. Chandler, 'London's urban climate', *Geog. Journ.*, 1962, pp. 279–86; J. Paszynski, 'Investigation of local climate in the Upper Silesian industrial district', pp. 83–95 in *Problems of applied geography*, Geographical Studies No. 25, Institute of Geography, Warsaw, 1961; and J. W. Reith, 'The Los Angeles smog', *Yearbook Assocn. of Pacific Coast Geographers*, 1951, pp. 24–32.

In Paris, where in the mid-nineteenth century urban planning envisaged the presence of 50,000 vehicles, there are now over a million and the total rises by 140,000 a year. At pavement level along the main roads, the air in Paris now contains 92 litres of carbon dioxide per 100 cubic metres, and the 'safe' limit is 100 litres per 100 cubic metres. Blood samples taken from individuals who have been stationed for several hours at a great traffic junction have shown the danger to health of such an atmosphere. Calm days are especially dangerous. The presence of trees and grassed open spaces is a help, because of the action of chlorophyll in reducing carbon dioxide, but such things are often absent just where they are needed, in proximity to the most intense traffic streams.

Another disadvantage of town air is the presence of dust. In Paris, between 60 and 70 grammes per square metre are deposited every year, and this figure is much lower than in London (118 grammes/square metre/year) or in industrial cities like Leeds (210), Newcastle (232) and Pittsburg (404). The microbe content of the air varies considerably according to the degree of confinement, the traffic intensity, and the part of the city chosen; in some of the big shops there may be 4 million microbes in a cubic metre of air, but 575,000 on the main boulevards and only 88,000 on the Champs-Elysées.[1]

The effect on the human organism of permanent contact with polluted air is difficult to establish in precise terms. However, certain statistics are sufficient to cause concern. Between 1950 and 1955, the mortality rate from bronchitis fell by 18 per cent over the whole of France but rose by 38 per cent in Paris. Medical specialists have noted that throat and respiratory infections increase in districts where oil heating is accompanied by sulphur fumes in the atmosphere. In London, on the occasion of dense fogs, people are advised to wear simple masks, and there is considerable fear of a sudden rise in the death-rate of elderly people from respiratory diseases during periods of winter fog when the air is dense and stagnant.

At all events, it would seem advisable not to allow towns to grow indefinitely in an un-planned manner.

URBAN RENEWAL

Consciousness of this double danger of unlimited and chaotic growth has led to an increasing preoccupation with town-planning. It must be stated at the outset that this is a difficult task, in which the expertise of the specialist and the goodwill of the administrator all too often encounter the indifference or the selfishness of the ordinary citizen and the more dangerous activities of the speculators.

[1] E. Sabourin, *L'espace vert au service de la santé des Parisiens*, Paris, 1955.

In Europe, it was in the United Kingdom and Germany that ideas on town-planning first began to take shape. We may take densely populated England as the type. Here in 1845, the first parliamentary Act was passed to safeguard the ancient common lands, which were considered as the lungs of the towns; then between 1848 and 1875 a series of laws was passed, regulating the manner of urban growth and designed mainly to protect public health. The idea of garden-cities was put forward in 1898, far in advance of its time, by Ebenezer Howard; this envisaged the association of residence and work in a town consisting of three concentric zones, a central core or business quarter, an inner ring of housing (with gardens, and no longer in the endless terraces of the mid-Victorian period), and an outer ring of industry, with a green belt beyond. Howard's ideas were put into practice in England's first garden city, Letchworth, in 1902.[1]

In 1909 planning was officially recognised and legislation was passed to ensure that the public interest came before individual projects. Several other Acts were concerned with various points of detail such as ribbon-development along main roads, but the next major event was the publication of the Barlow Report in 1940,[2] which emphasised the vital necessity of halting the disordered growth of the great cities, of London in particular. In 1943 a Ministry of Town and Country Planning was created and Acts were passed in quick succession in the years following the war. A comprehensive and coherent policy of ordered urban development and the reorganisation of industrial location has been developed, many aspects of which provide models for the rest of the world; in particular the planning of New Towns.[3]

In the United States, the problem of urban zoning began to be attacked in 1916, and after 1937 housing and slum clearance came to the fore; the law has been strengthened since the war. Progress in the various States varies enormously, and Illinois has been in the forefront of the planning movement.[4]

In the light of research undertaken in various countries, of communist experiments in urban planning, and of the progressive legislation that has been passed in many countries, we may classify town-planning problems under three headings, the creation or maintenance of a balanced urban network, the replanning of urban

[1] E. Howard, *Garden Cities of Tomorrow*, 1898; republished London, 1946 (ed. F. J. Osborn).
[2] *Report of the Royal Commission on the Distribution of the Industrial Population*, Cmd 6153, H.M.S.O., 1940.
[3] M. P. Fogarty, *Town and Country Planning*, London, 1948; *Town and Country Planning in Britain*, C.O.I. Reference Pamphlet No. 9, H.M.S.O., 1959.
[4] M. J. Proudfoot, 'Public regulation of urban development in the United States', *Geog. Rev.*, 1954, pp. 415–19.

land with due regard for open space and harmonious architectural development, and the improvement of the relations between the various functional zones of the town.

Under the first of these headings, one fundamental task is clear: at all costs certain already grossly overgrown urban agglomerations must be prevented from expanding still further. In Europe, London, Paris and Moscow fall into this category, and for these three capitals a policy of decentralisation is envisaged.

The problem of London was already, before the last war, beginning to alarm responsible people in England. The size of the built-up area and the lack of available space within a reasonable distance of the city centre were the most serious problems; space for all the schools that were necessary, for example, could only be found at the expense of housing areas containing 173,000 people. In 1943–4, the Greater London Plan, prepared by Sir Patrick Abercombie, and the County of London Plan, attempted to put forward solutions to the problems of what more than a century before William Cobbett had called 'the Great Wen'. Abercombie distinguished four zones within the area studied: first the County of London, from which 1,125,000 people were to be dispersed; secondly, a belt of mainly interwar development that might be allowed to remain more or less unaltered; thirdly, a green belt that must be preserved at all cost, and, fourthly, an outer zone in which a number of new towns (originally called 'satellite towns') could be built to house the population displaced from the central areas. Abercrombie proposed a number of sites for new towns, but not all were found to be acceptable, and eventually a dual policy came to be adopted, first of building new towns from scratch or around quite small village nuclei, with industries to occupy the inhabitants—such as Crawley, Bracknell and Harlow; secondly of 'exporting' Londoners to selected small market or industrial towns in which large new housing estates and accompanying industries were created—such as Swindon, Bletchley, Haverhill and Thetford.[1]

In the case of Paris, industrial decentralisation has been the expedient by which an attempt has been made to check the demographic growth of the city, which was endangering the economic equilibrium of the nation by its overwhelming superiority. Since 1955 several laws have been passed to forbid the construction and enlargement of factories, and later of major office buildings as well, within the city limits. Between 1950 and 1960, 590 decentralisation operations were carried out, involving the movement of 140,000

[1] S. W. Wooldridge, 'Some geographical aspects of the Greater London Regional Plan', *Trans. Inst. Brit. Geog.*, 1946, pp. 1–21. The latest developments in this field are foreshadowed in '*South-East Study*' prepared by the Ministry of Housing and Local Government, H.M.S.O., 1965.

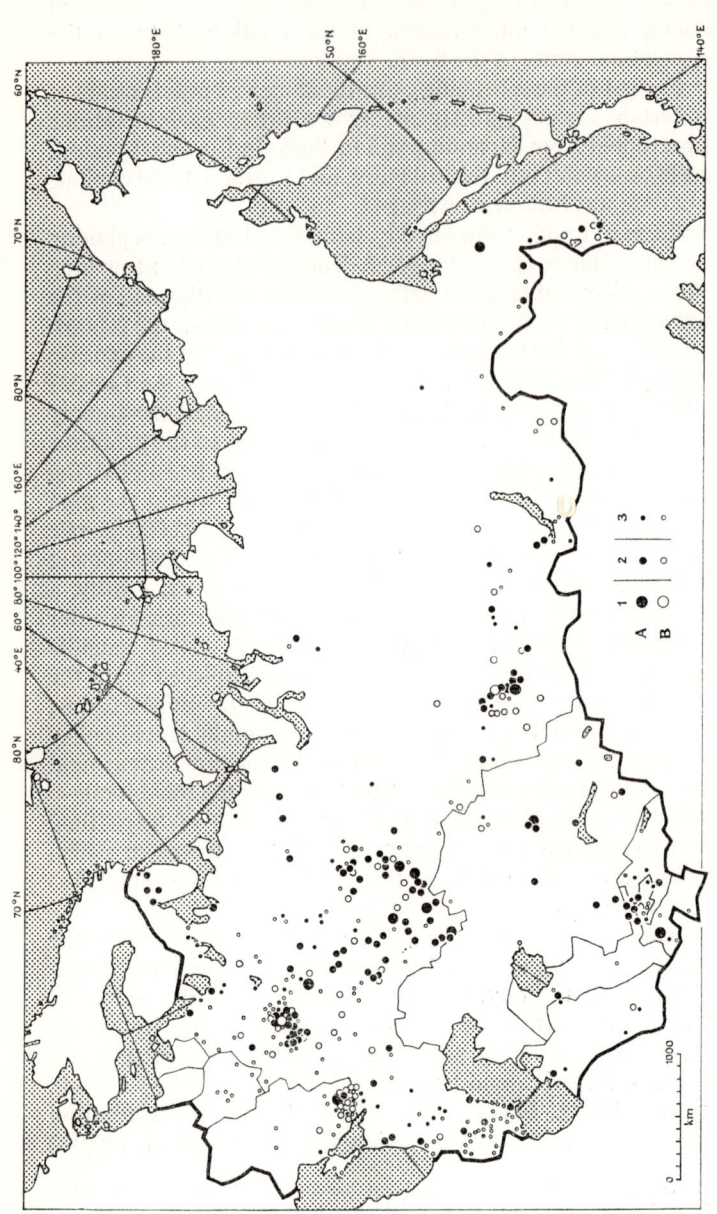

Fig. 31. Development of Soviet towns
A. Towns of recent creation. B. Old towns newly developed.
1. Important towns. 2. Less important towns. 3. Small towns.

employees, or with their families, something like half a million people, from the Paris region. It must be admitted that this did not prevent the population from increasing by 150,000 during the same period. The plans for the future of the French capital foresee 12 million inhabitants in 1985 and more than 14 million at the end of the century.[1] It is not all that easy to modify current trends.

In Moscow the 1935 plan forbade the construction of any new enterprise that was not directly concerned with serving the capital city's population, the limit of which was set at 5 million. This also has been less successful than was hoped, and in 1960 the population was 6 million.[2]

Apart from such attempts to control these monstrous urban growths, the authorities have been preoccupied with the task of improving the lot of the urban dweller, and experts have been trying to find a formula for the ideal town, and for the establishment of new towns where none existed before. English sociologists and town planners have been concerned to discover the optimum size for a small town in which people can live comfortably in the kind of social and professional milieu that is appropriate to their needs and aspirations. The Germans have been working along similar lines in their creation of small centres of rural industry in agricultural towns as a means of unobtrusively introducing urban life into the countryside. The Russian planners have been concerned not only with the establishment of a reasonable balance in the urban structure of industrial districts, but also with the penetration of urban life into the countryside through the creation of 'agricultural towns'.

In France, attempts have been made to fix an optimum level of urbanisation and to outline its ideal structure. Towns are no longer considered simply as a mass of houses, people and occupations, but as places in which to live, work, move about, and enjoy relaxation of mind and body. The concern is no longer, as in former times, with the question of how cheaply can the town be run, but with how can it offer its people the maximum opportunity of leading a full and useful life.[3]

Thus throughout western Europe and the areas of the world which are influenced by western European civilisation there is now implicit recognition of the fact that towns can no longer be allowed to grow spontaneously in an unplanned and uncontrolled fashion, as in the late nineteenth and early twentieth centuries. These ideals, however, cannot always be put directly into practice, for, except in the still

[1] *Schème directeur de la Région parisienne*, Paris, 1965.
[2] *Goroda sputniki*, p. 46.
[3] P. Pinchemel, A. Vakili and J. Gozzi, 'Niveaux optima des villes', *Cahier du C.E.R.E.S.* No. 11, Lille, 1959.

vacant parts of underdeveloped countries, the map is not a blank sheet. Not all countries have the magnificent town-planning laboratory possessed by the U.S.S.R. in its Asiatic territories. But if most countries have no opportunity of creating an urban network from scratch, all are confronted with the problem of the rehabilitation of the older quarters of their towns and the construction of new ones. The main concern is the availability of the necessary space; town centres must be opened out and congestion eased, and the standards seem to get more and more ambitious as the years go by.

English town-planners reckon 13,000 people to the square mile as an ideal density; and a glance back at the examples given on p. 266 will show the difference between what exists and what may be hoped for. In the replanning of Birmingham, the following proposals have been made concerning the amount of space to be allowed to different aspects of city life:

Land use	Existing (per cent)	Proposed (per cent)
Dwellings	47·0	27·4
Industry	21·8	21·2
Public open space	0·5	15·6
Schools &c.	1·7	8·6
Roads	1·3	6·7
Other uses	27·7	20·5
	100	100

Factories that have recently been reconstructed in Birmingham occupy from three to five times their former space, but they are surrounded by more open space and even sometimes by lawns and flower beds. In Manchester the housing areas built since 1918 occupy five times the area of the pre-1918 residential quarters, but they only house 25 per cent more people.

In Paris the slum areas lying between the right bank of the Seine and the Gare de l' Est will gradually be rebuilt; a beginning has been made near the Hotel de Sens, replacing tall tenement blocks and narrow streets by fine blocks of flats separated by wide roads and small gardens. In many places within the capital, small houses and half-ruined hovels are being replaced by tall residential blocks rising amidst green open spaces.

In new towns the residential quarters exhibit the same two characteristic features, the small area occupied by the buildings and the large open spaces that surround them. In Brasilia, dwellings occupy scarcely one-quarter of the total area, but they are surrounded by lawns, by wide streets with shopping centres, by playgrounds and pools. The same spirit, so far as one can judge from such published information as is available, clearly animates the new cities of Soviet Asia, with their wide roads, individual and public gardens, and adequate provision of sanitation, and educational, social and cultural facilities.

These policies are costly, both in respect of new towns and reconstructions; land prices, and the amount of equipment necessary put them beyond the reach of many countries, however great the willingness to adopt them. The State or public authorities must stand a large part of the cost. In Birmingham an inner ring road was a vital necessity—but it took fifteen years to achieve, involved the destruction of several thousand houses and had cost £20 million when it was completed in 1960. The prospect of opening new roads across Paris is sufficient to daunt even the most optimistic planner, when he recalls the expense and the complication that arose in driving new motorways through the suburbs. In Ottawa a large-scale reconstruction plan involved the suppression of two railway lines that had been engulfed in the urban expansion, and these were replaced by a wide transverse boulevard.[1]

In some countries it has been possible to go even further: the development of the various parts of a town is no longer left simply to the laws of economic competition and the forces of social segregation. The organisation of the urban area is conceived with the object of providing the people with the most practical and agreeable way of life. We may examine several examples of this new approach.

The New Towns of Great Britain represent a revolution in town-planning policy. They are entirely artificial creations, resulting from the New Towns Act of 1946. They aim to be complete in themselves, to permit their inhabitants to live within a closed circuit, with shops to provide for their material needs, schools and hospitals for their children and themselves, factories in which to work, and religious and cultural buildings to satisfy their spiritual, intellectual and social needs, all located within the town. In all, twenty-one New Towns are planned or in course of construction, sixteen in England, one in Wales and four in Scotland. Eight of them are designed to take London's overspill population, and two fulfil the same function in relation to Glasgow.

[1] J. Greber, 'Un example d'aménagement urbain: Ottawa', *La vie urbaine*, 1958, p. 226.

Sites for the New Towns have generally been chosen with regard to main lines of communication and the possibility of adequate water supply, and in most cases almost virgin sites were selected so that planning would not be hindered by pre-existing settlements. The areas delimited are variable, from 865 acres at Newton Aycliffe to over 10,000 acres at East Kilbride, and populations of 20,000 to 50,000 are envisaged, at densities varying from 6,200 per square mile at Crawley to 14,700 at Aycliffe. The planning of the layout avoids monotony and terraced housing and aims to give life to the town by the creation of a centre in which each functional building has its own style.

Land use at Crawley

(per cent)

Residential	44·0
Town centre	2·5
Industrial zone	7·6
Schools	9·3
Open spaces and sports grounds	25·0
Communications	8·3
Other uses	3·3

Source: P. and G. Pinchemel, 'Les villes nouvelles britanniques', *La vie urbaine*, 1958, no. 4.

In these spacious towns, in which each family generally has its own house and garage, the provision of commercial and educational facilities is carefully planned; the number of schools of different types is related to the population structure of the different neighbourhoods. In Stevenage for example there are eight primary schools, three secondary modern and two grammar schools. The residential quarters are so situated that a working mother can deposit her children at a creche or at school in the morning on her way to work; and pick them up later in the day after having passed through the shopping district en route. Or the children may go unaccompanied to school, by routes that lead across grassy spaces and through passageways devoid of vehicular traffic, which is confined to the perimeter of each small residential unit. Lastly, the town offers employment on the spot; it is estimated that there should be a 1:4 ratio between the number of industrial jobs and the number of inhabitants. At first it was thought that there would be only one industrial job for each one in the tertiary sector, but the progress of industry seems to have been rather more rapid than this. The industrial zones of the first fifteen towns have about 600 factories, employing more than 100,000 people, whilst the total population of the same towns is slightly under half a million.

Housing priority in these towns is given to those who take up employment in the local industries, but people who commute to neighbouring centres of employment are also accepted. The population includes a very high proportion of young children under five years of age, actually twice the national average. The majority of the adults are young married couples, and there are very few old people (only 2 per cent over sixty-five, compared with an average for England and Wales of 11·6 per cent). A few houses for old people enable grandparents to live near their families. It is clear that the peculiar demographic makeup of these towns is due to their essential characteristic of newness.

The construction of such towns has rendered necessary a considerable application of public funds, and their administration is in the hands of Development Corporations whose plans are subject to review by the Ministry of Housing and Local Government. Most of the houses are built to let, but latterly some have been made available for sale as well.

The New Towns are experimental and it is perhaps too soon as yet to pronounce as to their success and their future; but they certainly represent a remarkable effort to solve current problems of urban development. The close association of residence and workplace is interesting and so is the character of the residential neighbourhoods, for the absence of social segregation is noteworthy; at Crawley, about 33 per cent of householders are so-called middle-class, 58 per cent skilled workers and only 4 per cent unskilled, whilst for East Kilbride the figures are 23 per cent, 68 per cent and 3 per cent. The poorer classes are but feebly represented, for the rents are among the highest in the country.[1]

France too has entered this field, with plans for New Towns round Paris. Eight will be built during the next twenty years and they will house about 1·5 to 2 million people.[2]

The communist countries have also set in motion new policies for urban development and no better example could be found than Warsaw, which was so completely destroyed during the war that it was almost literally a case of building a great city from scratch.

[1] P. and G. Pinchemel, *loc. cit.*; A. Trintignac, 'La planification en Grande-Bretagne: le développement des villes nouvelles', *Et. et Doc. du Centre de Recherches économique et sociales*, 1960, No. 3; *The New Towns*, C.O.I. pamphlet; J. H. Nicholson, *New Communities in Britain*, London, 1961.

[2] *Schème directeur, op. cit.*

The ideas behind it were twofold, first the organisation of the different parts of the city, with rapid radial communications from the centre to the periphery, and secondly the juxtaposition, on the outskirts, of industrial and residential quarters with adjacent lateral links.

In the centre, on the left bank of the Vistula, the heart of the old city (Stary Miasto) has been reconstructed exactly as it was before the war, brick by brick and with all its plaster and paint decoration faithfully reproduced—the old market place, the churches, the narrow streets and even the walls and entrance gate; around this core are the administrative buildings, the university, the banks and large shops, whilst movement is made easy by wide roads crossing at right-angles, with a road tunnel leading to one of several bridges across the river. The whole of this section of the city is surrounded by a green belt of parks and gardens attached to palaces and embassies. Beyond this belt and on the other side of the river is a succession of industrial suburbs along the main arteries of communication (chemical and metallurgical industries along the Vistula downstream from the city, along the main western railway line and motor road, and across the river at Praga); between the industrial blocks are residential quarters; these are separated from the factories by open spaces and consist of four- or five-storey blocks of flats housing the workers in the neighbouring industries.

The residential structure is pyramidal—for every 100 flats there is a children's playground and a creche; for every 500, a primary school; for every 2,000 a more wide-ranging school accompanied by a commercial and cultural centre. Such a group forms one of the elementary 'cells' or 'neighbourhoods' of the city, housing some 10,000 people within a radius of about 550 yards. Clusters of such units form city sectors with about 100,000 inhabitants, and these have larger and more important cultural, sporting and administrative centres.[1] The construction of large blocks of flats reduces the journey to work and hinders the excessive spread of the city. The same principle has been adopted in the new town of Nowa Huta, near Krakow, and in the reconstruction of Wrocław, which now houses half a million people.[2] The rents are very low, and there is no social segregation within the residential quarters. Moreover, all the services are arranged so that each inhabitant has at his disposal an equal share of all the facilities, administrative, scholastic, cultural, medical and so on.

In quite another part of the world, the Japanese are also preoccupied with urban development, and their solutions greatly resemble those adopted in Britain and Germany, for example in the construction of

[1] P. George, 'Varsovie 1949', *Population*, 1949, pp. 713–26.
[2] I. Rutkiewicz, 'Wrocław en 1963', *Pologne-Allemagne*, 1963, pp. 42–9.

satellite towns such as Sagamihara on the edge of the alluvial plain of Kwanto, the finest plain of its kind in Japan but already invaded by industry, at an equal distance from Tokyo and Yokohama. The site was used for military and industrial buildings during the war, and

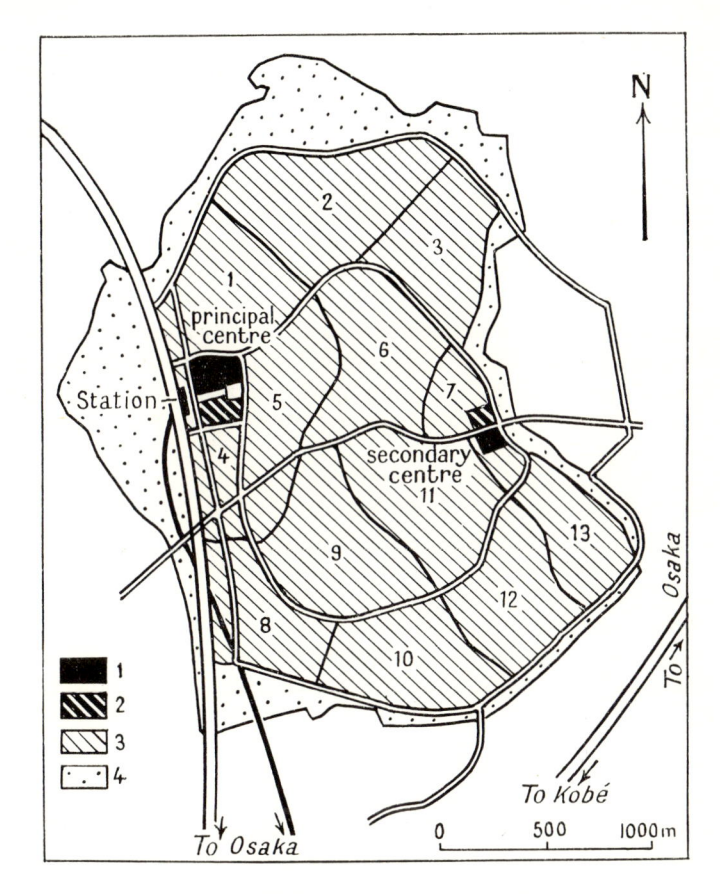

Fig. 32. The new Japanese town of Senri

The figures represent the different quarters of the town. **1.** Administrative centre. **2.** Commercial centre. **3.** Residential areas. **4.** Open spaces.

now a town of 200,000 people is planned, in which residential quarters and industrial zones will lie side by side; in 1959 there were 95,000 people, mostly working in vehicle building and consumer goods industries, but it is hoped to attract mechanical and electrical engineering. It will not be a well planned industrial town, but rather a collection of nuclei clustering round the main lines of communication, each having a small business centre and residential quarters surrounded by large industrial areas.

More original in conception is the large dormitory town of Senri, in the hills north of Osaka; here, on an area of 3,200 acres, it is proposed to house between 150,000 and 200,000 people, and the work of construction began in 1959. The town will have no industry at all, and will simply consist of high-class residences and such business establishments as are necessary to serve their needs. Some 45,000 to 50,000 commuters will go to Osaka every day by two underground railways and a motorway.

The whole town, which will be moulded round the hilly landscape, will form a single unit for the provision of services, but will be divided into thirteen administrative divisions, each having its own schools, shopping centres, park and public buildings; two ring roads will link the residential zones, which will also be crossed by a transverse road, and a green belt will surround the entire town. There will be a major administrative centre on the western side and a secondary centre in the east. The whole project is due for completion in 1967.[1]

Finally, we may return once more to the example of Brasilia, where, up to the present, there are only an administrative centre and residential quarters. The administrative area contains the equipment of a capital city, assembly halls, ministries, large public bodies, embassies, and a group of banks that are essential to the heart of a city whose influence will extend throughout the country. The residential zones consist partly of flats arranged in parallel rows, with lower buildings in between, housing the commercial and social functions. So far only civil servants, shopkeepers and their employees live in the city, and the rents are so high in relation to the normal Brazilian salaries and wages that many people prefer to remain in the wooden prefabricated housing where they were accommodated during the construction period.

We may venture to hope that these numerous experiments, and the research that has accompanied them, will lead to the creation of a new urban environment that will make the towns of this world, which bid fair to house nine-tenths of the population during the next century, places in which happy and fruitful lives may be lived.

[1] P. Schölleu, 'Wachstum und Wandlung japanischer Stadtregionen', *Die Erde*, 1962, pp. 202–34.

CHAPTER 27

THE URBAN POPULATION

The urban dweller is one among many. The countryman can feel solitary; even if he returns to the village at night, which need not necessarily be the case, work on the land offers him individual employment, in which he can toil in his own way and at his own pace; he is bound to stop at night and in bad weather, and he can never be certain that the result of his labour will be all that he desires.

The townsman works at someone else's elbow; he does the same daily journey at the same time as hundreds of his fellows and he works during hours over which he has no control; he can plan and control his activities, for his income depends on other men and not on the vagaries of nature. He is simply a cog in an enormous machine.

The town imprisons man behind its screening walls and within the rectilinear pattern of its streets; under its influence citizens are assimilated and moulded into a uniform pattern. It is no good looking for the character and the soul of a people in the government offices, in the concrete blocks that make up the European quarter, from which will emerge females of all colours, dressed, hair-styled and made-up after the fashion of Paris or New York, and men dressed in the lounge suits that have become an international uniform.

THE URBAN MELTING-POT

Most towns of the world have heterogeneous populations. They are centres of attraction and of immigration, acting as a magnet for individuals or crowds drawn from many places both near and far. By their very size they offer possibilities of employment and shelter that are regularly sought by those who are dissatisfied with their lot or have uprooted themselves from their rural surroundings. Because of the diversity of their population they can offer a variety of employment; the well-to-do seek various types of personal servants, the industries and businesses are on the lookout for labour. And finally, because of their adequate policing, they appear in the eyes of persecuted minorities as a kind of refuge where security can be sought under the protection of the law and the constable.

It is not surprising, therefore, that the greater part of the population of any town was not born within its confines, and that the urban population often differs appreciably in its racial, religious or even national composition from that of the country as a whole or even from that of the neighbouring region.

In the United States, at the census of 1950, 64 per cent of the population was classed as urban, but of the foreign-born population 83·5 per cent were town-dwellers. As far back as one can go with the United States censuses, it is the same story; and it is not only the European immigrants who are urban, but also the Japanese and (particularly) the Chinese, 93 per cent of whom live in towns. The Negroes uprooted from the South are also town-dwellers; over the whole country 62·4 per cent of the Negroes live in towns, but in New England, the central Atlantic States, and the states east and west of the Great Lakes, the urban percentage rises to 90.[1]

The same phenomenon is found in areas such as the Middle East, where various religious groups live side by side. In Syria, for example, apart from a few thousand foreigners, the population is 85 per cent Muslims of various sorts, 14 per cent Christians of several denominations, and 1 per cent Jews. The towns contain a much smaller proportion of Muslims, and the sunnite Muslims, who form 69 per cent of the whole population, are nothing like so numerous in the towns. At Latakia, the Alaouites are most numerous, in Damascus there is a large colony of Druses, whilst at Aleppo the three main groups are in the proportion of 64·2 per cent Muslim, 32·5 per cent Christian and 3·3 per cent Jewish.[2]

This heterogeneity of urban populations is a fundamental characteristic that recurs everywhere. In Pakistan, after the partition, there were 9·5 per cent of refugees from India in the population as a whole, but 48 per cent in the population of Karachi. In south-east Asia, the Chinese are almost all in the towns. In Asiatic Russia, Slavonic people have become numerous in all the new towns, bringing with them the science and technology of the west and acting as teachers and instructors of the native peoples.

In a town the exile feels less exiled, for he finds others like himself, and having left his native country he endeavours, in company with his compatriots, to recreate a small piece of it in his new surroundings. He makes a beeline for the main centres of human activity, where he feels less cut off from his native country; ports and communications centres are thus favoured localities.

In the United States the three main regions in which one finds the highest proportion of foreigners in the towns are the entire east coast (with 80 per cent, rising to 90 per cent in the New York sector), the region east of the Great Lakes, and the Pacific coast, especially California which is undergoing an astonishing demographic expansion. In Great Britain the Irish are most numerous in London,

[1] J. Bogue, *The population of the United States*, New York, 1959.
[2] J. Beaujeu-Garnier, *Géographie de la population*, vol. 2, p. 190; A. R. Hamide, *op. cit.*, p. 12 ff.

Liverpool, and Glasgow. Four-fifths of the Irish immigrants in England, and three-fifths of those in Scotland, are town-dwellers. Chicago, the most important focal point in the interior of North America, offers a vast range of employment and so shows this trend to perfection. It contains more than 600,000 negroes and 700,000 foreign-born whites in its population of just over 5 million. What is even more significant is the distribution of these 'outsiders' within the city:

Composition of Chicago's Population, 1950
(percentage of total)

	Negroes	Foreign-born whites
City proper	13·5	14·5
Rest of urban zone	5·1	9·1

We have already mentioned the peculiar features of the structure of Chicago and have called attention to the distribution of different categories of immigrants (see Fig. 23). The situation is very complex since there is a great variety of Europeans (17·9 per cent Poles, 10·8 per cent Germans, 10·4 per cent Italians, 10·1 per cent Russians, 5·9 per cent Swedes, 5·8 per cent Irish and also Czechs, Lithuanians, English, Austrians, Hungarians, Greeks, Jugoslavs and Norwegians) and also immigrants from neighbouring North American countries (7·7 per cent Mexicans, 3·2 Canadians).[1] The attraction remains powerful, for between 1940 and 1950, more than 180,000 Negroes entered the central part of the city and a similar movement seems to have taken place in the following decade. The location of the Negro quarter remains the same, but since 1950 the white newcomers are proportionally more numerous in the outer districts while the Negroes are six times as heavily concentrated in the centre. We are dealing, then, with a permanent feature of urban geography.[2]

In Africa, where tribal and racial clashes may still be quite dramatic and where people tend to remain obstinately in the land of their forefathers, the town is often the only place where they meet and get to know and understand each other—at least until some sharp conflict ranges them against one other and perhaps almost literally sets the town alight. At Conakry, for example, Susus from the coast,

[1] M. J. Proudfoot, 'Chicago's political structure', *Geog. Rev.*, 1957, pp. 107–17.
[2] W. Petersen, *Population*, New York, 1961, p. 204.

Fulani from the mountains and Malinkes from the interior savanas dwell side by side, thus enabling the national government to embrace elements that in the rest of the country and in day-to-day life are strictly separated. At Dakar nearly 160 different ethnic groups are represented in the city, of whom a dozen are of some importance, forming 95 per cent of the African population. The Wolofs are the most numerous, coming from all over Senegal; since they have recently been entering the city in increasing numbers they are found more or less throughout the urban area, but earlier groups such as the Lébou, the Serer and the Tucouleus, have a tendency to remain in homogeneous blocks, where they have established what are really urban villages, each with its traditional chief.[1]

Although perhaps this clustering of small groups within the urban environment would seem to be perpetuating the existence of isolated nuclei, the influence of the town is important in the development of the population. People get accustomed to seeing each other and get to know each other; the solidarity of the ethnic group is replaced by that of the occupational class. A narrow outlook is broadened by education, which is much more readily available in the town than in the rural areas; and finally, individuals may marry outside their original group.

Towns are thus melting pots primarily because they attract those who are on the move; the larger and more accessible they are, the more composite their population; but they act as crucibles also because they are places of assembly, of mingling, of contact and so of synthesis. Who can deny the overwhelming part played, in the past and the present time, by the towns of the United States, in the assimilation of the masses of immigrants from across the Atlantic?

SEX COMPOSITION

All towns do not have the same proportion of men and women in their population, for they are composed of diverse elements and are set in demographic environments that vary widely in character, and moreover they have not all reached the same stages of development. It is possible to distinguish three broad types: (i) towns in countries where rural-urban migration has virtually ceased or at least has slowed down; (ii) towns, which though perhaps falling into the first category, have some special function that gives them a special demographic character; (iii) the vast ranks of towns in countries in which urban evolution is still taking place.

In the countries of north-west Europe, and in particular the United Kingdom, town growth is so slow that the composition of the urban

[1] *Recensement démographique de Dakar*, 1955.

population depends more on normal demographic factors—natural increase or decrease—than on migration. In the population as a whole females are slightly more numerous than males—in England and Wales where there are 92·4 men to every 100 women. The towns have similar proportions—90 men to 100 women. The situation is slightly more ill-balanced in the large cities: in the seven conurbations the male proportion falls to 89·2 and in London alone to 88·4. This British example is typical, for it is confirmed by other European countries:

Number of Men per 100 Women in Urban Population

England and Wales	90
Sweden	92
Spain	86·5
Iceland	89
Canada	97·8
United States	94·9
Brazil	91·4
India	116

Source: *United Nations Demographic Yearbook*, 1960.

What this amounts to is that in developed countries women prefer urban residence to the comparative discomfort and thraldom of life in the country. In Ireland, for example, many young girls leave their country homes and go to the towns or even abroad, in order to get away from the chores of farming and find lighter work elsewhere, with the prospect of a less austere existence. The case of Spain, like that of other Mediterranean countries, is easily understood when one remembers on the one hand the essentially manual and unmechanised character of the agriculture, which keeps the men in the countryside, and on the other the importance of personal service as an occupation in the towns.

In North America the proportion of men to women in the urban population is appreciably higher than in Europe, but this reflects the composition of the population as a whole, for Canada has 97·2 men to 100 women overall and 97·8 in the urban population. This is a relatively recent phenomenon, for in the United States it was only between the 1920 and 1930 censuses that the preponderance of women in the towns began to appear. Until that time the main influence had been the successive waves of immigrants that broke upon the Atlantic shores during the nineteenth and early twentieth century; the majority of these immigrants were male (68 per cent in 1900, 72 per cent in 1907, and 56 per cent after the postwar restrictions in 1925), but with the slackening in the tide of immigration, and even the reversal of the previous trend (there were only 48 per cent men in 1957), the natural tendency has begun to make itself felt. In former times there were actually more men than women in the whole population (106 per cent in 1910, 100·7 per cent in 1940), but now the balance has been reversed, and in 1960 there were only 97·9 men to 100 women. This explains the urban percentage, to which also the reasons adduced in the case of European towns apply.[1]

A second type is also distinguishable in advanced countries, namely the towns that have a specialised function that is reflected in the demographic composition of the population. In industrial towns with mainly male occupations, the population naturally has a masculine bias; as in Sheffield, where there is relatively little employment for women, or in Bristol in the 1930s, with an industrial expansion based largely on the aircraft and engineering industries and the trade of the port.

This factor of industrial employment is much more marked in countries with rapidly-developing towns. In Brazil the overall proportion of men to women is 91·4:100, but in São Paulo it is 98, and there are almost equal numbers of both sexes; this indeed is true of all the Latin American states.[2] In India the disproportionate number of males in the industrial cities is very striking—175·4 per cent in Calcutta, 162·6 at Howrah, 130·7 at Ahmedabad. We might be tempted to query the accuracy of these figures, because the suppression of women in the censuses of eastern Asia is well known, but if one compares, for example, the Howrah figure with its neighbouring rural district which has 108·1 males to 100 females, it is clear that there is in any case a preponderance of the male element in the population.[3]

Conversely, some towns are occupied mainly by well-to-do people, or are chosen as places of retirement, and here, since women commonly live longer than men, the female element predominates. At Bournemouth in England there are only 71·6 men to 100 women, and whilst the predominance is noticeable at all ages above fifteen, it becomes much greater in the over sixty-five age groups, in which there are nine times more women than men. The record in this direction is held by Worthing, another retirement centre on the south coast, which has only 66·4 men to 100 women. The same thing

[1] J. Bogue, *op. cit.*, pp. 358 and 158–59.
[2] *São Paulo*, p. 210 ff.
[3] A. B. Chatterjee, 'Demographic features of Howrah city', *Geog. Rev. India*, 1958, pp. 150–69.

is to be found in certain residential London boroughs, which have large numbers of domestic servants and also many unmarried females who are employed in the offices and shops of the city; thus Hampstead and St Marylebone have only 71 men to 100 women, and Kensington about the same proportion.[1]

The French Côte d'Azur provides another example of the same phenomenon. Nice, except for the under fifteen group in which there are slightly more boys than girls, is largely a city of women, and of the over sixty-fives, 62·2 per cent are females.[2] In the United States, Florida and California show the same tendency.

The third type comprises towns in countries that are rapidly becoming urbanised; in short, the sex ratio depends on the nature of the masses of people who are crowding into the towns, and two subtypes are recognisable.

In Africa and Asia, the men often migrate alone to the towns; often they are bachelors, but if they are married they leave their families in the village and visit them periodically. The women accept the situation, either because they are frightened of the town or because they are bound by tradition. In any case, they are but poorly represented in the growing towns of Asia and Africa, and the disproportion between the sexes increases with the size of the town. There appears, however, to have been a slight tendency recently for the disequilibrium to diminish; in the largest Indian cities, the figure of 170 men to 100 women, which was characteristic a few decades ago, had fallen in 1941, and by 1951, for all the towns with over 100,000 inhabitants, it was only 127.[3]

Number of Men per 100 Women in Indian towns

Census	5 to 10,000 population	50–100,000 population	100–500,000 population	Over 500,000 population
1931	111	120	128	173
1941	—	—	127	161

It seems possible that there is some correlation between the tendency of men to migrate to the town alone and the distance from which they come. If the journey to the town is short, the man may consider that he can easily travel to and fro and leave his family behind, but at the

---

[1] United Kingdom census, 1961.
[2] R. Blanchard, *op. cit.*, pp. 158 ff.
[3] *Census of India*, 1951, vol. IX part IIA, p. 26.

same time young women, who have learnt something of the possibilities of urban life, also leave home more readily. Thus in and around Hyderabad, the new immigrants include 86 women to every 100 men; but in the case of long-distance movement the migrants are mostly unmarried males, and amongst the migrants from other provinces of India arriving in Hyderabad the proportion is only 72·5 women to 100 men.[1]

In Black Africa the same thing happens: the proportion of males at Abidjan is 140, at Brazzaville 124, at Duala 120, at Lagos 118 and at Dakar 106. A detailed study of the sex composition of the Dakar population leads to certain tentative conclusions: first, the importance of the migratory element is emphasised by the fact that there is not the same discrepancy between the sexes amongst the people who have been in the town for a long time, such as the indigenous Lébou; secondly, men are much more predominant in the over thirty age group (147 per cent), whilst women are more important in the fifteen to thirty age group—perhaps as a result of a change in the composition of the immigrant masses, for with the social evolution of the African continent, young women are now much more disposed than their elders to migrate on their own to the towns.[2]

The traditional social structure and the behaviour of the female element seem to exercise a decisive influence on the townward migratory movement. In Latin America, for example in northeastern Brazil and in some of the Andean republics, where the relations of urban and rural development do not differ greatly from those found in India and Africa, there is a much larger proportion of migrant females. Thus Brazil has a predominance of women in the towns, just as in Europe and North America, and the same thing occurs in all the other South American countries for which adequate statistics are available, such as Argentina, Chile and Venezuela.[3]

There are insufficient data extending over a long period to enable us to come to more precise conclusions or to analyse the relation between the attractive power of towns and the sex composition of their populations. But the latter is not the only aspect of urban demography in which an important part is played by the migratory element.

AGE COMPOSITION

Here again we are confronted with statistical data that show wide variations depending on a number of factors. In the first place, the many types and stages of immigration into towns make a direct

[1] G. Haberland, *Gross Haiderabad*, Hamburg, 1960, pp. 53 ff.
[2] *Census of Dakar*, pp. 37 ff.
[3] United Nations *Demographic Yearbook*, 1960.

contribution to the urban population, and the immigrants also contribute indirectly through their reproductive capacity; secondly, the immigration has an important influence on the conduct of individuals, that we shall analyse later on. As a result of these two factors the age composition of urban population may present wide differences.

In general the town, as the focus of immigration, attracts adults at a time when they are not held back by family ties or occupations and so can the more easily uproot themselves and start a new life in a new environment; in other words, between the ages of fifteen and thirty-five. Amongst the migrants who move for purely economic reasons more than 50 per cent are generally under thirty years of age, and more than one-third are between twenty and thirty. As example we may quote Poto-Poto, in former French Congo, where 86 per cent of the newcomers into the town are under thirty, or Salvador in Brazil, where migration studies have shown that 36 per cent of the new arrivals were between twenty and thirty years of age. Such immigrants are young individuals who enter the town just at the time when they are ready to raise families; and since they are young and active the town has many employment opportunities to offer them. The abundance of adults is thus one of the fundamental characteristics of urban populations.[1]

In the United States in 1950, 63·8 per cent of the urban population was aged between eighteen and sixty-five, but only 53 per cent of the rural population. The average age of the urban population was 31·6 years, that of the rural non-agricultural population 27·9 and of the agricultural population 26·3.[2]

If we examine and compare the age composition of the population of several countries, at the same time placing in juxtaposition the figures for the total population for towns and for the large conurbations, certain features stand out. It is necessary to present several tables since all countries do not adopt the same age-groupings in their census publications.

A comparison of these three tables confirms the peculiar character of the age composition of urban populations: the number of adults is everywhere greater in the towns than in the country as a whole, whilst the proportion of children and old people is appreciably less. That these characteristics are accentuated in the larger towns is shown by the comparison of São Paulo with the sum total of Brazilian towns, or by an examination of the age composition of the twenty-five largest

[1] M. Soret, *Démographie et problèmes urbains en A.E.F.*, Brazzaville, 1954, J. Beaujeu-Garnier, Les Migrations vers Salvador. *Cahiers d'Outre-mer*, 1962, pp. 291–300.
[2] J. Bogue, *op. cit.*, pp. 99 ff.

Age composition of Population in United States
(percentage)

Age groups	Urban population	Rural, non-agricultural	Whole U.S.A.
0–8	16·7	20·5	18·0
9–17	11·3	14·6	13·2
18–64	63·8	56·3	60·6
Over 65	8·2	8·6	8·2

Source: J. Bogue, *The Population of the United States*, pp. 96–9.

Age composition of Population in Paris and France
(percentage 1962)

Age groups	Paris	Paris Region	France
0–20	21·6	28·5	32·8
20–40	30·1	29·7	27·2
40–60	28·1	24·6	22·9
Over 60	20·1	17·2	17·1

Source: *Annuaire statistique de la Région parisienne*, Paris, 1965.

Age composition of Population in Brazil
(percentage 1950)

Age groups	São Paulo	Towns	Suburbs	Whole Brazil
0–19	39·8	44·3	49·9	52·5
20–39	37·7	33·7	41·6	29·7
40–59	18·0	16·5	14·1	13·5
Over 60	4·5	5·4	4·4	4·2

Source: *Contribuçoes para o estudo da Demografia do Brazil*, pp. 139–50.

cities in the United States. On the other hand, moving from ordinary urban populations to those of the larger conurbations, we note the increase in the number of children and a corresponding reduction in the proportion of adults. This is shown in the 'rural, non-agricultural' category in the United States (which in actual fact largely covers the urban fringes that are not yet classed as urban), and in the suburbs of Brazilian towns; it also appears in the comparison of the city of Paris and the whole Paris conurbation.

A more detailed study of age groups enables us to trace the correlation between the intensity of the migratory flow and the growth of particular age groups in the urban population. Thanks to the excellent statistical material which is available over a long period

THE URBAN POPULATION 365

for the United States, we can trace the consequences of the slackening in townward movement on the age pyramid. During the period 1900–20, the urban population increased by 5 or 6 per cent per decade, and in 1910 the towns contained 59·2 per cent of adults aged twenty to sixty-five, with those in the twenty to thirty age group being more numerous than those aged forty-five to sixty-five; this was the great surge of urbanisation. On the contrary, since 1940 the urban population has only increased by 2 per cent per decade, and the urban population now includes 61·1 per cent of people aged twenty to sixty-five, with the forty-five to sixty-five group more numerous than those between twenty and thirty.

Age composition of United States Towns
(percentage)

Census	0–4	5–19	20–29	30–44	45–64	Over 65
1910	9·9	26·8	20·9	23·1	15·2	4·0
1950	10·1	20·5	16·7	23·0	21·4	8·2

Source: J. Bogue, *op. cit.*, p. 103.

In African towns that are still being swelled by migration, the proportion of adults between twenty and forty years of age is especially marked, whilst the ten to twenty age group is somewhat deficient. Dakar illustrates this perfectly:

Age composition of Population of Dakar
(percentage 1955)

Age group	Total	Male	Female
0–9	30·1	15·0	15·1
10–19	13·7	5·9	7·8
20–39	38·1	19·5	18·6
40–59	14·3	8·8	5·4
Over 60	3·8	2·1	1·7
	100	51·4	48·6

Source: *Recensement de Dakar*, p. 14.

In comparing the male and female age groups, the comments made above are illustrated: young women migrate with greater willingness than older women.

The statistics for São Paulo also bear out the correlation between the strength of the migratory movement and urban growth on the one hand, and the proportion of young adults on the other.

Besides the high proportion of adults, towns are also characterised, as we have seen, by the relatively small proportions of young people and old. However, at the two extremities of the scale there are important variations. The number of children may be as large in the

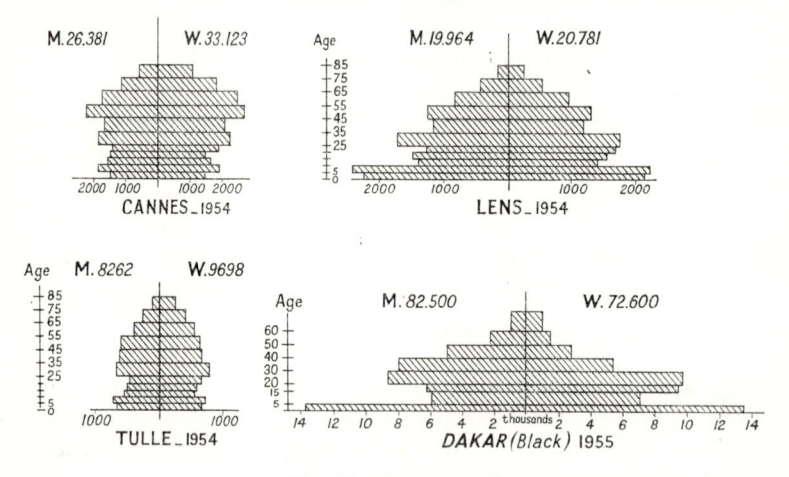

Fig. 33. Age pyramids

Cannes (a 'retirement' town)—Lens (working-class town)—Tulle (rural centre)—Dakar (large tropical town).

towns as in the rest of the country if the population as a whole is very prolific and demographically immature, and if the migrants are but recent arrivals and equally balanced as to sex. Thus in Hyderabad, the proportion under fourteen is 37·3 per cent or more than double that of a city like Paris, whilst the corresponding figure for the whole of India is 38·3 per cent: the proportion of adults is slightly higher, at 55·8 per cent, than in the country as a whole (53·4 per cent), and that of old people distinctly lower, the over-sixty-fives being 6·9 per cent in the city and 8·3 per cent in the whole country.[1]

A similar situation is found in the British New Towns, where a large proportion of the population are families with young children. At East Kilbride, for example, there are 39·6 per cent under nineteen

[1] G. Haberland, *op. cit.* pp. 61 ff.

and 49 per cent between twenty and forty-four, whilst in England and Wales these percentages are 28·6 and 34·2 per cent respectively; and in this town there are only 2·2 per cent over sixty-five as against 11·6 in the country as a whole. The same situation is found in the new towns of Soviet Asia, which are mainly peopled by young workers.

In the towns of the United States, the proportion of children depends largely, in towns of equal importance, on their situation within the country: towns with a large proportion of Negroes, and those in the mixed population belt of Texas, have many more children. New York has 26·9 per cent under twenty, but Houston 31·7, New Orleans 31 and Louisville 30·5. On the other hand, the historic towns of the north-east, which no longer have much power of attraction, and towns that have a high proportion of retired people, have many more in the older age-groups. The suburbs of Boston have a higher percentage of over-sixty-fives (9·9 per cent) than can be found in

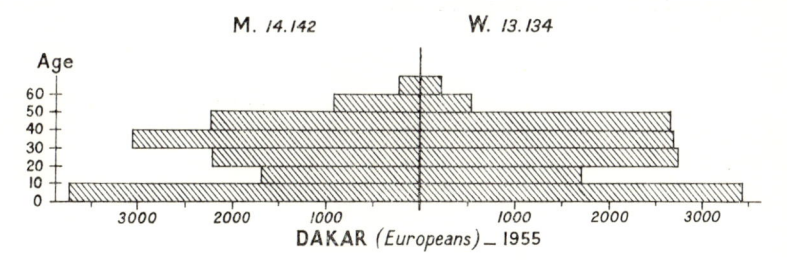

Fig. 34. Dakar. Age pyramid for European population

any of the twenty-five largest cities of the United States; and second on the list comes the suburban zone of Los Angeles with 8·7 per cent.

In order to represent graphically the variation in age composition, we can employ two kinds of diagrams, population pyramids and triangular diagrams. The first of these enables the age groups to be seen at a glance, either in terms of the actual numbers of males and females in each group or in terms of the percentage of the whole population (Figs. 33 and 34). The second enables rapid comparisons to be made between the demographic characteristics of a number of towns, but only in terms of three age groups, children, adults and old people (Fig. 35).

Finally, we may calculate the 'dependance rate', which is the ratio between the number of individuals under twenty or over sixty-five, and the twenty to sixty-five age group, multiplied by 100; we can also add the two extreme categories together and compare the total with the middle age group. The calculation has considerable economic significance, for with their high proportion of working adults, towns are obviously favoured localities. Some sample figures from the United States are given below:

Dependence Rate in 1950

	Total	In relation to under 20s	In relation to over 65s
Towns	63·3	50·0	13·2
Country	99·9	84·7	15·2
Total U.S.A.	72·7	58·5	14·2
New York	52·8	41·1	11·7
Boston	63·7	47·6	13·1
Houston	57·9	50·0	7·9

Source: J. Bogue, *op. cit.*, pp. 92–120.

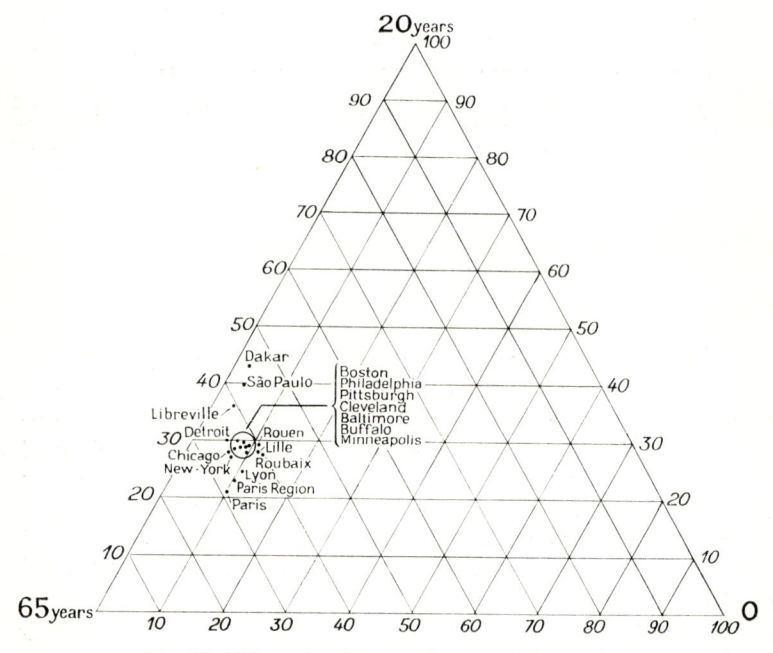

Fig. 35. Triangular diagram for certain large towns

The comparison of New York, Boston and Houston is made more significant when we add these figures to what has been said above about the demographic characteristics of these three cities.

28 Bombay: the English quarter

29 Bombay: old Indian houses

30 Bangkok: the floating quarter

31 Kankan: hutments

32 Labé (Guinea): market place

FAMILIES IN TOWNS

Urban populations contain large numbers of adults of reproductive age; but town life tends to reduce human fecundity. The town, whether large or small, offers distractions and a liveliness that are not found in villages; the individual is continually having his attention attracted or monopolised. He has friends who live on his doorstep; every day he sees the advertisements of the cinemas and the travel agents; he is tempted by the shop-windows and by itinerant salesmen who offer him the latest products of modern domestic technology—on hire-purchase terms; he can pay in cash or run a credit account. All this occupies his time and his money, and it is little wonder that many young married couples prefer a car to a cradle.

Another aspect of the urban way of life is the opportunity that it offers for female employment. Not that the country-woman does not work; she does, and often much harder than her sister in the town, but the work is done in limited surroundings, often at the side of her husband in the fields or in the farmyard. On the contrary, work in the town lies outside the home, and takes the young girl or married women away from the domestic hearth; she is cut off from her household during the day and returns in the evening to irksome domestic chores that are merely an addition to her daily job and not an essential part of her life. Thus absorbed and emancipated—the two things go together—she develops a greater than normal tendency towards the practice of contraception.

It is possible that this is the only reason for the notable diminution in feminine fecundity in the towns in comparison with the countryside; but some medical experts believe that diet also plays a part. The consumption of protein-rich foods like meat, milk and eggs, reduce the possibility of conception. In Brazil we find that 54·2 per cent of town-dwelling women have children as against 60·1 per cent in the rural areas[1]; and in some North African towns about 37 per cent of the women are childless whilst the percentage for the same racial groups in the country is 27.

The second factor in the reduction of the urban birth rate comprises the material conditions of life. Dwellings seem to be designed to limit the growth of families. The houses have no gardens, the streets have narrow pavements and few open spaces, and in the central parts of towns the flats, especially those of the lower classes, are small both in total area and in number of rooms. In France, present day apartments designed to let at a moderate rent to the working and

---

[1] B. Hutchinson, 'Fertility, social mobility and urban migration in Brazil', *Popul. Studies*, 1961, pp. 182–90; W. C. Robinson, 'Urban—rural differences in Indian fertility', *Popul. Studies*, 1961, pp. 218–34.

lower middle classes, are designed for childless couples, or at a pinch for families with only one or two children. Only the more luxurious flats have six or more rooms. For a large family to live in such limited accommodation is a danger to health and a severe test of the mental equilibrium of all concerned.

It costs a great deal to raise a child in a town; the infant cannot be allowed to play out of doors, to run around scantily clad and to indulge in his natural instincts; he must be watched at all times and decently clothed, and when he goes to school there may be fares to pay; and even those families working on the most modest budgets must take at least an annual holiday. When the child grows up he does not take his natural place at the side of his parents as he would working in the fields, in many country districts. He must be educated for a career; unlike the young farmer's son he does not render services which help to balance the family budget, on the contrary he is an expense to his parents throughout his whole childhood.

Lastly, psychological factors help to keep down the size of families. In the town education is better developed and more universal; people acquire new desires and discover new horizons, and the town may offer them one means, amongst others, of satisfying some of their aspirations. Birth control information and the means of applying it are available, why should they not use them?

The example of Japan, where propaganda in favour of family limitation has been officially undertaken since 1948, is very striking. Numerous studies of the results have been made, and in 1954, only six years after the official adoption of the policy, 37·2 per cent of married couples in towns practiced contraception and 30·4 per cent in the rural areas.[1] In almost all countries of the world, private enquiries will yield the information that educated and well-to-do couples practice birth control, whilst in the more advanced countries of western Europe, and over most of the United States, there is little difference between town and country in respect of the voluntary restriction of births—it is a family matter that has no sociological or geographical differentiation.

One other factor may in exceptional cases, but very powerfully, help to limit urban fecundity, and that is the unequal sex ratio. When a town has many more men than women, large numbers of men are condemned to celibacy, and this makes for female habits that are not conducive to the raising of families: prostitution is rife and divorces numerous, for the women, being scarce, are subject to many solicitations.

Two examples may be taken from the former French Equatorial Africa; the first is Dolisie, where there are only 593 women of

[1] Publications of the Japanese committee for planned births.

marriageable age to every 1,000 men, and 63 per cent of the marriageable males are bachelors; the second is Bacongo, where the figures are 747 and 40 per cent. In these towns even the married women are hard to please—they do little work, have much liberty and much money, and have frequent recourse to easy divorce, in which a quarter of all the marriages end. Further, the situation is most unfavourable to polygamy, for wives are scarce and expensive and bring in no return as they would in rural area where they would do most of the agricultural work; no wonder the men hesitate. In the two towns named above, the number of women per family is only 1·2 and 1·06 respectively.[1]

At Léopoldville (Kinshasa), where the female deficit is also considerable, 44 per cent of the men are bachelors and 8 per cent have wives in the country; conversely, 88 per cent of the adult women in the town are married.[2] Such towns, however, in which the inequality of the sexes is so marked, are very exceptional and we can draw no general conclusions from them.

It is nevertheless worth noting that urban life seems to favour marriage rather less than that of the countryside, particularly in the case of women:

Percentage of Married Persons in White Population of United States, 1950

Age group	Towns Men	Towns Women	Country districts Men	Country districts Women
20–24	38·7	61·3	34·7	72·9
40–44	87·3	81·1	86·9	92·0

Source: J. Bogue, *op. cit.*, pp. 224–6.

There are, however, variations which are related to the size of the town, and it is in the cities of over a million people in the United States that the deficiency is most marked; in relation to what would be regarded as normal, given the age composition of the population, it is 2·5 per cent in the case of white males and 4·2 in the case of white females. In towns of medium size, the proportion of married males is slightly higher than 'normal', and appreciably so in the case of small towns with 1,000 to 10,000 people. The deficiency of married

[1] M. Soret, *op. cit.* pp. 95 ff.
[2] J. Denis, *op. cit.*, pp. 217 ff.

women is noticeable in all towns of over 10,000 population, whilst on the contrary there is a surplus of 8·4 per cent in the agricultural zones. It is clear that women who find abundant employment opportunities in the town are capable of living more easily on their own.

A second variation results from differences in the age of marriage; young men get married at a later age in towns than in the country, and it is only in the thirty to thirty-five age group that the proportion of married men becomes larger in the country than in the town. Girls however, marry earlier and in large numbers, and it is noteworthy that the proportion of girls of under twenty who have married is higher in the town than in the country.

If it is more difficult to achieve, urban marriage is also less stable; it seems, at least according to the United States statistics, that the proportion of divorces is between three and five times greater in towns than in country districts. Within the forty to forty-five age group, 3·4 per cent of the men and 5·1 per cent of the women living in towns are divorced, whilst for the same group in the country the percentages are 1·4 and 0·9 respectively. The instability is even greater amongst the Negro population.[1] In France, the highest proportion of divorces is found in the *département* of Seine, which forms part of the Paris conurbation, and in the industrial *départements* which are heavily urbanised.

The effects of urban life are equally important in respect of the most direct result of marriage, namely the birth of children. It is well known that the birth rate is lower in towns than in the country, and that the larger the town, the lower the birth rate. In France in 1954 the birth rate in the city of Paris was 15·7 per thousand, in Seine *département* excluding the city, 16·5 per thousand and in the whole of France 18·8 per thousand.[2]

Naturally enough the birth rate varies widely according to the general circumstances of the country in question, and whereas in western Europe the great metropolitan cities like London and Paris have birth rates of about 16 per thousand in the Far East we find Chinese towns with rates of over 40–45·7 per thousand at Shanghai (in 1957) and 42 per thousand at Pekin.[3] In Brazil the birth rate in the State of Rio de Janeiro was 44 per thousand in 1957 whilst in the capital city itself it was only 25 per thousand.

The normal calculation of birth rate, however, is made on the basis of the number of births per thousand of the population. This is

[1] P. Glick, 'Marriage instability: variations by size of place and region', *The Millbank Memorial Fund Quarterly*, 1963, pp. 43–55.
[2] J. Bastie, *loc. cit.*
[3] S. Chandrasekhar, *China's population*, Hong Kong, 1960, p. 54.

a disadvantage in trying to compare the country districts with the towns, for the former, as we have seen, have a population whose age structure is markedly different from that of the towns; we must therefore adopt a different calculation that takes account of the number of women. Unfortunately the experts differ in their methods of calculation, and true comparisons are therefore rarely possible.

We may take as an example some calculations of 'fecundity rate' made by demographers in the United States; this is the ratio between the number of children under five years of age and the number of women aged twenty to forty-four. It shows clearly the increase in fecundity as one passes from the largest cities to the rural zones:

Fecundity rate in the United States
(per thousand)

Cities of over 3 million inhabitants	433
Cities of 1 to 3 million	478
Cities of 250,000 to 1 million	503
Cities of under 250,000	510
Towns of over 25,000	522
Towns of over 10,000 to 25,000	525
Towns of over 2,500 to 10,000	570
Towns of over 1,000 to 2,500	609
Towns under 1,000	629
Agricultural population	766
Average for whole United States	587

Source: W. Petersen, *Population*, 1961, p. 191.

In Argentina in 1957 the number of children under five for every 100 women aged fifteen to forty-four was twice as large in the rural areas (52·9) as in the towns (24·8). In Germany in the late 1930s 28·1 per cent of the families living in towns of over 100,000 people had no children, while in the small towns of under 2,000 the figure was only 13·5 per cent.

We have already noted the difference in the birth rate between the city of Paris and the suburban fringe, and a detailed examination of the available statistics shows that a similar contrast is to be found in the United States between the crowded cities and their more open suburbs. The number of children under five per thousand women aged fifteen to forty-four rises from 404 in the central areas of cities with more than 3 million inhabitants to 491 in the suburbs.[1] In Brazil

[1] J. Bogue, *op. cit.*, pp. 305 ff.; T. L. Smith, *Fundamentals of population study*, 1960, pp. 309 ff.

the number of children born to 1,000 mothers aged fifteen and over is 447·9 in the towns, 503·7 in the suburbs and 563·6 in the surrounding rural districts.[1] This is but the normal and natural consequence of the fact that the suburbs offer young families more favourable conditions; there are more houses with gardens, more open space and rents and land values are lower. Many city-dwellers with several children leave the town for the comparative comfort of the suburbs; and furthermore, part of the population of the suburban fringes of large cities consists of new arrivals from the countryside and from small towns who have preserved, at least for the present generation, the habit of having large families.

A similar situation can be found in India.[2]

Rural and Urban Fecundity in India

(*Number of children under five per 1,000 women aged fifteen to thirty-nine*)

	1921	1951
Whole of India	647	712
Towns of over 500,000	388	632
Towns of 100,000 to 500,000	544	726
India excluding towns of over 100,000	651	712

The figures for 1921, as for 1931 and 1941, are what one would have expected, and they confirm the observations of Kingsley Davis on the subject of the demographic evolution of the Indian population.[3] But 1951 is quite a different matter, and one of the suggested explanations for the difference is the intensity of rural–urban migration. In 1951 70·8 per cent of the population of Bombay was immigrant, 39·8 per cent of that of Madras and 20·2 per cent of Hyderabad. It must be assumed that these immigrants are for a time keeping up their rural fecundity, and so blurring the distinction between town-dwellers and country-folk. The population growth curve for Hyderabad, for example, shows a flattening towards 1931 and then a sharp rise, with 58 per cent increase between 1931 and 1941 and 53 per cent in the following decade.

---

[1] *Contribuçoes para o Estudo da Demografia do Brasil*, pp. 320–5.

[2] W. C. Robinson, 'Urban-rural differences in Indian fertility', *Pop. Studies*, 1961, pp. 218–34.

[3] K. Davis, *The Population of India and Pakistan*, Princeton, 1951, pp. 70–1.

The same phenomenon has been noticed in Europe. Around Milan, in the quarters inhabited by new arrivals from southern Italy, the immigrants retain for one generation at least a birth rate substantially higher than that of the genuine Milanese. This of course applies to only a small part of the urban population, and has little or no effect on the fecundity rate of the whole conurbation, but the circumstances are very different in the towns of those underdeveloped parts of the world where population is increasing rapidly.

Other anomalies are to be found in countries well developed demographically, in which one would normally expect the large industrial towns to be islands of low fecundity; these are newly created urban centres, like the British New Towns, where the bulk of the population consists of young families at the peak of their reproductive period, or the new towns of the Communist world in which the proportion of young adults is high. An example of the latter type occurs in the 'recovered territories' of western Poland, which were completely repopulated after the war; the colonists came from the Russian occupied areas of the east or were uprooted from other parts of Poland, and their birth rate was much higher than in Poland as a whole, reaching 34.2 per thousand at Szczecin as against 26·3 per thousand for all Poland. In actual fact, 130 towns in this western region have now surpassed their prewar population totals.[1]

Lastly, among those countries of advanced demographic development in which the country-folk practice birth control as much as the town-dwellers, there is the curious example of France. Here, the birth rate continues to decline, even in certain rural areas which, because of their strict observance of Roman Catholic prescriptions, or from sheer force of habit, had maintained until recently a rate that was above the national average; on the contrary, the decline has been arrested in the towns, where new legislation on the subject of families has caused a slight rise; the wage-earning classes benefit appreciably from family allowances and other concessions, and this has tipped the balance in favour of larger families.

This analysis of the family in the town does not lead to simple conclusions. It is clear that just as there are towns of different sizes, ages and functions, so there is very great variation in the attitude of their inhabitants towards family life. Nevertheless, we may tentatively suggest that towns fall into three groups:

1. Towns in demographically well-developed countries in which there is little difference in fecundity between town and country, the

[1] L. Kosinski, 'Les problèmes démographiques dans les territoires occidentaux de la Pologne et les régions frontières de la Tchécoslovakie', *Ann. de Géog.*, 1962, pp. 79–98.

fecundity rates vary but little, and if anything there is a slight tendency, under certain circumstances, for the town rate to increase.

2. Towns in countries of active demographic development, in which the urban population is more educated and informed; birth control is widely practised in the towns but not in the country and there is a large difference in the fecundity rates between the two environments.

3. In countries where the demographic evolution has hardly begun, masses of rural people are converging on the towns, which are incapable of absorbing them, assimilating them and employing them; here there is little difference in fecundity between town and country, and the rôle of the town in reducing the birth rate has not yet commenced.

Once established, the urban family has a tendency to form an isolated cell consisting simply of parents and children, thus differing profoundly from the situation in the rural areas where families of two generations often live together, even in western countries. In regions where tribal life still exists in the countryside, as in Africa and the Middle East, the change is fundamental and farreaching; the transition from a collective existence to a family unit is often difficult, and it creates in the young people feelings both of pride and disquiet. It is true that this 'independence' often does not last long; it is counteracted by what one might call family parasitism. When an African man finds paid employment in a town he soon brings a relative from his native village or perhaps one or two members of his tribe, to accompany him.

In the housing accommodation built for the workers on the Fria project in Guinea (a new bauxite mine), it was not uncommon to find ten or twelve adults in the space provided for a family with two or three children; in Brazzaville, anything from two to ten people live on one person's wages. In Latin America, too, the same custom exists, but the 'family' is rather more limited in scope.[1] We might be justified in assuming that this is a passing phase; but Italian films have portrayed the reunion, in the working-class quarters of Milan or Turin, of all the members of a Calabrian or Sicilian family after the immigration of the eldest son.

In general, setting up a home in a town marks an important transition for the individual, for he passes from the traditional, restrictive and often narrowly conservative atmosphere of the tribe or the family to the comparative liberty that the town offers.

---

[1] M. Santos, 'Quelques problèmes de grandes villes dans les pays sous-développées', *Rev. Géog. Lyon*, 1961, pp. 197–218.

THE HEALTH OF THE CITIZENS

Urban concentration is both good and bad for the health of the citizen.[1] Its worst aspect is the indiscriminate mixing, the crowds, the involuntary contacts, that favour the spread of epidemics and the development of illnesses due to poverty and lack of adaptation to a somewhat artificial mode of life. Its best feature is that it permits, especially in large and old-established towns, the installation of excellent sanitation, the presence of many medical specialists and the existence of efficient and rapid means of medical attention. We must not be surprised, then, if towns in their early stages are human death-traps, and later on become privileged localities. As always, there are examples of both stages in the presentday world, but if we glance back to nineteenth-century Europe, we shall find that the growth of the industrial towns was accompanied by a cruel slaughter of humanity.

Descriptions of the birth of industry and towns in the United Kingdom or in France, in the early nineteenth century, are quite shocking: at Manchester and at Lille, between 60 and 80 per cent of the babies born died before reaching the age of five; at Roubaix, where industrial workers from neighbouring villages or from across the Belgian frontier were crowded, 34·9 per cent of the children died before reaching their fifth year. Whilst births regularly exceeded deaths in rural districts, about 1840, the town of Lille scarcely achieved equilibrium and in some years deaths actually exceeded births. Industrial diseases made their appearance and the poor workers were grievously affected. At York, in 1839–41, the average life-span of the workers was only 23·8 years, as against an average for the whole population of forty-eight years and for the 'gentry' of fifty-six years. About 1860 the death rate in the new mining and manufacturing towns was higher than in most rural districts. But the transformation came about gradually: wages rose, sanitation improved, legislation took more care of the workers' health, and towards the end of the century the sum total of progress neutralised the deplorable affects of the early urban growth.

It is not surprising that similar conditions are to be found today in the urban growths that are springing up in the underdeveloped countries. In India tuberculosis has spread appreciably as a result of urban overcrowding, poverty and undernourishment; in the rural areas it is reckoned that about 20 to 35 per cent of the people are affected, but the proportion rises to 80 or 90 per cent in the poor quarters of the towns, and the latent infection develops into the characteristic disease more often in the town than outside it, in 7 per

[1] On this subject, cf. many papers in *Population Studies*, March, 1964, especially: B. Benjamin, 'The urban background to Public Health Changes in England and Wales 1900–1950', pp. 225–48.

cent of the urban population of Bengal and only 0·6 per cent of the rural population.[1]

In Black Africa the situation is no better. The low huts, damp in the wet season and badly aired, and the deficiencies of the diet favour tuberculosis (0·8 per cent of the illness and 12·4 per cent of the deaths), rheumatism (11·5 per cent of the illness), respiratory diseases that cause almost one death in every five, intestinal parasites, and skin diseases. At Brazzaville in 1948, with an African population of roughly 50,000, the hospitals and clinics dealt with 53,000 cases of illness during the year.[2]

Quite apart from offering concentrated liablity to infection, town life also disturbs traditional habits. This results in the appearance of deficiency diseases, as the medical profession has for long recognised. Africans who have just arrived in a town tend to adopt the dietary habits of the whites, but as they have little money they can only afford the cheapest food, and instead of the natural products and tasty sauces of their accustomed diet they eat bread and canned foods. Dental decay rapidly follows. An investigation in Salvador showed that the urban immigrants drink 30 per cent less milk and eat only half the fruit and 60 per cent of the vegetables that they consumed in their native locality.[3]

Coming as they do from all directions, the urban immigrants convert the towns into breeding-places of infection, and they are the more susceptible if they are weakened by dietary changes and unaccustomed habits. As long ago as 1902 it was noticed, during a cholera epidemic in India, that more than half the cases were attributable to Puri pilgrims. Pilgrimage shrines are often centres for the spread of disease. The mass movement of people in Africa renders almost impossible the complete suppression of malaria; amongst the crowds of workers who seek employment in Southern Rhodesia, numbering 120,000 a year, from Northern Rhodesia (Zambia), Nyasaland (Malawi) and Portuguese East Africa, countries where the medical battle against this scourge is not very strong, between 20 and 30 per cent are infected—and they go into areas from which malaria has been eradicated or in which it is under control.[4]

Even in the well-equipped urban environment of the United States it is recognised that country-folk who migrate to the towns often go through a difficult period of adjustment, during which they are a ready prey to certain ailments and especially to lung troubles.

[1] J. Beaujeu-Garnier, *Géographie de la population*.
[2] M. Soret, *op. cit.*, pp. 114 ff.
[3] J. Beaujeu-Garnier, 'Les migrations vers Salvador (Brasil)', *Les Cahiers d'Outre-mer*, 1962, pp. 291–300.
[4] R. M. Prothero, *Migrants and Malaria*, London, 1965.

However, we must not paint too black a picture of the consequences of urbanisation on the health of citizens, for the statistics indicate that human beings find the town a very favourable environment. Transformation proceeds. At Libreville, a health campaign was undertaken in 1946, and thanks to the use of insecticides after 1952, to the drainage of malarial swamps, to systematic vaccination since 1953, and to the distribution of preventive medicines in schools, excellent results have been obtained; the malaria index has fallen from 70 or 80 just after the war to about 25 in 1954. Libreville, which had the reputation, like many other African towns, of being a white man's grave, is no longer so; white families live there in good conditions and the improvement of the health of the African is very marked.[1]

A similar transformation is affecting Brazilian towns; for in the eight largest towns the death rate is only 16·7 per thousand compared with 22 per thousand in the country as a whole. There may, of course, be errors in the compilation of the statistics, but it is indeed remarkable that São Paulo, the richest and most advanced of all the Brazilian cities, situated on a high and healthy plateau, with good medical services and a high average level of earnings, should have a death rate of only 11·5 per thousand. Rio de Janeiro, second city in economic importance but with a coastal situation, a slightly less favourable climate, and a more mixed population, occupies second place in the list with a death rate of 16·4 per thousand, but mortality from tuberculosis is two-and-a-half times greater than in São Paulo. Amongst the least favoured cities are those of the north-east, like Recife and Fortaleza with a rate of 27 per thousand—the result of a hot climate, of the interpenetration of land and sea, of low wages and inadequate urban services, and of a continuous stream of immigrants from the dry hinterland.[2]

During the second stage of urban development, therefore, the town appears as a more favourable environment than the neighbouring countryside. It provides better economic conditions, better cultural facilities, better sanitation and better health services. In Brazil, the death-rate in São Paulo, Rio or Bahia is lower than in the states of which these cities are the capitals. In the Guinea Republic, an accurate survey made in 1955 showed that infant mortality was 202 per thousand in the interior, and expectation of life only thirty or thirty-one years, whilst at Conakry, the capital, the figures were 142 per thousand and thirty-five to thirty-six years. People living in the immediate hinterland could make use of the doctors and hospitals in Conakry, but for those living in the interior there was but one military doctor to 40,000 widely scattered people, and the native nurses,

[1] G. Lasserre, *Libreville*, Paris, 1958, pp. 260 ff.
[2] *Contribuções para o Estudo* . . ., pp. 722 ff.; and Brazilian *Annuaire Statistique*.

despite their devotion, simply could not cope with all the cases of illness.[1]

Similar evidence is forthcoming from other countries. In Yugoslavia, for example, in 1950, the officially recorded death rate was 13·2 per thousand, while the large towns had rates between 8 and 10 (Belgrade 8·6, Zagreb 9·1 and Ljubljana 9·8).

Women and children are the principal beneficiaries in this intermediate stage. Pregnancy and childbirth are a severe test of the female organism, especially in primitive countries where women have to work hard in the fields and where, at the moment of parturition, they may be subject to certain rituals and traditions. In Black Africa the countrywoman must undergo her accouchement alone in a hut or subject to the ministrations of an untrained and possibly unhygienic old woman, and if she experiences any difficulty she will be plied with unguents and potions the composition of which is best left undescribed. As soon as medical assistance is available and hygienic precautions are taken, the death rate of females aged between twenty and forty drops appreciably.

Infant mortality likewise falls. In Brazil during the years 1948–50, the north-eastern towns had an infant mortality rate of 240 per thousand, but in São Paulo the rate was only 86 per thousand. In the towns of Black Africa and south-east Asia that have some medical services infant mortality is only half what it is in the country districts where such services are lacking. In China, where remarkable efforts have been made to preserve child life, urban infant mortality rates have fallen remarkably; between 1952 and 1956 the rate fell from 65·7 to 35·1 per thousand in Pekin, from 81·2 to 31·1 in Shanghai and from 47·7 to 25·1 in Canton.[2]

Even in advanced countries there is still a lower infant mortality rate in the towns than in the country districts. In France, in 1953, the overall rate was 33·7 per thousand, but for rural *communes* it was 38·2, for towns of under 10,000 inhabitants 34·5, for towns of 10–50,000, 31 per thousand, and only 26·8 in towns with over 50,000 people. Clearly the larger the town, the better the health services and the greater the possibility of preserving infant lives. The rate also varies with socio-professional class, and French statistics for the same year showed rates varying from 17 per thousand for the upper classes, 22 per thousand for the middle classes (business and professional people) and 54 per thousand for the lowest working classes.

Thus the urban infant mortality cycle consists of two periods. In the first, all the unfavourable factors that we have described above in rapidly growing and poorly equipped towns contribute to raising

[1] *Enquête de Guinée*, 1955.
[2] S. Chandrasekhar, *op. cit.*, p. 56.

the mortality rate; in the shanty towns on the outskirts of Bombay or Calcutta rates of 666 per thousand are recorded, and 348 per thousand even in the textile town of Ahmedabad. But from the moment when the advantages of proper sanitation and health services begin to overcome the inconveniences of overcrowding, infant mortality is sensibly lower in the town than in the country.

The same thing is not true for the adult population, for it is well known that in highly developed towns the death rate is higher than in the country districts. What is the cause of this inversion? American sociologists have put forward five reasons why the advantage lies with the countryside: the smaller degree of concentration, which hinders the spread of infectious diseases; greater maternal care; open-air work and physical activity that are conducive to a more stable physical equilibrium; a less hectic life; and the easier adaptation of the human organism to rural life, which is more natural in contrast to the artificiality of urban existence. Furthermore, in advanced countries sanitation and health services are more uniformly distributed, and the towns can no longer derive from these two factors the decisive advantages which tipped the balance in their favour in the intermediate stage.

The United Kingdom provides us with an excellent example. Taking the mean mortality rate for England and Wales as 100, the figure is much less in the rural districts (90 for men and 97 for women) whilst all the large conurbations except London have figures in excess of 100. Mining and industrial towns appear to be particularly lethal, especially for the male population: the county of Glamorgan holds the record with 116 for men and 113 for women, in its towns, and the highest male figure is for Liverpool (123) with working-class towns next on the list (Manchester, Middlesbrough, Merthyr Tydfil and Oldham). The same towns are also high on the list in respect of female death-rates, as are the textile towns like Rochdale and Burnley. On the other hand the mainly rural counties appear privileged, even in respect of their towns. The London conurbation is the one exception; its death rate is slightly below the national average, but there are great variations within the urban complex, for the index rises to 120 in the densely peopled East End where male mortality is high, whilst in the open suburbs of the west it is below the national average. This anomaly presented by the capital city is to be correlated with the intensity and quality of the means available for the battle against disease and death. The case of Paris is very similar.

Urban death rates are in any case related to the general state of the country and so to the national rate. In France, the general death rate is 12 per thousand, and in Paris it is 10·2 per thousand. In the United States, the general rate is 9·6, with 10·5 for the urban population and

8·3 for the rural. In India, it is difficult to establish the death-rate with any precision (it was officially 16 to 17 per thousand about 1950), but fragmentary data for the same period gave figures of 14·2 per thousand for Great Bombay, 13·6 for Calcutta, 9·5 for Delhi, but 28·4 for Madras.[1] In China published figures show a striking fall in the death-rate, from 9·9 per thousand to 5·9 per thousand between 1953 and 1957 in Shanghai, for example, with a similar drop in Pekin.[2]

Urban health services are clearly always superior to those of the countryside, and are generally better in the larger towns. But the quality varies, and certain specialist services and equipment are only available in very large centres—to the great benefit of the inhabitants of such towns.

SOCIAL BEHAVIOUR

The citizen has a higher personal income, in general, than the countryman. In the United States, an urban family had an average income of $5,221 a year in 1956, whilst the agricultural family had scarcely half this, only $2,371. There are also some correlation between the size of the town and the average income:

Average Income in the U.S.A. 1956
(dollars)

	Per family	Per man	Per woman
U.S. average	5,953	4,807	2,935
Towns of over 1 million	6,480	5,057	3,134
Towns of 250,000 to 1 million	5,836	4,647	2,969
Towns of under 250,000	6,012	4,852	2,866
Towns of over 50,000	5,819	4,775	2,757
Towns of under 25,000	5,241	4,384	2,531

Source: J. Bogue, *op. cit.*, p. 659.

In general, the income for a white farmer was only 49 per cent that of a citizen of a town of over 100,000 inhabitants; and in the case of Negroes the disproportion is 1 to 3.

In Brazil, to take an example from quite a different region, the minimum monthly wage in Rio or São Paulo is 15 to 30 per cent higher than in surrounding rural districts and two or three times higher than in the countryside of the north-east.[3]

[1] G. Haberland, *op. cit.*, p. 52.
[2] S. Chandrasekhar, *op. cit.*, p. 54
[3] *Annuaire Statistique*.

The citizen, moreover, has more opportunities than the countryman; his tastes develop through what he hears and what he sees; he is tempted by a variety of things and so creates new needs. The town has better appointed streets, and more comfortable housing. To take the telephone service as an example, we find 7·6 telephones per thousand people in the whole of France, but 60 in Lille, 88 in Lyons and 200 in Paris.

The town dweller not only goes himself more frequently than his country cousin to schools and other places of culture but he appears more willing to send his children there as well. In the United States in 1957 the urban population aged twenty-five and over had completed on an average more than ten years' schooling, whilst the rural population had had only 8·5 years. Contrary to what one might expect, the urban–rural difference is greater amongst the younger folk than in the case of the more elderly; for men of over seventy-five the difference in length of schooling was only 1·3 years, but for those aged twenty-five to thirty it was 3·3 years.

If the young citizen gets a better education than the rural child, this is partly due to the existence of better facilities, but it also results from the higher family income—itself a consequence of the smaller number of children per family—and from the more ready availability in the town of jobs for people with university or technical college qualifications. Amongst the Negro population of the United States, the urban children have a longer education than rural whites, and nearly twice as long as rural Negro children.

There are also many distractions for the town dweller, both cultural (cinemas, theatres, museums, and libraries) and sporting. As he gets older, the accent is very much on sport, as a necessary compensation for a sedentary life. All these activities make great demands on the town's facilities; every small town in France has its cinema, and even medium-sized towns have their theatre, museum, library and several cinemas—and France is by no means exceptional in this. In the U.S.S.R. the building of new towns is always accompanied by the creation of a social centre where cultural facilities are available for the people, together with meeting rooms and exhibition hall; the community centres of British New Towns and new estates in older towns are smaller-scale examples of the same trend.

The sporting facilities of French towns are very poor indeed, but as soon as one crosses the Swiss or German frontier there is an abundance of swimming pools, tennis courts, athletic grounds and sports stadia. In Canada the general rule is that there should be a sports ground of $7\frac{1}{2}$ acres to every 13,500 people; it must be accompanied by a park and must provide for spectators. In England the minimum suggested size for sports grounds is 11 acres. Yet a city the size of Lille, with 372,969 people, has only 42 acres devoted to sport and it is by no means the worst off French town in this respect.[1]

In the old colonial towns, there is a great contrast between the indigeneous quarters where sporting facilities are meagre or non-existent, and the European quarters with their numerous and often magnificent polo grounds, golf courses and the like, that are unavailable to the mass of the population.

Even if the citizen is not himself a performer, he likes to watch the exploits of others, and hence the vast stadia built to hold a huge army of spectators, which at weekends attract thousands or tens of thousands to witness athletics, football, horse or dog-racing. A large stadium and a race course are essential components of a large town.

Lastly, the citizen, harassed and crowded all the week, deafened by noise and suffocated by petrol fumes, escapes whenever he can to the countryside. Many large towns thus have their ring of second homes in the green belt. Around Paris some villages have been almost completely abandoned by the local agricultural population and have been transformed into collections of reconstructed cottages and new houses that are occupied from Saturday to Monday during the summer half of the year. Around Moscow there is a belt of villages—the famous 'datcha' zone—in which during the summer, some 300–400,000 Moscovites spend their nights.[2] The wealthier Londoners have their country or seaside cottage in Kent or Sussex, or in the Cotswolds, whilst in parts of Canada and the United States the weekend cottage or 'summer home', preferably on the shores of a lake, is a very widespread amenity.

All these elements combine to give urban life a particular rhythm and to introduce some remarkable features into the urban landscape, such as large buildings for performances and exhibitions, grounds of different shapes and sizes for various sports, and the convergence or divergence of the citizenry towards the displays and sporting contents or out of town to their country residences.

Contrary to rural life which responds to climate and the seasons, urban life obeys a man-made rhythm. The streets are thronged in the morning, at midday and in the evening, transport is full at the same hours; food shops and markets are busy in the mornings, other sorts of shopping are done mainly in the afternoons. The appearance of a departmental store and of certain streets changes remarkably during the course of the day.

Over most of the world the urban atmosphere on Saturdays and Sundays is different from that of the rest of the week, but in Jewish

[1] A. Millon, 'Les espaces verts à Lille', *Bull. Soc. Géog. Lille*, 1962, no. 5, pp. 5–41.
[2] *Goroda sputniki*.

areas the religious day is Saturday, and in Muslim countries, Friday. Saturday afternoon, being free for most office and factory employees, is the time when shopping is at its height and the activity of the central business district reaches its climax. On Sundays, in western countries, other than in the places where performances of one kind or another may be staged, and in the promenades of the public parks, life moves at a slower pace; all those who can leave the town do so, the rest, apart from their church attendance and perhaps an afternoon stroll, stay at home. In the U.S.S.R. the shops are open on Sundays, thanks to a rota of assistants, so the citizenry can make its purchases or gaze into the shop windows.

One could make, for any given town, large or small, a series of maps indicating the concentration of people at different times of the day and week. This would indeed be a living picture of the town, in which the people move like the blood-stream in the human body.[1]

SOCIO-PROFESSIONAL CLASSES

The citizen is not merely in the town to live and to die, to move around and to enjoy the facilities offered by the health and education services, the sport and the culture; he is there to work. The variety of employment opportunities offered and the wide range of possible earnings are characteristic features of the urban environment.

In the traditional village, life revolves around the local environment. The farmer exploits his land, and the craftsmen—who still exist in the developing countries—are spinners, weavers, and carpenters for their village neighbours, while the very small number of shopkeepers merely cater for the local demand. On the contrary, town life is characterised by the wide range of its external relations, for the export of manufactures, the import of raw materials, and trade of all kinds.

The prime characteristic of the occupation structure of urban populations is thus the growth of that part of the tertiary sector that comprises business, finance and administration in all their various forms, education and entertainment. This part of the tertiary sector must be distinguished from another part which is officially included in the same category, namely domestic service and petty trading. In the towns of the world's underdeveloped countries, the man who sells the odd pound of sugar or packet of cigarettes, and the woman who hawks foodstuffs by the roadside, are called 'traders'. If we were to take the official statistics at their face value we should end up by

[1] F. S. Chapin and S. H. Stewart, 'Population densities around the clock', in H. M. Mayer, *op. cit.*, pp. 180–4.

regarding Bahia as a more advanced city than São Paulo, because the proportion of its people employed in the tertiary sector as a whole is higher.

In the second place, the urban population includes a variable number, according to the function of the town, of employees in the secondary sector, which includes all branches of industry. There is here, it is true, another awkward point of classification, as to whether or not extractive industries such as mining should be included. Some authors assign mining to the primary sector, and this enables certain towns to be distinguished, in which this sector plays a large part in the occupation structure. Thus in the northern coalfield of France, a town like Lens has 72·5 per cent of its population engaged in industry (*sensu lato*) but of these 50·1 per cent are in mining. It is a matter of convention; here we shall regard extractive industry as falling within the secondary sector.

If to the best of our ability we compare occupation statistics which are not only different in their method of compilation but also in the information they provide, and frequently imprecise and difficult to interpret, we may follow many previous workers in attempting a classification of towns.[1]

1. First, those in which there is a substantial amount of primary employment remaining: these are towns that have barely emerged from the rural environment, either because they have evolved slowly from local market villages into towns, or because they have grown spatially by the immigration into their fringes of new arrivals from the countryside. We may also add the rural centres of the future that may be developed to house the agricultural workers. There are thus three subtypes:

(*a*) European country towns which are incompletely built-up so that substantial farms persist on the fringes or perhaps specialised market gardens or orchards. In northern France, a small town like St Omer has 9 per cent of farm-workers amongst its population, and a sub-prefecture like Hazebrouck has 7·3 per cent. This persistence of farming employment is generally accompanied by produce markets and sometimes by food preparation industries.

(*b*) Towns in the underdeveloped countries where there is no definite boundary between the town and the villages of the rural fringe. Libreville is an example: only a small part of the population is officially recorded as active—only a third of the African population, indeed—but many women work in the plantations that surround the

---

[1] Compare for example P. Pinchemel, *Niveaux optima des villes*; C. D. Harris, 'A functional classification of cities in the United States', in *The Sociology of Urban Life*, New York, 1951, pp. 84–96; F. Carrière and P. Pinchemel, *Le fait urbain en France*, Paris, 1963.

town and thus maintain the link between urban activities and the primary sector.[1]

(c) The 'agrovilles' that the Soviets have planned for their rural population, represent in evolutionary terms a return to the close relationship of town and country. A somewhat similar idea lies behind the rehabilitation of 'rural centres' in France.

2. In the second group are those towns whose occupation statistics show a predominance of the tertiary sector. But though the statistics may put them all into one group, qualitative considerations enable us to distinguish several very different types.

(a) In the underdeveloped countries, where the towns are poorly developed economically, a large part of the population lives by its wits, as poor artisans, or as petty traders selling articles made at home by the women, or else in domestic service. All these occupations are classed in the tertiary sector. Amongst the active African population of Libreville, mentioned above, there are 13·8 per cent household servants and 17·2 per cent artisans and self-employed traders, of whom four-fifths describe themselves as 'employers'; together these groups comprise nearly one-third of the active Africans. But if one takes the statistics without further comment, one finds that 25·9 per cent of the Africans in Libreville are employed in industry and all the rest in the tertiary sector.

Similar conclusions can be reached with regard to a town like Dakar, which contains 7 per cent of farmers and fishermen, especially in the old Lébou villages; 39·5 per cent of the active population finds employment in the secondary sector—in building, engineering, local crafts and transport; whilst 53 per cent are in the tertiary sector, of whom 8·5 per are domestic servants.[2]

Certain Indian towns present a similar picture: in Madras, 73·2 per cent of the active population are in the tertiary sector, with commerce employing 22 per cent and 'various services' 41·9 per cent; the historic town of Vellore, the most famous hospital centre in India, has 26·3 per cent in the secondary sector (weavers and other craftsmen) but 70·2 per cent in the tertiary, of whom 45·6 per cent are in 'various services'.[3]

(b) A well-developed tertiary sector is also found in some ancient historic towns in the old countries of Europe; these towns have retained an administrative function or have become regional commercial centres, but they have not been able to attract modern industry. So they are generally quiet middle-class towns, with many monuments

[1] G. Lasserre, *Libreville*.
[2] *Census of Dakar*, 1955.
[3] J. Dupuis, *Madras et le Nord du Coromandel*, Paris, 1960, pp. 380 ff. and 511 ff.

attesting to their ancient importance. Amongst French towns in this group we find Arras, with 61·1 per cent in the tertiary sector, of whom about half are in commerce, banking and insurance and 22·8 per cent in other services, and Vannes, with 66·3 per cent in tertiary occupations.

(c) Lastly, there are included in this group certain well-developed towns that have specialised in one or other technical branch of modern civilisation, or in finance. They may be based on a number of different activities. Transport, for example in a great port, carries with it, as well as the personnel employed in the port itself, a whole train of specialists in insurance, commerce and banking. Thus Dunkirk, with 56·6 per cent of its active population in the tertiary sector, includes 21·8 per cent employed in transport and 26·1 per cent in insurance, banking and commerce.

Another dominant activity may be shop-keeping. The town of Kearney, Nebraska, has 36·2 per cent of its active population employed in retail trade, and 25·6 per cent in other specialised services.[1] In the United Kingdom, Crosby in Lancashire has the highest proportion in retail trade, with 23·9 per cent; a small residential town with a high proportion of 'white-collar' workers who commute to Liverpool, it has two-thirds of its active population in services of one kind or another.[2] Phoenix, Arizona, has the highest proportion in the United States of males employed in wholesale and retail trade, whilst Washington holds the record for employment in public administration (27 per cent) and Hartford, Connecticut, the record for employment in finance and insurance (7·6 per cent). In these towns, apart from the speciality named, over half the active population finds employment in the tertiary sector.

3. A third category of towns comprises those in which employment in the secondary sector is dominant. These are the industrial towns. Here too, however, there is no uniformity; we may follow C. D. Harris in including in this category towns in which more than 60 per cent of the active population works in industry.

(a) Some towns are dominated by a single industry, these are mainly coal-mining towns or based on some other extractive industry. In the northern French coalfield there are six towns, the best known being Bruay, that have over 63 per cent employed in mining.

In the same group appear towns that have a more complex industrial structure but still depend largely on one industry. In England and France, the car-building towns are examples, Dagenham having

---

[1] T. L. Smith, *Fundamentals of population study*, pp. 245 ff; H. J. Nelson, 'A service classification of American cities', in H. M. Mayer and C. F. Kohn, *op. cit.*, pp. 139–60.

[2] C. A. Moser and W. Scott, *British Towns*, Edinburgh, 1961.

73·7 per cent in the secondary sector, and Coventry 69·4 per cent; Sochaux-Montbéliard relies on the Peugeot factories, and Le Creusot on the Schneider works. In the U.S.A. Flint, Michigan (65·2 per cent) is also dominated by the motor-car industry.

(*b*) Other industrial towns are highly specialised in one industry and its associated trades; into this group fall the textile towns such as Roubaix (with 67·4 per cent in the secondary sector including 64·4 per cent in textiles) and Armentières (with 63·4 and 47·4 per cent respectively), and also metallurgical towns such as Valenciennes, with 62·2 per cent in the secondary sector, including 31·3 per cent in metallurgy and engineering.

(*c*) Towns with varied industries provide a more varied pattern. In northern France St Amand has 65·3 per cent in the secondary sector, including 20·5 in metallurgical and engineering industries, 17·5 in chemicals, 10·5 in textiles, together with building trades, coal mining, and food preparation. Somewhat akin to this are the towns in which the establishment of an 'industrial zone' or 'trading estates' has favoured the growth of diversified industry.

4. Finally, there are some very evenly balanced towns in which the secondary and tertiary sectors are more or less equal. Large towns are frequently in this position. In Paris, 55 per cent of the active population are employed in commerce and services, and 45 per cent in industry and transport. In Birmingham, whilst industry employs 62·5 per cent (37 per cent in metallurgy and 8 per cent in textiles), the tertiary sector accounts for 37 per cent (services 16, commerce 15, transport 6). In New York, the main occupations are in industry (25·8 per cent), retail trade (13·5 per cent), transport (7·8 per cent) and special services (7·3 per cent).[1] In Tokyo, industry employs 44 per cent and the tertiary sector 55 per cent (including 25 per cent in wholesale and retail trade and 17 per cent in various services).[2]

Matching these great cities are smaller ones that are similarly well balanced if to a solid industrial foundation they add administrative, commercial or other varied functions. Lille, for example, has 49·8 per cent of its active population employed in the secondary sector and 48·8 per cent in the tertiary; Douai, an old historic administrative centre that contains also the headquarters of the Nord and Pas-de-Calais coalfield, has 40·6 per cent in the tertiary and 58·4 per cent in the secondary sector.

Somewhere between all these extremes, taken from various parts of the world, can be found a mean for each country. The United States may be taken as an example: for the 897 towns of over 10,000 inhabitants, the mean percentage in industrial occupations is 27

[1] H. J. Nelson, *loc. cit.*
[2] *Census of Japan*, 1960.

(from 26·7 in the smallest towns to 31 in the cities of over a million), in retail trade 19·2, in special services 11 (the proportions in these two decreasing as one moves from small to large towns), in wholesale trade 3·8 (varying in the opposite direction), in personal service 6·2, in public service and administration 4·6, in transport and commerce 7·1, and in banking and insurance 3·2 per cent.

To this differentiation of towns by the proportions employed in the various sectors of activity we must add a classification based on income or social categories. Pinchemel has proposed for northern France a sixfold classification, that we can reduce to five.[1]

(a) Towns in which the middle and upper classes predominate. The type-specimen is Arras, which has 38 per cent working-class, 52 per cent middle-class and 10 per cent upper class inhabitants; it is an administrative town, with a large number of civil servants and business people; set in the midst of a rich agricultural district, it has little to do with manufacturing industry. Such towns are not very numerous; perhaps Strasbourg could be regarded as an example, with 43·4 per cent working-class and 10 per cent upper-class inhabitants.[2]

(b) Towns with an important body of middle- and upper-class inhabitants. These are very common, having a working class numbering over 50 per cent of the population but with a substantial middle class and a good representation of the upper class. Lille falls into this group, with 55 per cent in the working class, 41 per cent middle class and 4 per cent upper class. Other smaller regional centres are similar, such as Cambrai, a market centre well known for its long-established textile industry, and Béthune, one of the main towns of the coalfield; in Alsace, Mulhouse is an example, with the three classes in the proportion of 55·9; 36·8; 7·3.

(c) An intermediate group of towns, with smaller proportions of middle and upper class and a working-class proportion that rises to three-fifths or even two-thirds. Here we find specialised industrial towns such as Valenciennes, with 62 per cent working class, 35 per cent middle class and 3 per cent upper class, ports such as Boulogne and Dunkirk, and old historic centres with modern administrative and industrial functions such as Douai.

(d) Towns that are predominantly working-class, the proportion rising to two-thirds or even three-quarters of the total population, whilst that of the middle class falls below one-third and the upper class is between 1 and 3½ per cent. These are the industrial towns that have few other functions, like Armentières and Halluin, textile centres, Lens in the coalfields and Calais amongst the ports. The figures for Armentières are 65, 30·9 and 3·1 per cent and for Lens, at the other

---

[1] P. Pinchemel, *Niveaux optima des villes*.
[2] M. Rochefort, *Organisation urbaine de l'Alsace*, 1960, p. 258.

end of this group 75, 23·6 and 1·4 per cent. In the textile towns, the resident mill-owners help to raise the proportion of the upper class.

(e) Lastly, the mining towns, which are concerned with little else but coal and in which urban functions are reduced to the minimum necessary to sustain the life of working-class families—a few shops, and health and cultural services that are limited to the most elementary needs. The proportion of working class inhabitants may exceed four-fifths—in Mazingarbe there are 81·6 per cent working class, 17·7 per cent middle class, and only 0·7 per cent upper class. The structure of the coal industry and of the nationalised chemical industry does not require the residence on the spot of numerous members of the upper class.

In sum, for all the eighty-nine towns in northern France with over 5,000 inhabitants, there are two in the first group, eleven in the second, sixteen in the third, forty in the fourth and twenty in the fifth; the fact that in three-quarters of the towns there is a preponderance of working class inhabitants may be regarded as a regional characteristic. It would be interesting to apply the same classification to other regions of France, for the social structure of the urban network reflects the region that has given rise to it.

This is not the case, however, in all parts of the world; the colonial towns created by Europeans, and the cities of the Middle East, which have been called 'encysted nuclei' in the flat plain that surrounds them, differ profoundly from the general character of the region in which they occur. In the European towns in former colonial territories, there are often two social structures in parallel, the European with its predominance of managerial staff and the African which is mainly working class. Taking Libreville as an example, the European sector has 30 per cent managers and technical staff, 26·5 per cent salaried grades, 10 per cent workers and the remainder administrators and military; the African sector has none in the first group, 9 per cent in the salaried class, 14 per cent workers and 51 per cent labourers, the remainder being employed in public service. The same disparity exists in South Africa between the white and coloured populations, and there is also a great difference in rates of pay; a white worker and a black worker are only equal in the statistical classification.

SOCIAL MOBILITY

The urban environment presents to the individual and to the whole community an opportunity for development that the rural environment could never offer. For the individual, the countryside is a narrow and enclosed world, where everyone knows everyone else and where opportunities for social climbing are minimal. The small cultivator might, as a result of hard work and favourable circumstances such as a fortunate marriage, become a larger farmer, he might even become mayor of the village—but unless he quits the countryside what more can he hope for?

The town, on the other hand, is an environment in which evolution and transformation can more readily take place. Through the concentration of finance and intelligence that it represents, the town is the cradle of inventions and technical progress, and the broadcaster of ideas and innovations. It democratises science, fashion and taste, and in consequence creates new needs and stimulates activity. The urban way of life tends to create solitary souls, and to tear the individual away from his former habits[1]—or, in other words, to make new men.

The power and influence of the world's towns is immense. In 1929 it was estimated that the towns of the United States contained 64·7 per cent of the industrial establishments, 74 per cent of the salaried industrial employees, and 80·7 per cent of all salaried employees, paid four-fifths of all the nation's salaries, and were responsible for 79 per cent of the national output. The eleven largest urban agglomeration handled more than half the wholesale trade: 79 per cent of all rail traffic terminated in towns and the twelve most important cities were the starting point or the destination of half the railway passengers.[2]

The rôle of the town in regional development has often been stressed, and the 'poles of development' theory of certain French economists emphasises the importance of dynamic urban centres. Current planning in France is based on this dogma, as the effort to promote the growth of regional capitals shows.

But the town is also *par excellence* the place in which the individual can develop. The variety of employment, the opportunity for study, and the complexity of the socio-professional structure combine to offer the ambitious man the fulfilment of his hopes and desires. Moreover, the town offers greater anonymity and so greater freedom. The capable and hardworking newspaper-seller may become a Ford or a Rockefeller, and though such a destiny comes to few, examples of elevation in the socio-professional scale are to be seen all around us.

A new kind of human nature develops in the town, and it is because of this, as well as by reason of the economic and technical impulses, that the urban development which is proceeding in the world at such an accelerated pace is capable of transforming the lives of tomorrow's citizens.

[1] T. L. Smith and V. J. Parenton, 'Social cohesion and social control', *The Sociology of urban life*, p. 410.
[2] *Ibid.*, p. 81, 'City functions and perspectives'.

PART FIVE

THE TOWN AND ITS REGION

CHAPTER 28

THE NATURE OF URBAN INFLUENCE

One of the essential preoccupations of urban geography is to place towns in their setting and to relate them to that environment. Town planners themselves have increasingly recognised this point of view, in their enquiries into the ways in which a town and its surrounding region are interdependent.

This is what is sometimes called the regional function of the town, and we have said that this expression seems incorrect to us. In fact we are dealing with an application of urban functions in a particular case, that of the immediate environment. Within this environment, the town exercises a sort of tutelage over the neighbourhood; it is important to investigate just how far this tutelage extends, and thus to define the region round a town that, so to speak, belongs to it.

URBAN INFLUENCE ON AGRICULTURE

The citizen as country property-owner. It would seem that a town is, by definition, divorced from agriculture, leaving the neighbouring countryside to its fate, in order to devote itself to industry and commerce. In reality it continues to exert some sort of influence over the countryside.

First of all, the country is very often the property of the town. In every age townspeople have owned rural properties. Formerly many nobles who lived in towns were in fact landowners. The gentry only resided in London three months a year, and spent the rest of the year on their estates. It was not only the nobility who did this; many middle-class property owners did so too. The land was the soundest of financial investments. The estates round Metz belong to the bourgeoisie of Metz who bought them at a time when it was customary to invest in land, whereas the bourgeoisie of Nancy, dating principally from the industrial era, preferred transferable securities. And then land was a sign of power and renown, and it was a sign of good breeding to have an estate, a second residence or a hunting lodge. Lawyers bought land in order to acquire nobility. The middle classes on the other hand, particularly wished to acquire vineyards: these were properties of the highest value; the bourgeois could afford the risks of loss caused by an always uncertain harvest; and he was very proud to offer his own wine at his table. The vineyards of Burgundy exist thanks to the cities of that province.[1]

There were periods in Europe when the acquisition of rural property was a craze, as in the eighteenth century. Bourgeois property owners benefited from every political and economic unheaval, which presented excellent opportunities to buy at small cost. After the Hundred Years' War the bourgeoisie of Paris bought the lands of the nobles and peasants who had been ruined. The Revolution of 1789 and the sale of national property is another example of this in France.

More recently, the dividing up or the collectivisation of estates has been the ruination of city landowners in some countries, but the controlling power of towns has not been weakened by it. And, in capitalist countries, the rural exodus has increased the amount of land owned by townspeople.

Bourgeois landowners would often only let their properties for rent or on condition that part of the produce was given to the owner. This procedure allowed them to intervene directly in matters such as the choice of crops to be grown. Renting, while giving more latitude, still guaranteed the landowner a considerable influence.

This gave rise to many abuses, especially in districts where large properties remained intact, and town-dwelling landowners were not always the only people to blame. Round Homs in Syria, the land belonged to the middle-class inhabitants of Homs who kept it in a state of stagnation,[2] and it has been said that the town was sucking the country's lifeblood, and was in part living on it. The bourgeoisie, according to Étienne Juillard, is 'the *rentier* of the soil'.

Today, in almost every country, a large part of the land belongs to townspeople. Certain villages in France are almost entirely rented from absentee landlords. It is not only a question of rich landowners. Rural properties retained by people who have left the countryside are very numerous. Roger Facon[3] noted in the district of Charente that, the poorer the land, the more numerous are the city landowners. These landowners sometimes find it convenient to plant trees in order to avoid all the complications of letting and thus increase the agricultural recession irreversibly, as Denise Roumignac noted on the subject of the Millevaches plateau.

But it is not only family inheritance that is preserved. Townspeople are buying land in ever great quantities; it sometimes takes the form of a sort of vertical integration, as when town butchers buy fields on

[1] R. Dion, *Histoire de la vigne et du vin en France*, Paris, 1959; 'Métropoles et vignobles en Gaule romaine. L'exemple bourguignon', *Annales*, 1952.
[2] Nammar, *Le pays de Homs*, 1951.
[3] R. Facon, 'Villes et campagnes charentaises; problèmes de propriété rurale', *Norois*, 1958.

which to raise livestock. Often, too, these purchases are purely speculative: in countries which have suffered from inflation, the value of land remains stable; its exploitation can be supervised thanks to the motor car. And this movement has gone so far in France that it has given rise to a campaign of protest by agricultural workers.

In new countries there is no such thing as property passed down to heirs or patiently acquired over the centuries. The area of influence of a town is not small on this account, for colonisation was accomplished through the towns. The American farmer is in constant touch with the town and when the large-scale Canadian farmer lives in a town for three quarters of the year, one may well ask if he is not in fact a town-dwelling landowner. In any case new impulses always come from towns.

The town thus exerts an influence over the country thanks to the properties it retains there.[1] Most agricultural transformation has only been made possible through the funds of townspeople.

But this is not the only form of agricultural influence exerted by a town over its neighbouring regions.

SUPPLYING THE TOWN WITH FOOD

Even when a town seems unrelated to the country, the country is still working for it. A large part of the countryside is organised to feed the town, and the town depends on this agricultural production. Towns of the Middle Ages lived in constant fear of famine, a fear with which countries at war have since become familiar again. And Henri Pirenne[2] has described the rules which were anxiously applied to all agricultural produce until its arrival on the market. The ways in which food is supplied today are less restricted, but the needs of towns have increased; not only are populations greater; in the past some crops used to be grown inside the town itself; many townspeople had their own vegetable gardens, and in Paris the so-called Fermiers Généraux walls in the late eighteenth century enclosed some fields as well.

Fortunately, certain important goods can be brought from great distances; grain is not affected by long journeys; ancient Rome had

[1] We must not over-generalise. E. Juillard has shown that the citizens of Alsace have been quite indifferent to rural property-owning (*Publ. de la Soc. Savante d'Alsace et des régions de l'Est*, 1954), and though the citizens of Paris have acquired much land, the large landowning farmers dominate the countryside and have brought about a revolution in agriculture (P. Brunet, *Structure agraire et économie rurale des plateaux tertiares entre la Seine et l'Oise*, Paris, 1960).

[2] *Les Villes du Moyen Age*, Bruxelles, 1927.

granaries on every coast of the Mediterranean, and today grain can be obtained from Canada, Argentina and Australia. Supplies of meat are less easy to obtain; but it is above all the daily supplying of perishable goods which presents problems. Milk supplies, for example, call for the raising of dairy cattle within a greater or smaller radius, depending upon the speed of transport available.

Milk collection extends over great distances, for the thickly populated neighbouring regions use up all their own production, the area supplying Paris with milk extends for 250 miles from the capital; the entire English Lowland supplies milk for London and the rural areas of the north-eastern United States have been gradually converted into pastures to supply milk to the great cities of America.

The supplying of fresh vegetables demands a zone round every town devoted to their cultivation. Allotment-gardens form a part of the urban landscape in every country. But these are far from sufficient and, round every town, part of the country is organised to produce vegetables; these are the market-garden suburbs. They are generally found in the well-watered alluvial parts of nearby valleys: in Paris, the early market-gardens lay in a former loop of the Seine; they can be found on the terraces above the Garonne round Toulouse, in the upper valley of the Huveaune for Marseilles, on the terraces of the lower Dauphiné district at Vaulx-en-Velin for Lyons, and in the Lys valley for the towns of northern France. The market-garden zone for Hamburg is the Vierland, situated on fenland, the Marschen, at the mouth of the Geest. London was fortunate in having the loam-covered gravel terraces of the Thames extending both east and west, and as the urban area has expanded, the market gardens have moved further and further out into Essex, Middlesex, and South Buckinghamshire. Sometimes market-garden suburbs are created by the town itself: the soil used to be prepared with manure supplied by garrisons of cavalry, or by omnibus horses; today, sewage sludge is used or manure prepared chemically from the town's refuse.[1]

Ease of transport sometimes places market-gardens at some distance from the town. The horticulturists of the Somme have long been sending vegetables to Paris, and today the Saône valley, Brittany, the lower Rhône valley and the Aquitaine basin are all part of the zone supplying Paris. London gets produce from Bedfordshire, the

[1] On market garden suburbs see, for example: P. Barrère, 'La banlieue maraîchère de Bordeaux', *Cahiers d'Outre-mer*, 1949; M. Phlipponneau, *La vie rurale de la banlieue parisienne*, Paris, 1956; P. Gourou, 'L'agglomération bruxelloise', *Bull. Soc. belge de Géog.*, 1958; C. Christians, 'Problèmes de banlieue maraîchère: la culture fruitière d'appoint à Votten, près de Liège', *Bull. Soc. belge Et. Geog.*, 1960; E. C. Willatts, *Middlesex and the London Region*, Part 79 of *The Land of Britain* (ed. L. D. Stamp), London, 1937, pp. 192–206.

Fen district and even from as far afield as Cornwall and the Channel Islands. But transport costs are high; distant areas specialise in early spring vegetables which can stand the price; they cannot compete with the market-garden suburbs at periods of peak production.

So this special zone continues to exist near to the town. Its rhythm of life is dictated by daily needs, for supplies must arrive at the urban markets at dawn. And, in spite of the natural temperature, the maximum possible must be produced from this precious ground. Crops are grown under glass frames which are covered at night with plaited straw. In colder countries, and where coal supplies are abundant, the heating of greenhouses sometimes works out cheaper than bringing vegetables from further south: this is the case in Belgium. Thanks to the hot water geysers, bananas can be ripened in the greenhouses round Reykjavik, in Iceland.

Fresh flowers are more delicate than vegetables; and the high standard of living in towns necessitates a daily supply. They are brought by express train, and by air from Mediterranean countries; but, here again, it is useful to have some cultivated near towns as well. Floral suburbs adjoin the market; and, as flowers are expensive, special care is taken of them; glass-frame cultivation around Paris is principally devoted to flowers; glass-houses round Copenhagen usually provide flowers rather than vegetables.

The hinterland of the French Riviera has completely transformed its crop systems to make room for the cultivation of the flowers, fruit and vegetables which supply the coastal resorts.

Attempts have been made to define the roughly concentric zones in which the various crops are distributed in relation to the town. According to R. Dickinson and O. Jonasson,[1] there would be a first zone consisting of floriculture, and market-gardening. Then would come a zone of intensive agriculture represented by the rearing of dairy cattle, then the cultivation of sugar beet and cereals. Then comes the area of stock-rearing, and the whole is surrounded by forests. Of course this sketch plan is not universally applicable. Certain crops grown for oils and fats, for example flax, with its oily seeds, must take their place in countries where fats of animals origin are not used. In the United States, vegetables come from further off than milk, and the order would have to be reversed. Then again, this plan is only valid for humid temperate regions and it presupposes that the town is in the centre of a clearing isolated from the rest of the world. Nonetheless, this plan, in the authors' minds has the advantage of showing the extent to which the town dominates the life and economy of the region surrounding it.

[1] Agricultural regions of Europe,' *Econ. Geog.*, 1925, pp. 284–6.

SUPPLYING THE TOWN WITH RAW MATERIALS

The influence of the town on agriculture takes another form; it is seen in the outlet a town's industries offer to the products of the countryside. Often, moreover, these urban industries owe their origin to the raw materials found in the neighbourhood; they are only a consequence of the agricultural mode of life. The wool from the sheep of the Ardennes brought about the cloth-making industry at Sedan, the wool from the Champagne district caused the cloth-trade of Rheims; the glove industry at Millau is due to the sheep skins of the Causses; the box-wood of the Jura which is hard and rewarding to work on was the origin of the wood-turning workshops of Saint-Claude, even as the beechwoods of the Chilterns were responsible for the furniture industry of High Wycombe; the sugar-beet of Brie resulted in the sugar-mills of Meaux; and the flax grown in Northern Ireland gave rise to the linen industry of Belfast.

But more often still, the initiative comes from the town, which requires raw materials drawn from the earth; it is convenient if these can be found locally.

Products for dyeing used to be obtained from plants which were cultivated for the needs of the textile industry: Woad was grown round Toulouse, madder and indigo round Avignon. Mulberry plantations were developed throughout the entire Rhône valley as a result of the silk industry of Lyons. These influences continued to be felt even after the growing of woad and madder and the raising of silk worms were no more. Districts which practise commercialised agriculture in this way do not go back to polyculture but look for other forms of commercial cultivation.

To feed the canning industry during the off-season for fish, Nantes has developed market-gardening in all its surrounding districts. The viticulture of Champagne could not be properly carried out if it was not under the control of the great cellars of Rheims and Epernay. All the small landowners are dependent upon the companies which, alone, have the means of making the Champagne wine; therefore the large companies, while already possessing extensive vineyards, do not try to increase their extent; they are content to be in charge of production. Thus town and country are interdependent; the disappearance of a sugar-mill brings about the disappearance of sugar-beet cultivation.

It is not only through its own needs of food, or its factories' requirements of raw materials that a town represents a market for the products of the surrounding area. Its commercial organisation is also the source of agricultural work. Cavaillon (population 14,000) undertakes the dispatch of spring vegetables with a financial organisation and a fleet of lorries so that its surrounding areas have specialised in the production of certain early vegetables; the same may be said of Carpentras or Château-Renard-Barbentane. Narbonne cooperates with the rich farming land which surrounds it: it is a winemarket after having been a cornmarket.[1]

Thus a town, through its landowners, its requirements of supplies, and the facilities for export which it offers, exerts a very real control over the country surrounding it.

This is even more true in the case of the action of capitalist enterprises in underdeveloped countries. In countries which have developed under external influences, orders still come from towns. Thus plantations are arranged around towns. It is not a question of private properties (in Algeria the settlers lived on their land) but of large companies, which have their head office or representative in the town; it is there that the harvest is brought; sometimes it is from there that the labour force is recruited, by daily commuting in the opposite direction to usual, and Pierre Lasserre has portrayed the women of Libreville leaving every day to go to work on the plantations.[2] Dakar developed through exporting the products it had collected from the region.

So almost everywhere, and in all latitudes, the town exerts its controlling power over the countryside. However, this influence over agriculture was not always a matter of self interest. From towns, the physiocrats of the eighteenth century sought to improve the countryside, and admininstrators encouraged the introduction of new methods. Attempts at rational exploitation of new lands were initiated in towns. And lastly today the town is often the laboratory for agronomic research.

THE TOWN AS SUPPLIER OF THE REGION[3]

Towns in relation to their regions do more than supply employment and a market for country produce. Country people come to town to do their shopping and to choose from goods collected from all over the world. The more towns there are in an area, and the more urbanised the civilisation the greater is the quanity of foreign produce consumed.

The big shops of the town deliver goods to country customers within a reasonable radius; the traders also organise regular rounds. The town's area of influence has been substituted for the trade of the peddlers who used to go from village to village with their packs.

[1] G. Galtier, 'Une capitale languedocienne: Narbonne', *Ann. de l'Institut d'études occitanes*, 1948.
[2] G. Lasserre, *op. cit.*
[3] P. Claval, *Géographie générale des marchés*, Paris, 1962.

Local trade is often taken in hand by large companies who make contracts to supply the village grocer, or who run a chain of village stores. These companies buy in bulk and store and distribute the goods as required. Sometimes they practise vertical integration, marketing their own brand products. There is competition in this field, but monopolies exist in some districts. The head offices of these multiple companies are not necessarily in the largest towns; they may work from far afield through local warehouses and distributing centres; their zone of trade therefore cannot always be equated with the town's sphere of influence, but there is some element of unity where a region is entirely provisioned by one company.

This of course applies only to trade in capitalist countries; in some socialist countries the provisioning of local trade is controlled by the government. Towns still exert an influence, but as a form of administrative guardianship.

The extent of the market. The separate evaluation of sales to inhabitants of the town and those outside only expresses the strength of the area of influence, not the extent of the regional market. This can be studied in several ways.

Research can be done among customers outside the town, and enquiries made with the town's tradespeople. Unfortunately, tradespeople do not know all their customers by name, and cannot always say precisely where their out-of-town customers live. Also they do not know to what extent these customers go shopping in other centres. Direct enquiries from customers as they make their purchases on the spot is difficult; it has been tried in the United States. It can only be considered for very special articles which are obtainable at a limited number of places.

The only method of investigation which results in exact information seems to be inquiry from customers at their homes. It is unfortunately difficult to question all the inhabitants within a radius of 25 to 50 miles, and these investigations are always hazardous.

Oïva Tuominen, in Finland, made an inquiry in schools through the teachers; these teachers asked the children where their parents went to make certain essential purchases. A certain number of families were also questioned individually. The result was recorded in a book on the area of influence of the town of Turku[1] (Fig. 36).

The procedure of inquiry through teachers was also used by Sven Dahl in Sweden, and Yves Genet[2] in France (Versailles). Only the articles which strike the imaginations of children can be used (clothes, furniture, bicycles) or the obvious services (doctors, dentists, cinemas).

---

[1] O. Tuominen, *Das Einflussgebiet der Stadt Turku*, Helsinki, 1949.
[2] *Bull. Soc. Etudes de la région parisienne*, 1956.

Gunnar Arpi and Bengt Elfström have studied the little Swedish town of Växjö (population 22,000) in this respect.[1] The inquiry extended over an area with a radius of about 40 miles round the town, having a little over 100,000 inhabitants. As it was impossible to question them all, the researchers covered sections along the main roads. After compiling a detailed list of purchases and services, they

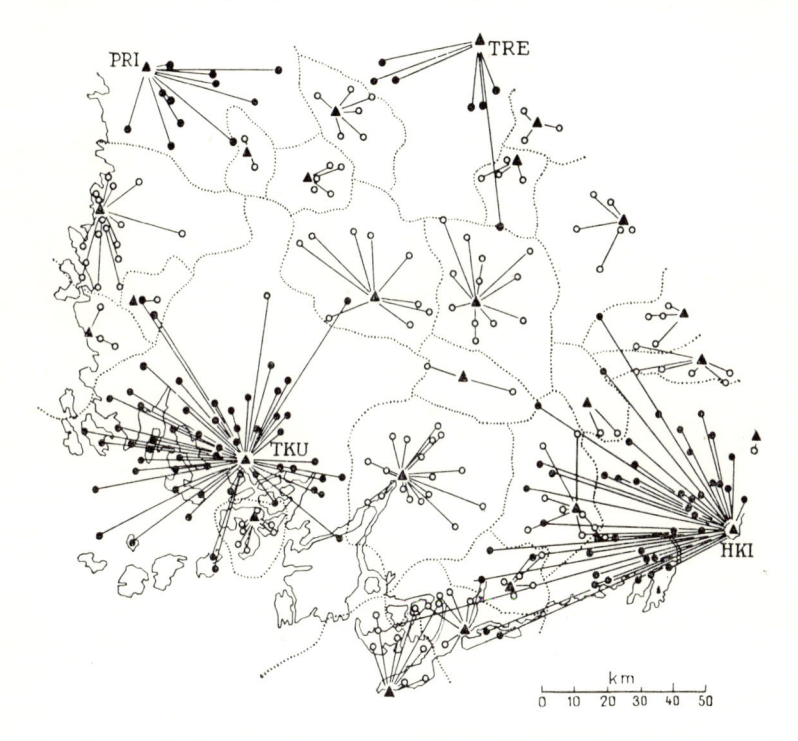

Fig. 36. Urban vegetable supplies in southern Finland

HKI—Helsinki; TKU—Turku; PRI—Pori; TRE—Tampere (from O. Tuominen, *Das Einflussgebiet der Stadt Turku*, Helsinki, 1949, p. 49).

decided in each direction upon the points at which people made 80 per cent, 60 per cent, 40 per cent, 20 per cent and 5 per cent of their purchases in Växjö. By joining these points, curves were formed which determined the town's zones of decreasing influence. This influence must be proportioned according to the importance of the goods involved. In fact, it is not enough to record that, in a certain parish, one product in four is bought in the town, or in another, one

[1] *Växjö, Detaljhandelsområde*, Stockholm, Helsinki, 1962.

product in ten, for there are key products which indicate a more direct dependence, and each product must be provided with a coefficient (Fig. 37).

Considering that even for a town with a population of 22,000 a sampling technique must be used, it will be seen that only a rough

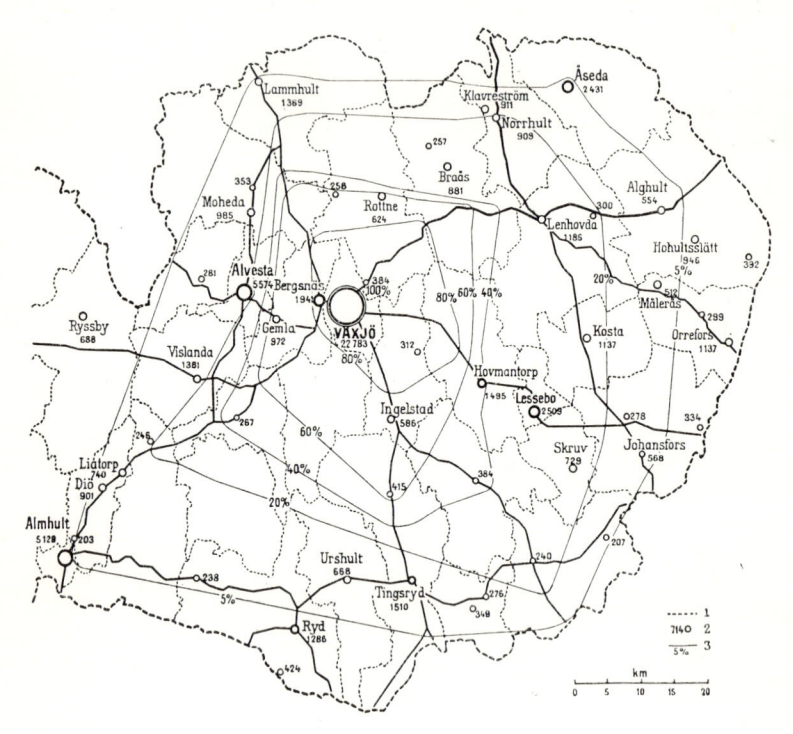

Fig. 37. The zone of influence of Växjö as a shopping centre

(from G. Arpi and B. Elfström, *Växjö Detalj Handelsområde*, 1962)

1. Commune boundaries. **2.** Number of inhabitants in 1960. **3.** Percentage of commercial relations with Växjö.

approximation of sphere of influence can be obtained. Special bus or train services on market days can also give a clue.

But the extent of the retail market is only one element of the zone of influence of a town, and we shall return to the subject of the determination of this influence.[1]

[1] See p. 428 for the studies that have been made to determine the sphere of influence based on shopping.

Financial influence. Townspeople have in the past invested in the country not only by buying houses and land, but also by leasing out livestock and equipment independent of land. The citizen leasing to the farmer thus to some extent becomes his banker and on occasion—such as a bad harvest—his creditor.

Neither farmers nor country-based industries can do without the financial organisations of the town. Conversely, financial houses endeavour to find in the country funds with which to feed other enterprises in which they are interested. The invasion of the countryside by the banks was one of the major economic changes of the nineteenth century, as Jean Labasse[1] has shown.

Financial cooperation is thus one of the closest links between the town and its region. In some countries financial domination by the town can become oppressive, sepecially if moneylenders are able to establish too firm a hold.

INDUSTRY IN THE COUNTRYSIDE

The countryside surrounding the town does more than cultivate fields and raise livestock. When the population began to increase, in modern times, it began to look for other resources. Work done at home gave many villages a way of earning a living, but was only possible in relation to a centre which had access to raw materials, distributed the work and collected and manufactured goods. As people went to work on foot, the centre had to be fairly near; hence the advantage of small towns, and even market towns. Thiers used to receive steel, passed it on to be worked in villages, each one of which had its own speciality, and then collected the knives and took charge of selling them. The villages of the Pays de Caux used to spin and weave on behalf of the merchants of Rouen. The peasants around Troyes and Aix-en-Othe used to come on Sunday to collect skeins of thread, made them up and the following Sunday brought back finished hosiery articles. In the mid-nineteenth century, Mauléon had sandals made in all the neighbouring villages.[2] Similar examples could be quoted from almost everywhere in Europe. The towns of Tuscany spread first the wool and then the straw industry thoughout the countryside.[3]

The weaving trade was to be found throughout the country, and not only to supply the peasantry. Members of a family would work in shifts from four o'clock in the morning to midnight. And the employers in the towns preferred this ill-paid labour, which competed

[1] *Les Capitaux et la région*, Paris, 1955.
[2] G. Viers, *Mauléon-Licharre*, Bordeaux, 1961.
[3] J. Sion, *Méditeranée. Péninsules méditerranéennes.* Vol. 2, Paris, 1934.

successfully with the workmen in towns. Hence the revolt of the silk-workers of Lyons who were deprived of their living.

In the case of many manufactured goods, workshops and factories supplanted work done at home; first spinning and then weaving were concentrated in this way, not always in towns, but in the country markets which depended upon large towns. Lyons is a particularly magnificent example. 'Manufacturers' have the raw material they have bought transformed by a series of operations (throwing, spinning, weaving, dyeing) in a network of small centres each of which has its speciality and which covers the whole south-east of France.[1]

Cottage industry has not entirely disappeared; it persists for some of the intricate work which requires individual attention, such as the gem-cutting around Saint-Claude in the Jura mountains, the manufacture of parts for clocks and watches around Morteau in France and Locle in Switzerland.

Cottage industry and factory industry have often appeared to be two successive forms of the life of a village working for a town. Daily commuting is another aspect of industrialisation round towns. This has been examined earlier, but it should be noted that it often constitutes the end result of the evolution of rural work. The peasant from the forest of Othe no longer puts on his Sunday best to bring his knitted goods to Troyes at the weekend as he used to; he goes there every morning to work in a factory.

A more recent aspect arises from those large factories which are planted in a rural area without any industrial tradition. It is one way of using the surplus labour force, and of making work for young people for whom mechanised farming can no longer provide a living. Often these are new industries—electrical or mechanical industries. There is thus a more malleable labour force, easier to house, and retaining certain rural advantages. Small industries strung all along the Seine valley downstream from Paris are clearly dependent on Parisian enterprises. Sometimes large undertakings are launched by a firm in search of land and workers: the Renault car firm opened a factory at Flins, near Mantes, which completely altered the life of the district.

It should be noted here again that contact with the town is necessary. Large firms are usually happy for their factories to expand at greater distances from the centre, but long-distance decentralisation is generally advantageous to other towns, and not to the countryside. When Parisian firms undertake decentralisation, going as far as the Alps, they usually place their factories in towns, as the firm of Gillette did at Annecy. The rural area influenced by industry is usually in the neighbourhood of a town.

[1] M. Laferrère, *Lyon, ville industrielle*, Paris, 1960.

The conditions which used to apply in Europe still persist in Asia and Africa. Small industries carried on at home presuppose urban control, whether in the case of carpet-making, tapestry and embroidery in North Africa, or lace in China. And if in the People's Republic of China village workshops have replaced cottage industry, these workshops are all the more dependent upon the neighbouring town because communications are more difficult with the more distant centres.

Thus in every country industry is connected with towns; the growing industrialisation of the countryside only reinforces urban influence.

ADMINISTRATION

There is a hierarchy of administration in which towns of varying status take their place. Status in this context does not necessarily coincide with size or economic importance, since a traditional administrative centre may in these respects have lost ground to an industrial upstart. The structure of the hierarchy varies from country to country, but in general terms the town serves its region as the centre of all administration, business and legal affairs, hospital services, education and so on (see below, p. 428 ff).

MEDICAL SERVICES

Country districts rely on their regional town for specialist and hospital services. Most towns have hospitals, though the extent of the services they offer is not strictly related to the size of the town; the status of the town in the administrative hierarchy, or the existence of a university with a medical faculty may be more influential.

It is interesting that the sphere of a town's influence through its doctors and hospitals is one of the most far-reaching. This is especially the case in underdeveloped countries.

CULTURAL INFLUENCES

Towns inevitably exercise a cultural rôle. It is not possible to establish facilities for advanced education in every village. Grammar schools and universities are sited in towns because it is there that their main catchment area is found and that the systems of communication converge. There are, of course, isolated boarding schools, but these are few in number and are reserved for children who need special care or who come from well-to-do families. As a general rule, the various levels of education follow the pattern of urban life.

It is important that educational centres above primary school level should be distributed as widely as possible: it is there that professional instruction is given, and children destined for higher education are selected. In general this is the rôle of the larger villages and small towns, and must be within the reach of everyone. Beyond that level, further studies leading to higher qualifications are available in the towns. These should be open to all who are capable of profiting from them. In fact the regional influence of the grammar school is very uneven: it depends on the social structure of the region and, to a certain extent, on the agrarian structure. American farmers, British landowners and the prosperous French farmers of Brie send their sons to boarding schools; whereas small farmers in the French countryside generally keep them at home to help on the farm.

The university attracts students from some distance away and this epitomises the cultural centre of the region to a greater extent than the more numerous grammar schools whose catchment area is often determined by administrative boundaries.

Special circumstances must be borne in mind, for example the superiority of certain laboratories, or the reputation of certain professors. The mobility of the students may be a factor against regional recruitment and this is even more true in countries where this mobility is greater, like the United States.

The cultural influence of the university may also be shown in the recruitment policy of industrial and commercial enterprises. This is particularly the case when the university specialises in the main activity of the town, mining, textiles, metallurgy or electricity. However this university influence (in the widest sense of the term) is limited to industrial undertakings within a certain social stratum. There is a form of cultural influence which permeates the whole region more deeply; this is the newspaper, especially the daily newspaper.[1] In the countryside of western Europe there is hardly a home where a newspaper is not taken every day, a paper in which world and national news are found as well as local news. In this way the town brings its influence to bear every morning. It expresses through the newspapers various points of view: one of them usually takes the lead, perhaps not because its political colour wins more votes (though it must avoid causing too much offence) but because it is the best informed about everything concerning the district. Thus the Parisian newspapers, less concerned with local news, often compete with local newspapers. In order to be able to give this regional news more space, some provincial daily papers allocate space to particular districts in order to cater for local readers. There are few criteria which demon-

[1] A. Chatelain, 'Le journal, facteur géographique du régionalisme', *Et. rhodaniennes*, 1948.

state better the extent of a township; it is a source of pride for a town to publish one or more dailies; sometimes the circulation hardly extends beyond the district; sometimes, on the other hand, it extends over a wide region, covering several thousand square miles. The

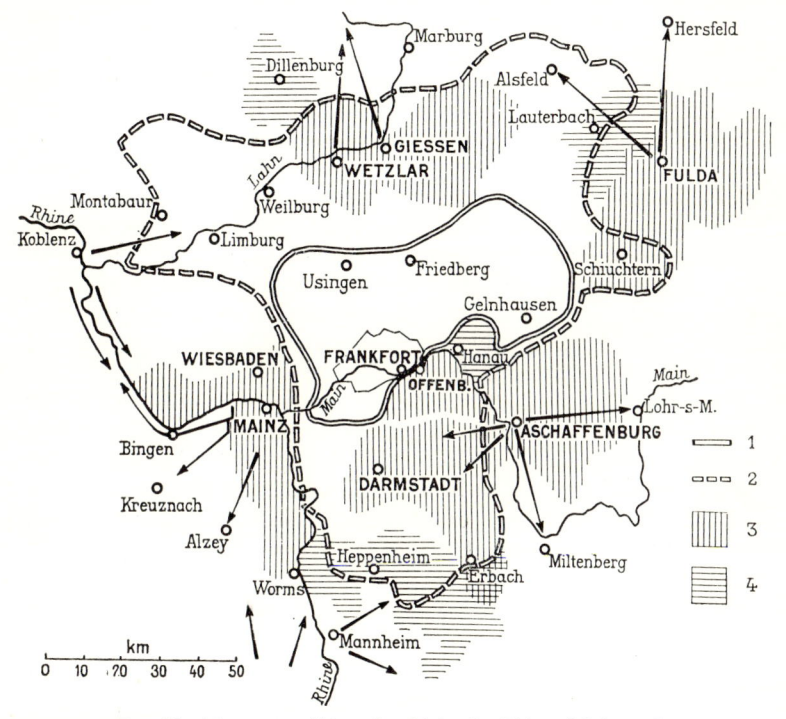

Fig. 38. Newspaper 'hinterlands' in the Rhine–Main region

(From W. Hartke, 'Die Zeitung als Funktion sozial geographischer Verhaltnisse im Rhein-Main Gebiet', *Rhein-Mainische Forschungen*, 32, 1952, p. 21.)

1. Major distribution area for Frankfort newspapers. **2.** Secondary limit of penetration of same newspapers. **3.** Main distribution areas of newspapers published in secondary centres. **4.** Areas strongly penetrated by newspapers from outside centres.

newspapers of Toulouse reach the whole of south west France.[1] the newspapers of Marseilles the greater part of the south east.

In England the 'national' papers are predominant, but there is a wealth of local papers, some of them of high standing; a map of their circulations would show a series of overlapping hinterlands.

[1] J. Coppolani, *Toulouse*, Toulouse, 1954.

There is no doubt that reading the same paper leads to a certain dependence on the town where the paper is published. The awareness of a certain feeling of centralisation is expressed in this way.

Unfortunately investigation is by no means easy. Newspapers appear in various forms, daily (morning or evening) or weekly (we can leave on one side periodicals appearing at longer intervals.) The small local weekly paper enables some country people to dispense with regular reading of the daily paper published in a larger town where the national weekly is not the main source of information. Research is often complicated by the presence of two or three dailies issued in the same town. Furthermore, the information given by the management of the paper, even if readily given, is often vague and sometimes deliberately inaccurate. Circulation is achieved in various ways by direct sale or by subscription. It is at the point of sale, in each village, that the information should be sought. J. P. Haughton, in Eire, has made an interesting analysis of the circulation areas of local newspapers, using the local advertisements and news from villages as the main lines of evidence for determining the areas served.[1]

The fact remains that this information can be a valuable source for determining the sphere of influence of a town. Wolfgang Hartke has done this for the Rhine-Main district. He shows the important part played by local newspapers in smaller centres (Fig. 49).

Finally it must not be forgotten that everywhere today television and the radio sometimes supersede newspaper reading, especially in the country.

However the daily paper which, at any rate in Europe and America covers all areas, still remains the most effective instrument of urban opinion and best expresses cultural influence.

CONSCIOUSNESS OF URBAN INFLUENCE

The town is thus connected with its region by all sorts of links, economic, demographic, cultural and social. One should not, however, overlook the psychological link which is often only an expression of the others.

The town represents for the region a well-defined concept; one knows immediately which town is referred to. Certain districts, moreover, have taken the name of the town round which they gravitate. In France there are well-recognised *pays* around Nantes, Metz, Toulouse and Lyons. Generally speaking this applies to connexions slowly forged over the centuries. Towns more recently developed have no such claim; there is a Metz *pays*, but one never speaks of the *pays* of Nancy. It is possible for this feeling to survive

[1] J. P. Haughton, ' Local newspapers and the regional geographer', *Adv. of Science*, **vii**, 1950, pp. 44–5.

new administrative boundaries; the capitals of the old provinces still keep their old prestige. Any organisation must take regional psychology into account.

This feeling of belonging to a town conveys not only the strength of the links connected with the town, but also the intensity of urbanisation in general. It has been observed that, in a fragmented district like Brittany, where the population live turned in on themselves, this feeling hardly exists, whereas it is very strong in other districts such as the one surrounding Toulouse.

RECREATIONAL LINKS

Privileged citizens have always indulged in periodic absences from town, to visit their country estates, to 'take the cure', to hunt, to climb mountains or to bathe. Nowadays all town dwellers seek to spend their holidays, and in summer their days of leisure, away from home, and, though their main holidays may be spent further afield, for weekends and short breaks they seek country or seaside near at hand. Indeed as a result of this habit, as we have already seen (pp. 182–187), recreational towns are in turn created. Luxury hotels, family hotels and guest houses, holiday camps and caravan sites, welcome a seasonal influx of visitors from the towns; weekend houses and well placed friends add to the quota of town and country connexions.

Local trade adapts itself to this temporary population, which spends lavishly. Moreover these seasonal visitors bring with them new customs, standards of comfort, outlooks and attitudes—in short, the urban mentality.

But recreational activities form a two-way traffic. The town is a great centre of entertainment. It is possible to go there on a day, or even a half-day excursion; or one can go for an evening up to fifty miles or so and back by car. The car may be a private one or a hired coach; there are travel firms who will arrange a theatre party and book the seats at the theatre according to the number of people on the coach.[1]

[1] L. Ducret, 'La banlieue est de Paris et l'utilisation des loisirs', *La vie urbaine*, 1952.

CHAPTER 29

SPHERES OF INFLUENCE

METHODS OF INVESTIGATION

Isochrones. The relations of a town with its surrounding region are to a large extent a function of the means of transport available. The town's influence will extend as far as communications permit ease of movement, and all progress in transportation brings town and country closer together.

It is of interest therefore to establish precisely the conditions under which the country can be reached from the town. This is the purpose of isochrones, that is, curved lines joining points which can be reached from the town in a given time.[1] Such isochrones would allow us to define the area of influence of markets at the time when people travelled on foot. The market could be reached from each village in about two hours, and the radius of active influence of the market town was therefore about 5–6 miles, and towns were about 12 miles apart.

With modern methods of transport, the problem is rather more complex, for one is involved not with personal means of transport (except in the case of the private car), but with public conveyances such as the omnibus, tramcar and railway train, which run to a timetable and have scheduled stopping places, and, in the case of railways and buses, often offer both express and 'all stations' facilities which complicate the drawing of isochrones. Clearly, in tracing the sphere of influence of a town the timing and frequency of the services must be taken into account as well as mere journey time. The 'workmen's train' may take as long to cover ten miles, stopping at all stations, as the businessman's express a couple of hours later in the morning takes to cover three or four times that distance.

Further, the isochrones must take into account the time taken to walk from the place of residence to the nearest railway station or bus stop, and, at the other end of the journey, the relation between the location of the terminal station and the central business district (which in some towns may be quite widely separated).

Clearly, too, the isochrone maps must show both rail and road services, for the buses frequently fill in the gaps between the railway

[1] G. Chabot, 'La détermination de courbes isochrones en géographie urbaine', *C.R. Congrès Int. de Géog.*, Amsterdam, 1938; M. Dumont, 'De isochronen Kart von Gent', *Naturwetensch. Tidsskrift*, 1943; *Atlas de France*, Plate 77; A. Aagesen, 'Över isochronic maps', *Geog. Tidsskrift*, 1943.

lines that radiate from a town (see Fig. 39), whilst in the case of large towns the railways may well have abandoned their inner suburban traffic to the roads, by the closing of stations.

An alternative method of studying the local sphere of influence of towns, which does not involve isochrones, is the analysis of the routes and frequency of bus services, without reference to journey times. This was developed by F. H. W. Green for England and Wales, and his map, produced after a very careful study of all the ramifications

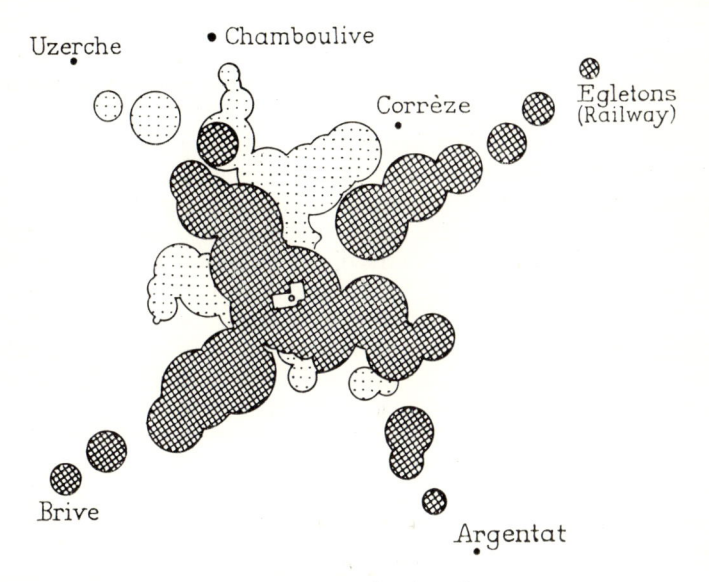

Fig. 39. An example of isochrones

(From G. Lafond 'Tulle et son environnement', *Urbanisme et habitation*, 1954, p. 269.)

Cross-hatched, isochrone of one hour from Tulle by train; dotted, by omnibus.

of the bus network, was subsequently published in the Ordnance Survey's Planning Series on a scale of 1:625,000.[1]

The advent of the private motor car has still further complicated the problem of the accurate delimitation of urban influence. The immense car parks round some American factories—a phenomenon that is becoming increasingly common in Great Britain and Western

[1] F. H. W. Green, 'Urban hinterlands in England and Wales: an analysis of bus services', *Geog. Journ.*, 1950, pp. 64–88; Ordnance Survey, *Local Accessibility Map*, 1954.

Europe—show that it is not only the managerial classes who travel to work by car. The speed of the cars will of course vary, but for the purpose of drawing isochrones an average must be struck—perhaps 25 m.p.h., depending on the traffic conditions.

Another aspect of this new complication is the use of the motor car to get from place of residence, not the whole way to the place of work in the city but simply to the local railway station, where car parks are increasingly being provided. In this way the 'catchment area' of suburban stations is greatly widened and the isochrones drawn on the basis of walking time must be redrawn.

Technical changes in the means of public transport may also materially alter the speed of travel and so the position of the isochrones. Electrification of suburban railways often has this effect, a consequence of which is to increase the commuter range. Electrification brought Brighton (63 miles) and Portsmouth (73 miles) within commuting distance of London, and the recent electrification from the Euston terminus has carried the possibility of commuting many miles further out in a north-westerly direction, as far as Northampton (66 miles) and Rugby (82 miles). In this direction, too, the M1 motorway has had a similar effect (which, however, increasing congestion at the London end may tend to nullify). If commuting is being considered there seems little point in drawing isochrones for lengths of journey time greater than one and a half hours.

Finally it may be suggested that cost as well as time should be plotted by isopleths, for the two are not necessarily closely related: for example, if two towns are competing for commuter labour it is the one that offers the lowest fares that will get the workers, even though the latter may have to spend more time in travelling.

THE TOWN IN REGIONAL DEMOGRAPHY

In a region, the demographic relationships between a town and the country surrounding it take on a double aspect; first, a town exerts a permanent or temporary power of attraction over the people occupying an area of varying size; secondly, it exercises a powerful centrifugal influence, throwing its inhabitants for one reason or another towards the periphery of the urban area. Attraction and repulsion, these are the two forms of movement whose different characteristics we shall study in turn.

Movement towards the town. An urban nucleus forms and develops by helping itself liberally from human reservoirs both far and near (cf. pp. 15–18). We shall not study here all the aspects of urban recruitment, but only the one concerned with the peripheral area where the town's power of attraction is considerable. It cannot be said that the recruitment area of Le Mans extends as far as the departments of Haut-Rhin and Gironde simply because, quite by chance, often in the appointment of certain officials, these distant departments are named in statistical information regarding the birthplace of the inhabitants of Le Mans. A certain Soviet town in Kazakhstan which (apart from the majority of its population which is of local or regional origin) includes 12 per cent of people from Moscow or the Ukraine who have come there as technicians or officials, offers another example of the same phenomenon.

Therefore the area dependent on a town's power of attraction, where permanent migration to the town is concerned, must be considered as that which is in immediate contact with urban territory and develops at varying distances from the town according to the attraction exerted.

This power of attraction varies considerably according to the town's size and dynamism. The two urban districts of Chicago and Los Angeles have populations of similar size, slightly over six million inhabitants, but the growth rate of Los Angeles is twice that of Chicago; the town of Chicago itself is declining slightly, while Los Angeles is increasing by 50,000 people per year; in fact, the latter is in that part of the United States which is at present undergoing the strongest economic and demographic development; the former is the capital of a large industrial area which is already well established. The same phenomenon can be seen in France with two towns such as Grenoble, which is developing at rapid pace and increased by 31 per cent between the years 1954 and 1962, and Rouen, established for a much longer time, where the region only added 16 per cent to its population over the same period.

It must be remembered that this 'dynamism' can be real when it represents the actual amount of employment available, or mythical when it only reflects the mirage of a possible improvement in people's daily lot: the Sicilians who crowd into the great industrial centres of the Po valley are anticipating the transformation of their lives, and quickly find jobs in the numerous factories, while those who reach the suburbs of Palermo may have to wait a long time for similar opportunities.

The extent of this power of attraction varies according to the network of communication routes, which is itself closely connected with natural and especially physical conditions.

The recruitment areas of Ajaccio and Bastia, the two principal towns of Corsica, are clearly separated by the mountain ridge which crosses the central part of the island: in the south-west, people are attracted to the capital (28·2 per cent of its population were born in the towns and villages of this region), whereas in the north-east people go towards Bastia (33 per cent of the town's population).[1] A study of the adult population born outside Arras leads to similar conclusions: the districts to the west, south and south-east of the town which lack easy communication with the local metropolis supply fewer new citizens than the districts immediately surrounding it, or those situated on the railways and roads, particularly those to the north-east and south-west.[2]

It is obvious, however, that the regional environment is of great importance: if there is only one well-known or prosperous town its influence is felt far and wide; if there are several close together, the intervening space is divided between them; the same applies to daily journeys to and fro. In this respect the situation is very different in highly urbanised and developed countries like those of Europe, and in territories where urban density is low and the flat country in between does not offer many possibilities, as in some parts of Africa.

A detailed study of the current of movement towards the towns of Central Africa brings to light the following observations: towns are centres artificially created by the whites, usually for the exploitation of mines, and bear no relation to the density of local occupation: Bukawu and Luluabourg have populations of under 50,000 although they are situated in very highly populated regions, whereas the two great centres of Léopoldville and Elisabethville were founded in the middle of what might be called a 'human desert'. Therefore, the recruitment patterns of these urban nuclei are very different:

Birthplace of Inhabitants

	Territory	District	Province	Exterior
Luluabourg	38·3	33·3	24·8	1·6
Elisabethville	3·3	12·6	31·9	52·2

Source: J. Denis, *Le Phénomène urbain en Afrique Centrale*, Paris, 1958, p. 141; (Elisabethville is now known as Lubumbashi).

The part played by main arteries of communication is very well defined: following the railway from west to east, the direction away from their home, Angolans are distributed as follows, 4,000 at Kolwezi, 2,000 at Jadotville and 2,000 at Elisabethville; inversely, following the railway from the south or the road from the east, there

[1] Y. Kolodny, *La géographie urbaine de la Corse*, Paris, 1962, pp. 164 ff.
[2] A. Cornette, 'Arras et sa banlieue', *Rev. du Nord*, 1960.

are 10,000 Rhodesians at Elisabethville, 3,500 at Jadotville and 1,000 at Kolwezi. Under these conditions it is very difficult to trace the limits of the regional influence of demographic recruitment with urban networks like these. The industrial towns of Africa have a quite special and individual regional influence.

Their rôle could be compared to that of the great mining centres of industrial countries where recruitment tends to follow purely economic factors rather than truly regional influences. In fact the coal and iron mining towns of France have taken a large share of their workers from the Polish countryside, and are at present using labour from southern Italy and North Africa; it is the same in the coal-mining towns of the southern Belgian coalfield, where 15 per cent of the total population are not natives of the Borinage or the centre of the country (even 25 per cent of the men between the ages of 25 and 40), while round Liège, in 1947, there were 46 per cent of foreigners, principally Italians.[1] A comparable phenomenon exists in the coal-ming areas of Donetz, which draw part of their labour force from the edges of the Caucasus.[2] What counts is the difference in economic standards between the area left behind and the point converged upon, or even the similarity of jobs in the two places in question: miners from a town which is in decline can move to busier centres, as was the case with the coal-miners of Alès who moved to the coal-mines of Lorraine.

When defining the permanency of a town's magnetic power, therefore, a clear distinction must be made between its local and regional rôle, and its more widespread and far less concentrated power. Only the first of these two aspects should be analysed in the 'regional rôle' of the town.

As well as being a place upon which people converge permanently, a town is also the scene of temporary gatherings: commuters come in great numbers to run its factories, and give life to its shops and offices, and their job done, return to their homes.

There are very few towns, large or small, which are without this daily coming and going. This commuting depends upon urban and peri-urban transport systems (cf. pp. 313ff.) but sometimes it engenders its own transport: coach rounds can be organised to pick up workers for a particular establishment. In the Liège basin, fifty-three bus services make a daily journey to fetch the workers for the coal industry; large suburban factories, like I.B.M. at Essonnes, help with the travelling arrangements of their non-motorised staff, at

[1] P. Lambert and J. Mineur, *L'économie de la région liégoise*, Paris, 1960; *Les Regions du Borinage et du Centre à l'heure de la reconversion*, Brussels, 1962.
[2] O. A. Konstantinov, 'Les villes de la République Socialiste Soviétique d'Ukraine', *Bull. Soc. Géog. de l'U.R.S.S.*, 1950.

418 THE TOWN AND ITS REGION

least in the matter of linking them up with public transport or reaching nearby residential centres.

In the case of small towns, it is possible to speak of a direct convergence on a single place; but in larger agglomerations movements

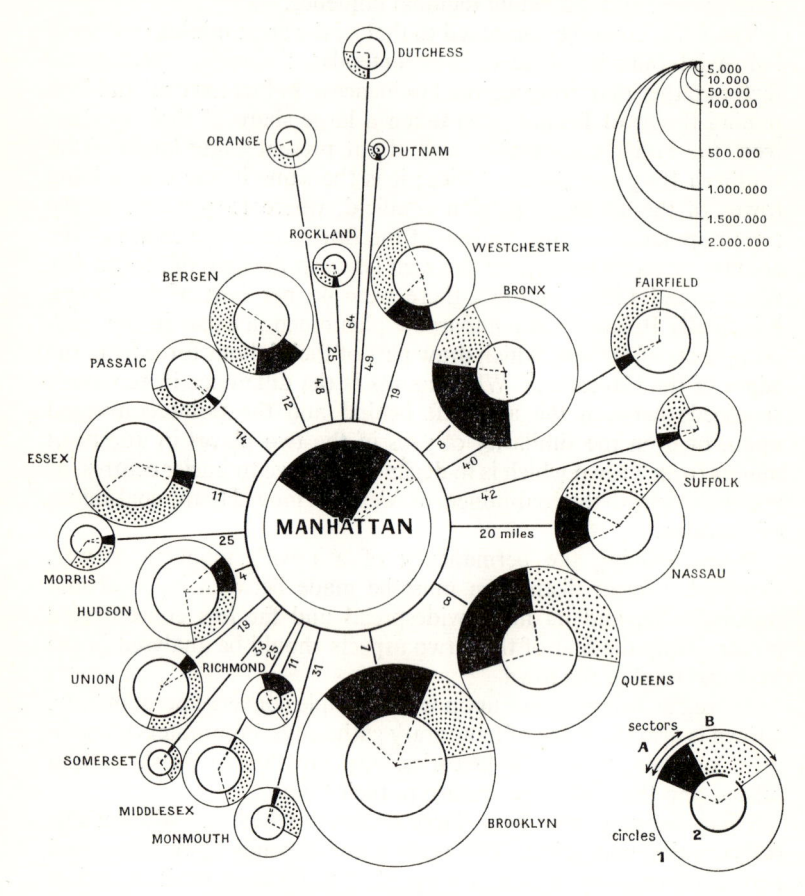

Fig. 40. Commuting to New York

Circle 1, total population. Circle 2, workers employed locally. Sector A, working population employed in Manhattan, Sector B, total working population. (Compiled from figures given in *Anatomy of a metropolis*.)

become complex, with intermediate areas of reciprocal employment formed by this more or less discontinuous peripheral industrial ring, or by districts scattered about the urban mass, the existence of which has already been pointed out (cf. pp. 288 ff.).

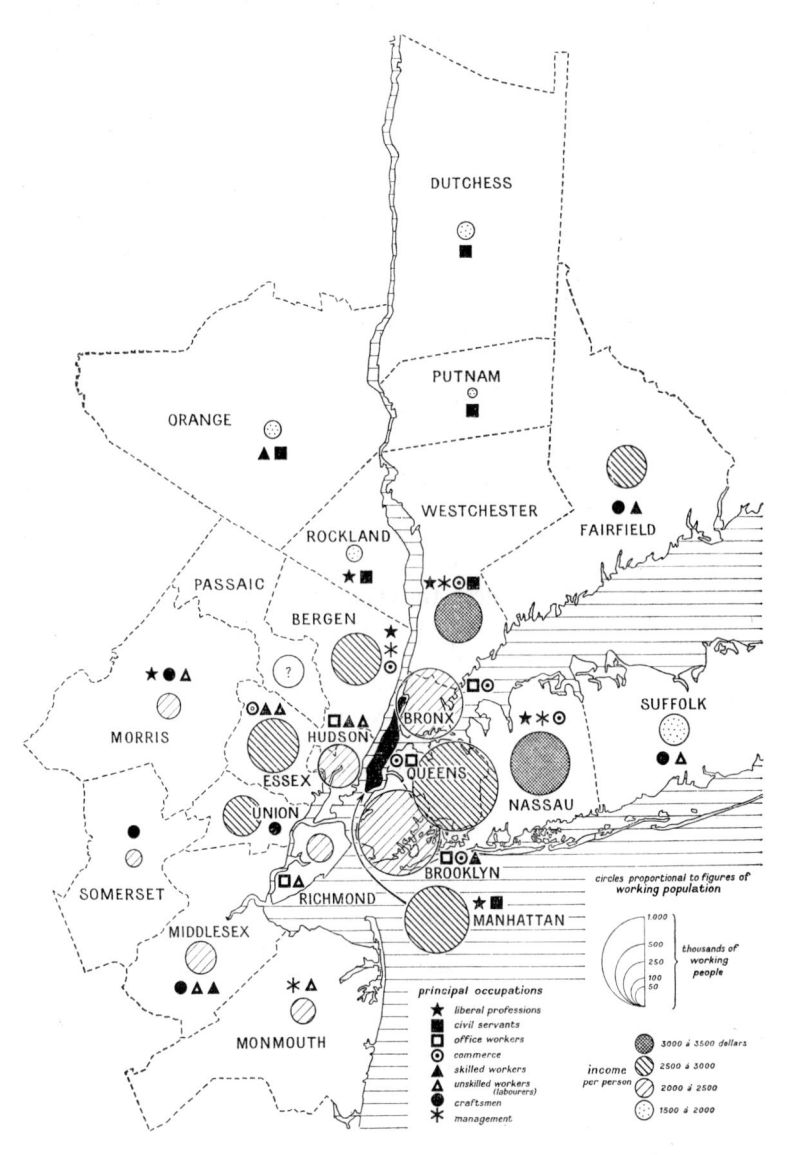

Fig. 41. Incomes and dominant occupations in the New York region (from figures contained in Anatomy of a metropolis)

The intensity of these movements depends in the highest degree upon the deficiencies in the labour force registered in the central urban organism, but also upon the relative levels of salary. A large influx of workers allows employers to keep wages at a lower level than if there were a shortage of manpower and competition between the various establishments. Its development is linked with the difficulties of finding housing on the spot, which becomes more acute the bigger and more densely populated the town. This form of movement contributes to the maintenance of a higher density over a larger surface area round towns, at the same time bringing extra resources to the surrounding countryside.

Here again, the question of defining the recruitment area arises: transport systems covering large areas, or special circumstances, cause certain commuters to travel large distances and seem to be exceptional. This is why Liège, which had great difficulty in recruiting labour for the coal mining industry, attracts it from as far off as Flanders where there is considerable unemployment; on the other hand, for its metallurgical and chemical industries which are far more sought after by employees, this same town seems to draw its labour from a more concentrated area nearer at hand; during the last few years the average distances travelled by mining commuters have increased while, on the contrary, those travelled by workers in other industries have been reduced.

It can therefore be stated in principle that, to define the area of daily commuting affected by an urban centre, we must consider only the space in which a certain percentage of the local active population comes to work regularly in that centre's enterprises. This percentage could be fixed at 5 or 10 per cent.

The organisation of transport systems is of considerable importance. Leningrad has 140,000 commuters, which is about 10 per cent of the city's labour force; 75 per cent of these come by train, but their recruitment area extends as far as twenty-five to forty miles to the north and twenty-five to fifty miles to the south, along the main arteries of communication. There is a noticeable difference in the development of the satellites of two towns like Kharkhov and Kiev because of the differences in conditions of transport: the satellites of Kharkov, which is better served, have a population of 200,000, and the town receives 75,000 commuters daily, while the numbers for Kiev are 65,000 and 30,000 respectively.[1]

Round Soviet towns, as collective transport usually predominates, regional recruitment areas take on the form of enormous stars. In the United States where private transport is much more highly developed the geometric shape of the area is much more compact. In

[1] *Goroda Sputniki, passim.*

the latter country the average town of the north-east or of New England where the urban network is very dense has a relatively modest percentage of commuters (between 8 and 12 per cent of the number of workers); on the other hand, in the centre of the country where industrial towns are often isolated crossroads, convergence on the town is much more massive and can reach 25 or even 30 per cent of the number of people employed.

In Europe there are inextricable cross-currents of commuting, with extremely complicated overlapping of recruitment areas in all the regions of intense economic activity: the provinces of the middle Rhine in Germany, in Switzerland, the Netherlands, Belgium and northern France.[1] Around large agglomerations however, the movements, while still remaining very complex, are more strongly concentrated, as in the regions of Paris, London or Milan.

Movement away from the town. This is a much more concentrated phenomenon, and one which is typical of towns which have already reached a certain level of development. It manifests itself in two ways: in connexion with daily movements to and from work and in connexion with residence.

In agglomerations of some importance, given the fact that industrial activities are located, as often happens, in specialised suburbs or along roads, rivers or railways, it is not unusual for the inhabitants of central districts or even of the principal urban nucleus, to take a centrifugal course in order to reach their places of work. Thus, 195,000 people leave Paris every day to go to work in suburbs at varying distances from the city itself.

This outward movement is generally made over shorter distances than the mass of centripetal movements. It is also much more limited, but very diffuse and difficult to study with precision. It is one of the contributions a town makes to the functioning of its region.

As for the residential aspect of 'movement away from towns', it corresponds partly to the spreading out of the suburbs and to what is frequently called 'suburbanisation'.

If a quite important town and the administrative districts which surround it are considered, it will be seen that the whole agglomeration develops demographically in the form of a series of concentric rings: first the central nucleus develops to saturation point, then the inner ring of suburbs, then a second ring.

Densities continue to increase over ever larger distances but still decrease from the centre outwards to the perimeter; on the other hand, relative development progresses from the interior outwards.

[1] R. E. Dickinson, 'The geography of commuting: the Netherlands and Belgium', *Geog. Rev.*, 1957, pp. 521–38.

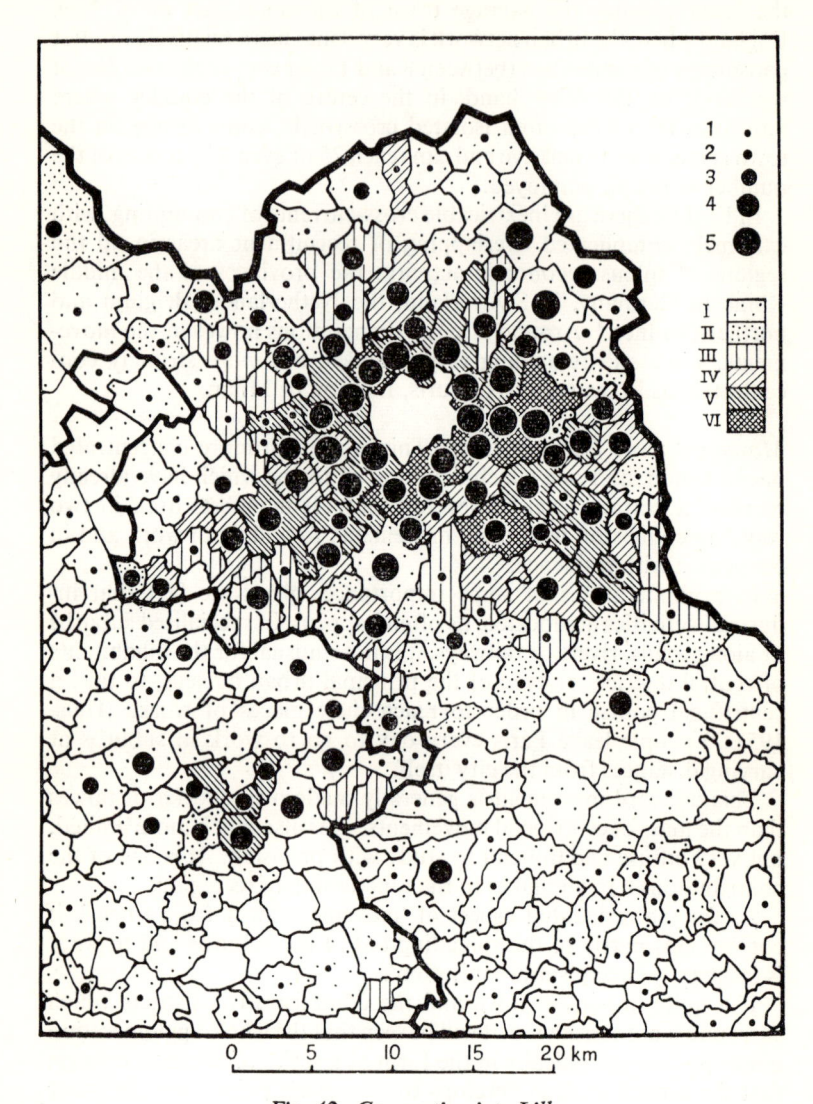

Fig. 42. Commuting into Lille

1. From 1 to 49 commuters. **2.** 50–99. **3.** 100–199. **4.** 200–399. **5.** 400 or more.

I. From 1 to 3·9 per cent. **II.** 4 to 7·9 per cent. **III.** 8 to 15·9 per cent. **IV.** 16 to 31·9 per cent. **V.** 32 to 64·9 per cent. **VI.** 65 per cent or more.

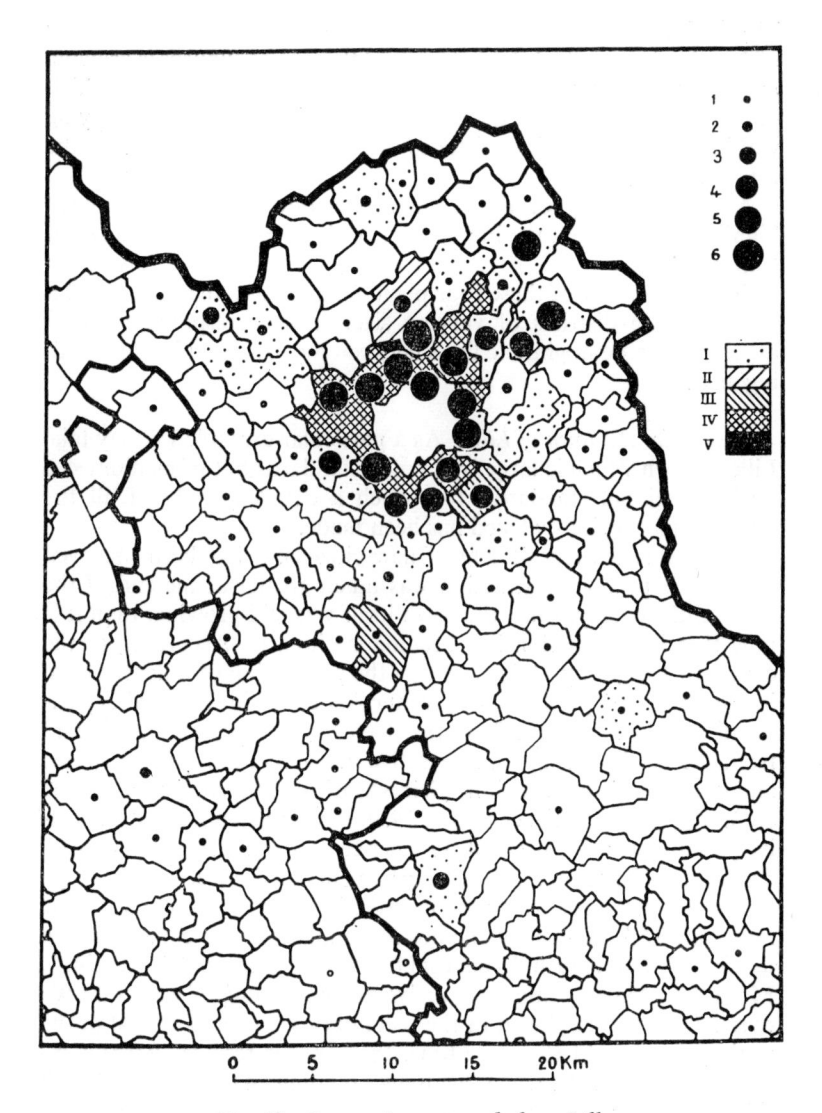

Fig. 43. Commuting outwards from Lille

1. From 1 to 49 commuters. **2.** 50–99. **3.** 100–199. **4.** 200–399. **5.** 400–599. **6.** 600 or more.

I. From 1 to 4·9 per cent. **II.** 5 to 9·9 per cent. **III.** 10 to 19·9 per cent. **IV.** 20 to 39·9 per cent. **V.** 40 to 80 per cent.

This is the case of the agglomeration of London in which the County of London has reached stagnation, and the ring of suburbs is growing only very slightly, while, between 1950 and 1960, the regions which have developed most of any in the United Kingdom were those lying within daily commuting distance by train from the heart of the capital.

A detailed study of the demographic development of suburbs highlights once again the importance of the localisation of means of transport.

An examination of these four phenomena allows the drawing of a map showing districts of varying size, each corresponding to an aspect of the demographic links between the town and its surrounding area.

An example: the town of Lille. As a particular example we may take Lille, a town which is large by French standards but of medium size by international standards, for which a detailed study of all the phenomena previously listed has been made.

As regards the recruitment of new citizens, it has been observed that Lille's power of attraction was much greater at the end of the nineteenth century, when the town itself was still growing, than it is now that its population has become stationary. According to the census of 1896, only 43·7 per cent of the population of Lille was born in the town, 21·1 per cent came from other towns in the Nord department, and 15·7 per cent from the department of Pas-de-Calais and from the neighbouring administrative divisions (Somme and Aisne) and the more distant divisions (Seine-Maritime, Seine-et-Oise, Marne etc.). According to the census of 1954, the percentages were respectively as follows: 57·1 per cent, 17·7 per cent, then 7·6 from the Pas-de-Calais and 11·9 from the other departments of France.

Lille and its suburbs attract a crowd of commuters every day: these represent 37·3 per cent of the active population (the corresponding figure is 17·8 for Rouen, 13 per cent for Nantes, 3 per cent for Rheims). More than 45,000 commuters arrive every morning of whom a little over 27,000 are employed in the secondary sector and 18,000 in the tertiary sector. At the same time, the suburbs receive 19,000 commuters from outside, while themselves sending 24,600 into the town: this is an example of the reciprocal commuting which was referred to earlier.

Out of the 1,572 communes of the departments of Nord and Pas-de-Calais, 551 send commuters to the chief town of the department of Nord: this expresses the extent of the phenomenon; but in order to limit the size of the concentrated area to be considered as typical, only the 136 parishes which send more than 4 per cent of their active workers to the factories and offices of the agglomeration need be included. The area of attraction of Lille extends further where the means of communication are more favourable: it extends along the roads and railways to the south-west, the north-west and the south-east. The average distance travelled by commuters is 12 miles: 54 per cent travel less than 6 miles, 30 per cent between 10 and 20 miles, and 16 per cent over 12 miles.

It is the key workers and the labourers who come from furthest afield while the skilled workers are usually recruited on the spot or from close by: paradoxically, it is those who are worst paid who travel the longest distances, but this is explained by the fact that the standard of qualification brings about greater stability of employment. The commuters are 67 per cent men and only 33 per cent women; 44 per cent are engaged in industries concerned with Lille and 51 per cent in industries concerned with the agglomeration, while the rest work in tertiary activities.

A very special group comprises the female labour force recruited in the coalfield to work in the textile factories: this force is usually collected by private coach services belonging either to the factories, or to specialised transport companies: therefore its recruitment area does not correspond to the main communication arteries.

In the reverse direction, 15,000 active residents of Lille leave the town every day to go to work, mostly in the highly industrialised communities or the near suburbs.

In short, the analysis of the demographic development of the town's neighbouring communities is typical: the town of Lille has not increased in size since 1911; the communities immediately surrounding it have themselves become stagnant or even started to diminish since just after the Second World War, while a quite extensive ring round them continues to develop vigorously.

The regional demographic phenomena of the town of Lille present as it were a striking example, epitomising urban influence. Cartography helps to emphasise all these subtleties (Figs. 42 and 43).

ZONES OF INFLUENCE

The combination of forms of influence. Many links are forged between a town and its region. They correspond with the different forms of communication and influences; they overlap and combine without necessarily all being represented. As a whole they represent the zone of influence which extends as far as the various forms of relationship with the town. Here the emphasis is not on the town but on the region which depends on it. The name *Umland* is given to this zone; it was, we believe, first used by André Allix and is in current usage among German and Scandinavian writers. It is less vague than the French *banlieue* which has many different meanings.

These relationships are expressed in many different ways and develop more or less freely according to physical or human restrictions. Even in countries with an advanced civilisation, rural–urban communications vary according to whether the country is rural or industrialised and whether the population is dense or sparse. The exigencies of physical surroundings are gradually being conquered by technical progress. Political restrictions are perhaps more hampering. Frontiers form a barrier even if no hostility exists between the nations; on the Franco-Swiss border the towns turn their back on each other to some extent, and face towards their respective countries.[1]

The influence of a town naturally varies with times and circumstances. Regions do not appear to be finally orientated around a town. The links which we have defined are mobile. Influence in medical matters can depend on the reputation of a famous doctor. Commercial enterprises may move from one place to another. Until the seventeenth century La Sologne was organised in relation to Orléans and included in that town's zone of influence; the evolution of transport, agricultural developments and the attractions of hunting orientated La Sologne towards Paris in the nineteenth century.

In western Europe the town has grown up with the help of the countryside; when this happened on a large scale in the nineteenth century, as Étienne Juillard observes,[2] the countryside was overpopulated; it supplied the town with workers and foodstuffs and a real symbiosis took place.

In underdeveloped countries urbanisation has been on a much smaller scale. Here the town is 'as though encysted in a countryside which presents a violent contrast to it'. The Lébou villages of the Cape Verde peninsula in the Dakar district still demonstrate the juxtaposition of a traditional village population with urban life. 'The psychology of the Negro and the special solidarity of the Lébou group both contribute in slowing up the progress of the social influence of the town.'[3]

Thus urbanisation is slow to establish itself. At one time the Lapps were content with one or two annual expeditions to the coast to exchange reindeer skin for needles, thread and tobacco. Under these conditions urban influence is very slight; the difficulty of communications makes penetration into the interior very arduous.

---

[1] S. Daveau, *Les Régions frontalières du Jura franco-suisse*, Paris, 1959.

[2] E. Juillard, 'L'urbanisation des campagnes en Europe occidentale', *Etudes rurales*, 1961.

[3] J. Gallais, 'Dans la grande banlieue de Dakar. Les villages lébous de la presqu'île du Cap-Vert', *Cahiers d'Outre-mer*, 1954.

However, the scarcity of towns makes the radius of this diffuse influence a wide one. The influence of the town is often determined by trading conditions. Abidjan seems to be above all the capital of the coffee and cocoa trade.[1]

Centrality. Thus the influence of the town on its region is very variable. There are regions which are closely bound up with their urban centre, others which are only loosely attached to it. So it is possible to speak of the centrality of each town, an expression which has proved very successful, and also to distinguish the degrees of centrality.

Attempts have been made to express this centrality mathematically. This is simply a matter of expressing to what degree the urban function is exercised within the region. All the functions cannot be taken into consideration: for example the industrial function extends widely over the region; the same is true of certain commercial functions. It is essentially ·a question of the retail trade and of services which, to the extent that they are used outside the town, express the connection with the region.

Sven Godlund,[2] in a study of Swedish towns, suggests a formula based on the relationship between the total population of the town (F) and the number of people actively occupied in retail trades and services (D); an index of centrality is defined thus: DF100. All towns whose index is more than 3·4 devote a proportion of their retail trade and services to the exterior. In regional centres, the index figure exceeds 6·5.

Thus one can attempt to observe to what extent the region utilises the services of its urban centre and to what extent it provides for its own needs. This was done by W. Christaller, who was the first to use an indirect procedure, based on the use of the telephone.[3] Then, as the telephone criterion raised numerous criticisms, Christaller adopted the following formula, based on the retail trade:

$$Ct = St - Pt \frac{Sr}{Pr}$$

---

[1] M. Santos, 'Quelques problèmes des grandes villes dans les pays sous-développés', *Rév. de Géog. de Lyon*, 1961.

[2] *Bus trafikens framväxt och funktion: de urbana influensfälten*, Lund, 1954.

[3] The telephone is nevertheless a useful element in statistical analysis. Arne Lysen ('Intensitet och distans', *Gothia*, 1955) takes up a formula applied by T. Hagerstrand (*Innovations förloppet ur korologisk synpunkt*, Lund, 1953), by which the intensity of telephone communications initiated by a town is given by

$$\frac{S}{a^1 \times a^2}$$

S being the number of calls during a given period, a^1 the number of subscribers in the originating town, and a^2 the number of subscribers in the town called.

Ct being the index of centrality for the town, St the number of inhabitants employed in the retail trade, Pt the population of the town, while Sr and Pr indicate the number of inhabitants of the region employed in the retail trade and the number of inhabitants of the region.

The greater the number of local inhabitants employed in the retail trade, the more these people have to work for the region as opposed to the town and the stronger the centrality. It is all the stronger if the number of inhabitants of the town in comparison with this number, is smaller (hence the *minus* Pt). On the other hand this centrality will be stronger if there is a smaller number of people employed in the retail trade in the neighbouring region, compared with the whole population, hence the formula:

$$\left(\frac{Sr}{Pr}\right)$$

It is thus possible to determine, according to the figures obtained, the different categories of centrality.

It must, however, be observed that the retail trade alone is taken into consideration. Many other elements also deserve to be considered, all kinds of administrative, medical and cultural services. Even in the retail trade is one justified, as W. William-Olsson asks, in including in the same diagram trades of a very different type, pianos and groceries, for example? One can in any case only obtain a convenient but rough estimate. Too many unknown circumstances escape these formulae and figures, and we have tried to show how varied and many sided are the various forms of influence exercised by the centrality of a town.[1]

Determination of the zone of influence. Various attempts have been made, making use of the criteria we have defined, to determine the zones of influence of certain towns and, in these zones, to distinguish the different forms of influence.

One of the earliest was that of R. E. Dickinson,[2] whose studies of Leeds and Bradford provided a model for much further work by British geographers; his delimitations were based on wholesaling, retailing, insurance, accessibility and other criteria. A. E. Smailes[3]

[1] It would seem that a town can hardly assume the rôle of regional centre if its population is under 6,000, and that it only effectively becomes a centre when it passed 20,000. G. Veyret-Verner, in *C.R. Congrès d'économie alpine*, Grenoble, 1960.

[2] R. E. Dickinson, 'The regional functions and zones of influence of Leeds and Bradford', *Geography*, 1930.

[3] A. E. Smailes, 'The analysis and delimitation of urban fields', *Geography*, 1947; *id. The geography of towns*, London, 1953; *id.* 'The urban hierarchy in England and Wales', *Geography* 1944; *id.* 'The urban mesh of England and Wales', *Trans Inst. Brit. Geog.*, 1948.

has discussed in general terms the analysis and delimitation of urban spheres of influence, and the 'hierarchy' of towns that has developed as a result of the differential growth of urban functions; whilst H. E. Bracey has studied especially the service areas of country towns and their influence on the rural countryside.[1]

Many similar studies have been made by American geographers, of which that by J. E. Brush on Wisconsin may be cited as an example.[2] In Germany J. H. Schultze has summarised the functions and spheres of influence of forty urban centres, and has commented on the theoretical basis of the 'central place' theory (see below p. 437).[3] And even the antipodes are not without their recent literature on this subject.[4]

In Finland Oïva Tuominen, as a result of several investigations made particularly in schools, has determined the limits of the zones from which people in Turku come to do their shopping, consult the doctor, or pursue their studies; and this pioneer work has remained a model of its kind.[5] Elisabeth Lichtenberger has depicted the zones of influence of various Austrian towns (in the Austrian Atlas). The comprehensive and imaginative study made by Sven Dahl for the Swedish town of Västerås (population 79,000) must also be noted.[6]

In France also attempts have been made to define the zones of influence. Jean Coppolani has done this for Toulouse[7] and Marie-Claire Berthe described the 'Area of influence of Toulouse' in her study of all possible forms of relationships.[8] In the same way Huguette Vivian studied the regional influence of Grenoble,[9] Blanche Monteux that of Valence and Claire Mangin that of Besançon; mention must also be made of articles published in the *Etudes normandes* concerning the definition of the Rouen[10] and Caen[11] regions, and the Atlases of eastern and northern France.

[1] H. E. Bracey, 'Towns as rural service centres: an index of centrality with special reference to Somerset', *Trans. Inst. Brit. Geog.*, 1953.
[2] J. E. Brush, 'The hierarchy of central places in South Western Wisconsin', *Geog. Rev.*, 1953.
[3] J. H. Schultze, 'Zur Anwendbarkeit der Theorie der zentralen Orte', *Petermanns Mitteilungen*, 1951.
[4] P. Scott, 'The hierarchy of central places in Tasmania', *Australian Geog.*, 1964; J. S. Whitelaw, 'The measurement of urban influence in the Waikato' *New Zealand Geog.*, 1962.
[5] *Das Einflussgebiet der Stadt Turku*, Helsinki, 1949.
[6] *Västerås kontakter med landet i övrigt*, Västerås, 1954.
[7] *Toulouse, étude de géographie urbaine*, Toulouse, 1954.
[8] *Rév. Géog. des Pyrénées et du Sud-Ouest*, 1961.
[9] *Rév. de Géog. alpine*, 1959.
[10] J. Matha and P-E. Perot, 1953 and P. Feuillet.
[11] M. A. Brier, 1959.

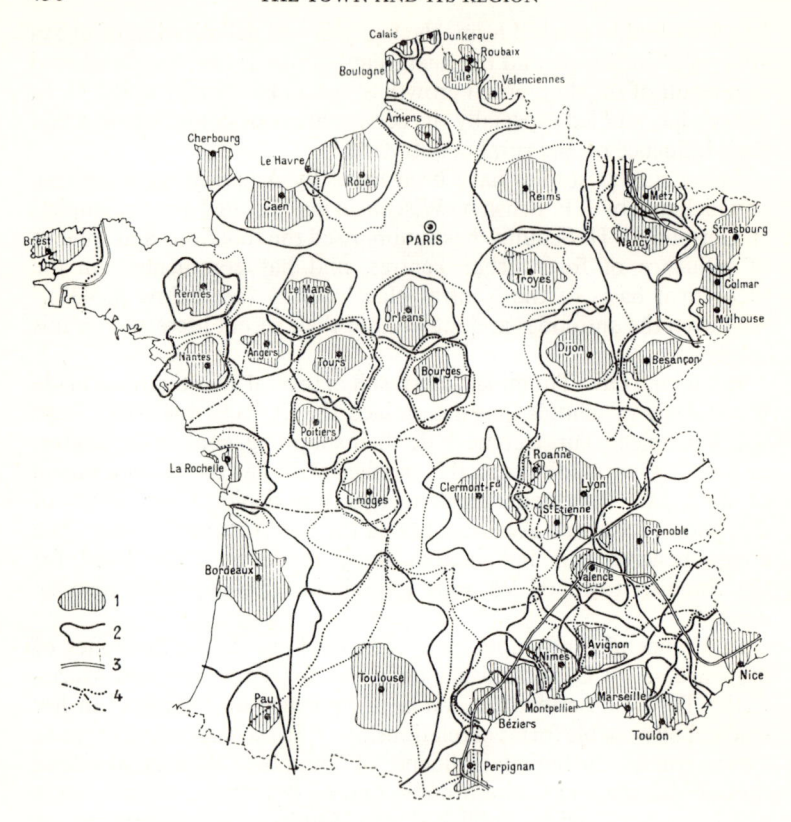

Fig. 44. Zones of influence of French towns with more than 50,000 inhabitants

1. Area having general and direct relations with the town: daily commuting of workpeople, regional recruitment of office staffs; fresh food supplies and general marketing (at least 50 per cent commercial relations with the town); main birthplace area of town's inhabitants; at least 50 per cent of all telephone calls made to the town.

2. Regional extension of main commercial relations: area covered by local banks: tributary area of local fairs; important commercial relations with town (at least 10 to 25 per cent); at least 25 per cent of telephone calls to town.

3. Areas in special commercial relationship with certain centres.

4. Extent of intellectual influence. University recruitment, local newspapers, hospital clientèle, sense of belonging to the centre. Each area has been determined by the reference to one or more of these criteria.

An attempt to determine the zones of influence of the large towns throughout France has been made recently.[1] Only towns with a population of more than 50,000, about fifty in all, were considered and Paris, whose complex influence permeates the whole country, was omitted. This research was undertaken by all the geographical institutes of the various universities. First of all an immediate zone of influence was determined, the activity of which was closely linked to the neighbouring town. Beyond this immediate zone various criteria were considered, permitting the delineation of a zone of economic influence and a zone of cultural influence (recruitment to the universities, newspaper circulation); naturally in the case of these two last zones, interference caused by the influence of a neighbouring town could be caused. It became apparent that a town's area of influence was very far from being in proportion to the size of its population; this depends very much on its position (thus Le Havre is blocked by Rouen). But it also seemed that many country districts do not appear to depend on any large town; the small towns play an important part, usurping the influence of one or more larger towns (Fig. 44).

On the other hand, a complete unit cannot be made out of this zone of influence, this *Umland*. Robert Dickinson has already shown that the region near Chicago is limited to a radius of fifty miles around the Loop, but that the zone of influence (which he calls *Hinterland*) extends as far as the city boundaries of Cincinnati, St Louis and St Paul.[2] With regard to the suburbs, we have already distinguished these as inner, near and outer[3] (see page 240). Peter Schöller distinguishes the *Umland* where communications with the town are close and constant, the *Hinterland* where these communications are less frequent, and the zone of influence where they are the exception.[4] This threefold division is generally accepted. Indeed it is possible to distinguish three categories of communication between the town and the region.

1. Basic communications, which are more or less essential and which provide close and permanent links with the neighbouring town: for example, those that carry agricultural produce to the urban market.
2. Occasional communications, which are orientated towards the town in a permanent manner but are much less frequent: these can be expressed by the purchase of manufactured goods in the town.

[1] Zones of influence of major French cities. Map made by G. Chabot, *Mémoires et Documents du Centre de Documentation cartographique et géographique*, 1961.
[2] Chicago, ville métropole et sa région', *La vie urbaine*, 1934.
[3] G. Chabot, *C.R. Congrès int. de Géog. Paris*, 1931 (vol. III, pp. 432–7).
[4] 'Aufgaben and Probleme der Stadtgeographie, *Erdkunde*, 1953; cf. E. Meynen and F. Hoffmann, 'Methoden zur Abgrenzung von Stadt und Umland', in *Geogr. Taschenbuch*, 1954–5; N. Lenort, *Entwicklungsplanung in Stadtregionen*, 1961.

3. Finally, exceptional communications, which certainly express the town's influence but in a much less regular way: itineraries of commercial travellers or the frequenting of certain medical or hospital centres.

Competing influences. In defining the zone of influence of a town one encounters the influence of the next town and the question of demarcation arises. J. Reilly has formulated a rule inspired by Newton's law and which, for this reason, has been called the law of gravitation[1] of the retail trade. Towns exert their influence in direct proportion to their population, but their influence decreases outwards in proportion to the square of the distance.

If two towns A and B have a population of a and b respectively, the amount of purchases made in these towns by a village C will be expressed by $\frac{a}{b} \times \frac{CB^2}{CA^2}$. Sometimes in making the statement $\frac{a}{b}$ a correction has been made, using logarithms. C will be placed over AB in such a way that $\frac{BC}{CA} = \frac{\sqrt{b}}{\sqrt{a}}$.

Oïva Tuominen has suggested another criterion: the distances from the limit of the two spheres of influence from their respective centres are proportional to the square root of the number of shops in these centres.[2] It must however be observed that the shops are of very varied importance and we cannot include in the same diagram a large departmental store, drugstore or supermarket, and a small grocer. The numbers should be weighted.

In any case this method clarifies only commercial relationships and these are not the only ones which express the influence of a town. In the undefined zone between two towns, the population may depend on one centre for commercial needs but the newspapers may be distributed from another centre. It is therefore impossible to draw a precise line and one can only indicate a zone of demarcation, often a fairly wide one, within which the various influences mingle (see Fig. 46).

This definition of zones of influence is particularly difficult, if not impossible, in the conurbations. We have seen that the characteristic of the conurbation is the urbanisation of the intervening countryside. The latter labours entirely to fulfil the needs of the neighbouring towns in supplying them with milk or vegetables. A high proportion

---

[1] *Methods for the Study of Detail Relationships*, 1929.
[2] *Marketing areas in Finland*, 1953; cf. also R. Ajo, 'Influence du réseau routier pour la lutte entre Turku et Tampere', *Fennia*, 1955; and I. Hustich, 'Divisions régionales de la Finlande', *Terra*, 1961.

of the country people go to work in the town, often, even those from the same village, in two different directions. It is possible, if the conurbation is not too congested, to try to define the respective influence of the towns by estimating the intensity of commuting.

This intermediate country also becomes a recreation zone, a playground; citizens of different towns mingle in the holiday resorts,

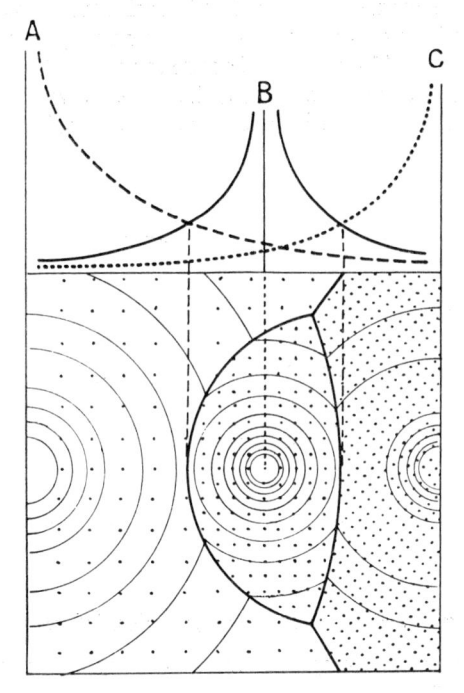

Fig. 45. Respective zones of influence of three centres

The attractive power of A is greater than that of C, which itself is greater than that of B. (From S. Godlund, 'The function and growth of bus traffic within the sphere of urban influence', *Lund studies in geography*, Series B, No. 18, 1956, p. 38.)

hotels and camping sites; the number of weekend residences centred on each of the towns can scarcely be a factor in demarcation.

The Megalopolis of north-eastern U.S.A. gives the best example of countryside which has been completely urbanised. One can no longer speak of the suburbs of this or that town; there is a common suburban area in which the zones of influence are very difficult to disentangle.[1]

[1] J. Gottmann, *Megalopolis*, New York, 1961.

The diagrammatic representation of these conflicting influences is complicated. An example can be found in an article by Anneliese Krenzin for the Rhine-Main district.[1] Sven Godlund, on the other hand, has established isodynes or lines of equal influence around towns situated close together.

If this influence extends equally in all directions, each town is found to be at the centre of a circle in which this influence is radiated. It is surrounded by other towns, each one having its own zone of influence, surrounded by a circle. Let us suppose that all these towns are of equal importance, possessing the same powers of attraction and that their surfaces are identical from the physical and human

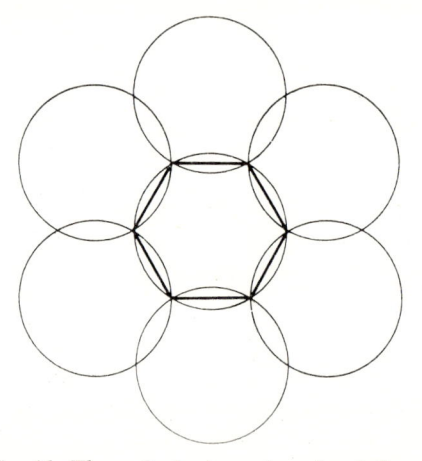

Fig. 46. Theoretical scheme for urban influence

The sphere becomes a hexagon through the juxtaposition of neighbouring spheres of influence.

point of view. We shall then have a group of adjacent circles, each of the same radius. The circles overlap so that no spaces are left between them. The chord joining the points of bisection limits the zones of influence. This chord being equal to the radius if there is a complete overlap between the circles, the town becomes the centre of a hexagon the vertices of which represent the furthest points reached by its influence (Fig. 48).

This was the starting point of the hexagonal theory initiated by W. Christaller thirty years ago[2] and taken up by A. Lösch.[3] We shall

[1] 'Werden und Gefüge der Rhein-mainischen Verstadterungsgebietes', *Frankfurter geographische Hefte*, 1961.
[2] W. Christaller, *Die Zentralen Orte Süddentschlands*, Jena, 1933.
[3] A. Lösch, *Die räumliche Ordnung der Wirtschaft*, Jena, 1944.

have an opportunity of discussing this again with regard to urban networks. It must, however, be noted immediately that these examples are taken from entirely theoretical ideal conditions. The area of influence is only hexagonal in form in cases where the regularity is not disturbed by a river, mountain or frontier and where the extremities of the hexagon are not pulled outwards by a stronger attraction.

Zone of influence and urban region. This determination of zones of influence makes definition of the urban region possible, that is to say, the region which, round each town, exists in symbiosis with it. The influence is, of course, reciprocal. The town is the reflection of its region, its population is to a large extent born there, and many of its workers continue to reside in it; its industries often find their raw material locally; its services are organised to fulfil the needs of the region and increase according to the size of the zone of influence; the tertiary employment sector expresses to a great extent the influence of the town on its neighbourhood.

A town like Béziers in Languedoc is simply the expression of a countryside almost entirely given up to viticulture, while Winnipeg is a response to the wheat growing prairies. Rodez, a sleepy little county town in the south of the Massif Central, in fifteen years has become an active centre, increasing in size by 40 per cent when the agricultural and transport revolution after the First World War made Segala which adjoins it, into a prosperous region.[1]

The urban region corresponds to this zone where the influences are reciprocal; this is also the zone where these influences are stronger than those of any neighbouring town and would therefore not spread as far as all the influences exercised by the town. The demarcation of the region must be fixed according to rival influences.

This demarcation has taken on a special importance with the realisation that the radiation of urban influence provides the only rational basis for a geographical division, the region, in other words, has become an urban or 'city-region'.

In France as long ago as 1910 Vidal de la Blache suggested the division of France into regions each one of which would be centred round a large town. Fifty years later, Etienne Juillard shows that we cannot imagine a region as other than centred round a town.[2]

C. B. Fawcett has also stated that 'each region must have a capital which must really be the centre of regional life'.[3]

[1] Y. Pomarède, *Rev. Géog. des Pyrénées et du Sud-Ouest*, 1954.
[2] 'La région: essai de définition', *Ann. de Géog.*, 1962.
[3] C. B. Fawcett, *The provinces of England*, 1919; see also E. W. Gilbert, 'Practical regionalism in England and Wales', *Geog. Journ.*, 1939.

Robert Dickinson has made various attempts to define this concept of an urban region, of a metropolitan city.[1] A town can be considered as metropolitan, that is as the capital of a region, when a high proportion of the region's products are concentrated there, when it pays for them by means of its own production and when it assumes the responsibility of the consequent financial transactions. In the end this metropolis becomes the cultural and administrative centre. Moreover it can happen that this metropolitan rôle may be divided between several towns in the same region.

In all countries it is by means of centring regions around towns that administrative organisation has been attempted. The plan for hospital organisation in Sweden ended in a demarcation of the country into regions, each one determined by the town centre round which it is grouped; the various choices of town centres determine the different solutions proposed.[2] Even in the emergent countries the region is organised round the town. Paul Pélissier observes that in Senegal regions are grouped around newly built towns, thus breaking up the old ethnic divisions.[3]

In order to define the organic regions considered in this way, the term *nodal* regions has been used, in regarding the region as entwined round the town; Hans Carol calls it the *functional* region. Robert Platt calls it the *organisational* region and François Perroux and Jacques Boudeville also incline to the latter concept, seeing in the region the framework of planned changes. Jacques Boudeville defines in this way the *polarised* region, emphasising the commercial influence of the agglomeration.

The region exists, then, with the town as its centre, among other regions centred round other towns. No doubt seen from an aeroplane, the small huddle of houses which form towns with well known names heard since childhood may seem insignificant, and yet it is around them that the whole area of the neighbouring countryside gravitates.

The urban network and the hierarchy of towns. Thus each town has around it a region of which it is the centre and which adjoins the neighbouring urban regions; but this town also depends on other centres that are better provided for. There is a whole hierarchy of

---

[1] 'The regional functions and zones of influence of Leeds and Bradford', *Geography*, 1930; 'The metropolitan regions of the U.S.', *Geog. Rev.*, 1934; 'Social basis of planning', *Sociological Rev.*, 1942; *City, region and regionalism*, London, 1947.

[2] S. Godlung, 'Befolkning—Regionsjukhus, Resmöjiligheten—Regioner', *Medd. från Lunds Univ. Geog. Inst.*, 1958; cf. O. Bonstedt, in *Stadtregionen in der Bundesrepublik Deutschland*, 1959.

[3] *Cong. Int. de Géog.*, Stockholm, 1960.

urban centres,[1] each one of them acting as a solar system with its planets grouped around it and itself part of a larger system.

There is the small town where people go to do their regular shopping for goods not to be obtained in the village; cars can also be repaired there when the local garage is inadequate. Beyond this is a centre where one is certain of finding more or less everything that one wants, cars, tractors, luxury goods. But the whole area is dominated by a metropolis which exercises a central control, houses the headquarters of large organisations and gives the tone to the whole region (see page 200). To borrow another image from feudal society, there are sovereign towns and vassal towns, which may themselves exercise sovereignty over other towns.

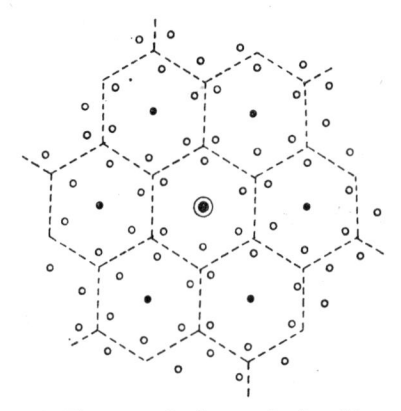

Fig. 47. Theoretical scheme of urban hierarchy

A major centre dominates six peripheral secondary centres

Here are some examples of towns which govern each other in this way: Die, Valence, Lyons or Beaune, Dijon, Paris.

W. Christaller, studying the towns of western Europe and particularly those of southern Germany, has expressed this hierarchy with the aid of a hexagonal diagram (Fig. 49). He groups the hexagons, each one representing the zone of influence of a primary centre, around a central hexagon containing the master town. By this method of grouping a secondary system is obtained, represented by a diagram of six primary peripheral hexagons with a seventh hexagon in the centre: this is the secondary figure the centre of which dominates $7 \times 7 = 49$ units.

In the same way a more advanced system can be imagined where a series of secondary figures is grouped around a central hexagon, the

[1] J. E. Brush, 'The urban hierarchy in Europe', *Geog. Rev.*, 1953; A. E. Smailes, 'The urban hierarchy in England and Wales', *Geog.*, 1944.

whole forming a tertiary figure. Such a tertiary centre would dominate 343 units.[1]

The theories of W. Christaller have met with great success; his diagrammatic representation is still used.[2] But while these theories have been supported by enthusiasts and have been applied to certain regions, they have also been subjected to numerous criticisms, and examples have been produced to invalidate his system. One could say that a large proportion of work devoted to the relationship of the town to its environment has been inspired by the ideas of W. Christaller, either in support of them or in opposition to them.[3]

It must be said immediately—and the author emphasises this—that it is a question of a theoretical and idealised diagram. One thinks of William Morris Davis, the great physical geographer, closing his eyes before looking at the countryside to see whether it corresponded to his mental image. The method has its advantages and sometimes permits anomalies to be more easily discovered.

At the same time the author has also perfected and complicated the system in order to meet certain objections. In its most recent presentation, at the Lund Symposium in 1960, the system of the perfect hexagon hardly applies any longer except with regard to administration; the supply of goods and medical services may disturb this hierarchy in the same way as they can be integrated within it. Above all traffic conditions provide a new factor in hierarchical organisation by placing the central towns on the important traffic routes, giving rise to a new and very different system. But it must, of course be recognised that, in allowing the possibility of choice between contradictory systems, much of their value is lost.

Attempts to adapt the theory, even to western Europe, are laborious. No doubt the French railway network, with six main lines out of Paris, can be invoked as support of the hexagonal theory. But these correspond more or less with the shape of France, which is not necessarily constituted to correspond with a hexagonal diagram. On

---

[1] W. Bunge, *Theoretical Geography*, Lund, 1962, p. 132.

[2] P. Zaremba, 'Functional divisions of suburban areas', *Czasopismo Geograficzni*, 1962.

[3] E. Neef, 'Das Problem der zentralen Orte', *Petermanns Geog. Mitteil*, 1950; J. H. Schultze, 'Zur Anwendbarkeit der Theorie der zentralen Orte', *Pet. Geog. Mitteil.*, 1951; R. Klöpper, 'Methoden zur Bestimmung der Zentralität der Siedlungen', *Geog. Taschenbuch*, 1953; L. J. King, 'Central place theory and the spacing of towns in the United States', in *Land and livelihood*, ed. M. McCaskill, Christchurch, N.Z., 1962. The most complete summary of the discussions to which Christaller's theories have given rise will be found in W. Bunge, *Theoretical Geography*, Lund, 1962. It must be added that the most severe criticism has been directed against the centrality of the smaller centres, with which we have not been concerned.

the other hand a diagram must not be too far removed from reality. In this case the physical reality is composed of mountains and plains, lakes and rivers. Diagrammatic figures can be rapidly mutilated, twisted or deformed.

From the beginning of administrative and commercial organisation, there are towns which have no satellites: this happens especially in underdeveloped countries with difficult communications; the influence of the town really only touches a narrow fringe and beyond this fringe remains extremely vague: the towns of Brazil have been described as an archipelago separated by vacant spaces.

The development of large centres may distort the hierarchy by attracting certain satellites out of their orbit, and W. Christaller himself admits[1] that the hierarchy may be profoundly modified by the growth of large centres which crush the intermediate centres. Épernay turns from Rheims towards Paris; Saint-Étienne from Lyons towards Paris.

Then again the theory presupposes the existence of centres spontaneously organised as a result of commercial or administrative links. But many towns—and they are becoming increasingly numerous—escape this definition. This is particularly true of industrial towns; established on the basis of mineral wealth, ease of communications or labour resources, these towns, although they have not risen as a result of regional needs, are to some extent superimposed on the neighbouring countryside and play the same rôle as the others. They have a monopoly of administrative functions and are tied up commercially with the small market towns: in their early years they are equipped with small shops for daily food supplies; later large shops, cinemas and hospitals follow, which confer on them a central position in the region; their regional rôle has been polarised to their own advantage. A similar destiny has already overtaken the old industrial towns. The forges of Bergslag in Sweden, grown up over the iron mines, have given birth to commercial centres exchanging manufactured goods with the neighbouring countryside against agricultural products.[2]

It might be thought that the development of urban networks in socialist countries and in countries with a free economy would be different. In countries with a socialist economy industrial affairs are bound up with the foundation of the towns; commerce is only a consequence of this, while in countries with a free economy it is the latter that is responsible for the origin of towns. Vadine V. Pokchichevski expounded this theme very strongly at the Stockholm

[1] 'Die Hierarchie der Städte', *Lund Studies in Geography*, Ser. B, No. 24, 1962.
[2] O. Nordström, *Die Beziehungen zwischen Hüttenwerken und ihrem Umland in Südschweden von 1750–1950*, Lund, 1953.

International Geographical Congress of 1960.[1] In fact, as Jacqueline Beaujeu-Garnier has remarked, the contrast is by no means absolute, for market towns organised for exchange arising out of their regional rôle are also known in socialist countries, while free enterprise countries are also experiencing the growth of more and more towns which, though of industrial origin, are becoming regional centres.

The objections, however, only hold good in a geometrical or mathematical approach to the phenomenon. The fact remains that, within the zone of influence of a metropolis, satellite towns are subordinate to it and themselves exercise an influence on a restricted zone. The urban network—from the moment that a network exists—is very hierarchical.

In considering the various forms of influence which we have classified into three categories, we can speak of centralities of first, second or third degree or, if preferred, of primary, secondary or tertiary centres. The majority of authors have supported this threefold division. Hans Carol[2] distinguishes thus (if we omit villages and small towns): towns, large towns and metropolitan cities. Jean Labasse[3] likewise suggests three degrees: 'the primary town is frequented daily or several times a week, according to the scale of the country or county. The secondary town has a zone of influence proportional to its monthly or fortnightly attraction. The regional metropolis is visited every few months, or once a year, for highly specialised services.' It might be simply remarked that the buying of food supplies does not necessarily involve a journey and that certain purchases in the large town can be made through the primary centres. Occasionally four degrees have been distinguished. Olaf Boustedt gives four degrees of centrality for towns in Bavaria.[4]

F. H. W. Green and R. F. A. Edwards (in the Lund Symposium volume, *op. cit.*) also distinguish the centres of districts, the regional capitals, the provincial capitals and the metropoles.

At times it seems as though the list of categories has been widened, but this is because villages and large towns are included. We do not consider that there is room to include in a general and diagrammatic manner more than the three or four categories mentioned above and all authors are more or less in agreement with this. One is forced to give to each of these centres a kind of status, to specify the qualities peculiar to each one of them: financial, medical, judicial, university.

[1] 'Types of localities (cities, towns, villages) in the U.S.S.R., and "central place" theory' (in Russian), *U.S.S.R. Acad. of Science*, Moscow, 1961.
[2] H. Carol, *op. cit.*
[3] J. Labasse, *Les capitaux et la région*, Paris, 1955.
[4] *Die centralen Orte und die Verkehrsräume in Bayern*, Munich, 1960.

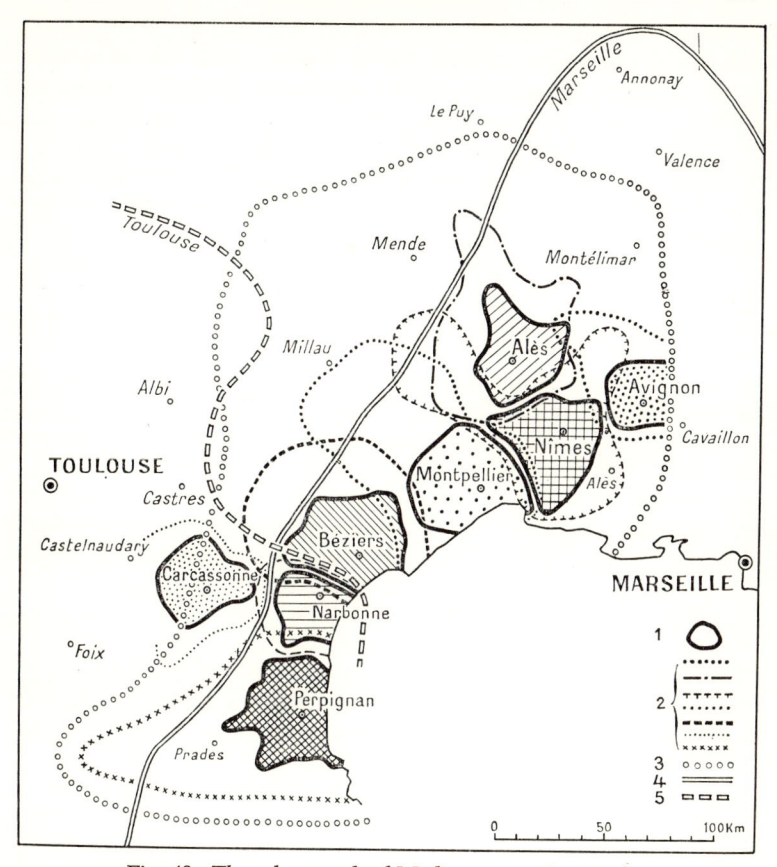

Fig. 48. The urban mesh of Mediterranean Languedoc

(From R. Dugrand, *Le réseau urbain du Bas Languedoc méditerranéen*, Paris, 1963.)

1. Zones with close relations with the town (daily commuting, market visits, etc.). **2.** Less close relations (luxury shopping, high schools, hospitals). **3.** Metropolitan influence—of Montpellier as a hospital centre. **4.** Zone of influence of an external town, Marseilles. **5.** Zone of influence of another external town, Toulouse.

It is according to these differences of function much more than to the size of the population that towns are classified within the hierarchy. And it is on this basis that we depend for equipping the various centres in a rational way.

We must be careful, however, to note that the criteria will not be the same in every country and to realise that, in geography, everything is regulated by the peculiar character of each region. Thus a more elaborate regional analysis may lead to a wider scale, valid only for a limited region. Michel Rochefort[1] in this way lays down the hierarchy of Alsatian towns: (1) the towns which are regional centres (Strasbourg, Mulhouse, Colmar); (2) the town centres of subsidiary regions; (3) small towns with a population of between 4,000 and 5,000; (4) industrial towns (between 5,000 and 8,000 inhabitants); (5) small market towns (between 2,500 and 3,000 inhabitants) to which the title of town can hardly be given. Even smaller towns play a transitional rôle between those which are centres of organisation and the villages they administer.

Within this network, each town has its part to play. Etienne Juillard suggests that the influence of the town and the importance of its regional rôle should be expressed by an index.[2]

One of the rôles which expresses the subordinate position of a town is that of 'relay station'. The town passes on the influence of the metropolis on which it depends; the surrounding countryside is mainly concerned with it and generally speaking the population does not travel further afield to find other supplies. And yet what is bought there comes from the metropolis which supplies its branches and undertakes to fulfil their needs. Chalon-sur-Saône appears in this way to be the relay-station of Lyons, and round Paris numerous towns play this part from the industrial as well as from the commercial point of view. It is on the margins of the zones of influence that these half-way towns play their biggest part; elsewhere one can go direct to the metropolis for special purchases.

So the urban network binds together the towns with many ties, placing them at the service of the regions and subregions and allows the full development of their urban function.[3]

Cartographic representation of urban influence. For cartographic representation on a small scale, it is sufficient to draw, by means of

[1] *L'organisation urbaine de l'Alsace*, Paris, 1960; cf. Dionnet and Frémont, 'La zone d'influence de Bayeux et le réseau des villes et bourgs du Bassin', *Norois*, 1962; L. Dethier, 'Contribution à l'étude du réseau urbain de la Belgique', *Bull. Soc. belge d'Et. Géog.*, 1962.

[2] Urban networks may be more or less closely woven, depending on the number of towns they contain. We are thus led to make maps showing town density (not to be confused with population density within towns). Such maps may vary in their basis, depending on the town population at which one starts—one could have a map showing the density of towns with over 2,000 people, for example, or over 20,000; and on a continental scale one could represent urban density on the basis of towns of more than 100,000. Considered thus, the urbanisation of a country is an expression of one aspect of its civilisation.

[3] 'Essai de hiérarchisation des centres urbains français actuels', *Centre d'études d'aménagement et d'urbanisme*, 1961.

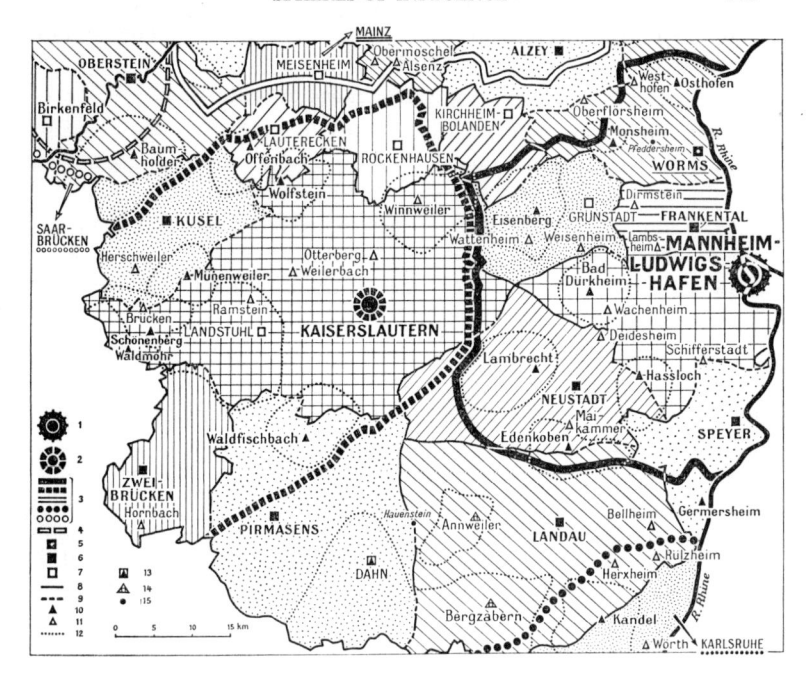

Fig. 49. The urban mesh of the Rhenish Palatinate

(From 'Rheinland Pfalz in seiner Gliederung nach zentralörtlichen Bereicher', compiled under the direction of E. Meynen by R. K. Klöpper and J. Körber, *Forschungen zur deutschen Landeskunde*, Bd. 100.)

1. Very large centre.
2. First-order centre (luxury trades, permanent threatre, museums, professional schools, higher administration).
3. Zone of undisputed extension of the influence of centre 2 (beyond these limits rival influences may operate).
4. Zone having predominant relations with a distant centre.
5. Zone having only moderate facilities but exercising a 'first-order' influence on certain regions.
6. Second-order centre (grammar school, hospital, specialist doctors, specialist trades, theatrical performances, scientific meetings, sometimes the administrative seat of its region).
7. Second-order centre but possessing a less complete range of facilities.
8. Limit of the zone of attraction of a single centre 6.
9. Limit of the zone of attraction towards several centres 6.

10. Third-order centre or lower (doctors, chemists, trade in local products, cinemas, finishing schools).

11. Low-order centre having a poor range of facilities.

12. Limit of zone of influence of a centre 10.

13. Centre having the range of facilities of a third-order centre but exerting the same influence as a second-order centre.

14. Centre having the range of facilities of a second-order centre but whose influence is that of a third-order centre.

15. Locality having the same range of facilities as a third-order centre but whose influence is negligible.

one or several curves, the limits of the zone of influence; a different line may be drawn for each aspect of the influence: this procedure has been used for the map of zones of influence of French towns with a population of 50,000, on a scale of 1/600,000.

On a larger scale it is possible to go into greater detail and to show the town towards which each village is orientated. This has been done by an arrow from the village pointing towards the town in question.[1] For the sake of clarity, a distinctive colour may be given to all the arrows pointing towards the same town. It is also possible to represent each town by a sign common to all the villages dependent on that town.

The difficulty is greater when it is a question of representing simultaneously the position of towns within the whole network and the regions or subregions which they dominate. The towns should not be differentiated according to the size of their population only, for a town of 100,000 inhabitants can play the part of a metropolis, while another with the same population is eclipsed by a neighbouring city. The primary, secondary and tertiary centres should be given special symbols. It is a more delicate matter to mark the limits of the regions and subregions. The entire region may be allotted a special colour on which are drawn two series of curves forming the primary and secondary subregions. The use of a mosaic of colours within the region seems preferable.

Olaf Boustedt, for Bavaria, marks the inside of the urban circles with lines showing four degrees of centrality and also, by tracing the circumferences with single or double arrows, the importance of commuting to the town under consideration.

One of the most remarkable examples of this kind of representation is given by the map drawn under the direction of A. Meynen, R. Klöpper and J. Körber for the Rhenish Palatinate.[1] Zones of medium

[1] O. Tuominen, *op. cit.*; J. M. Casas Torres, *Esquema de la Geographia de Jaca*, Zaragoza, 1942.

[2] Rheinland-Pfalz in seiner Gliederung nach zentralörtlichen Bereichen: 1/300,000.

or average centrality (*mittlere Stufe*) form the basis and they are distinguished by different plain colours. Greater centrality (*höhere Stufe*) is shown by drawing a bold line round the subregions, variously coloured, which look towards the greater centre. The smaller centres (*untere Stufe*), that is those equipped with doctors, pharmacies, by a cinemas, technical schools and agricultural markets, are marked green triangle and their zone of influence by a green line.

The optimal network. The urban network arises spontaneously and is consolidated by a slow evolution which may be disturbed by the creation of new towns, an alteration in the direction of trade or an administrative decision. But it is by no means always perfectly adapted and we are bound to wonder what would be the optimal urban network for a region.

It is at any rate certain that the crowding together of towns is damaging to their harmonious development and that of the region. This is the case with industrial regions where competition obliges each town to live its own life and does not allow for a sufficiently wide range of employment opportunities. There are also regions which suffer from the unbalance of their urban capital; a too large agglomeration has a weakening effect on the secondary centres. In Haute-Garonne there is no town of intermediate size between Toulouse (population 268,000) and Saint-Gaudens (8,000). Conversely, lack of urbanisation is a cause of stagnation. Poor distribution is a difficulty; it is regrettable that in France the largest towns are almost always on the periphery (eight out of nine towns with a population of more than 200,000). A proliferation of small towns is characteristic of the French Massif Central and there is a lack of real regional capitals.[1]

In the Mediterranean Languedoc the rivalry between neighbouring centres (Narbonne, Béziers, Nîmes) prevents Montpellier from playing its full part as regional capital.[2] Economic and political evolution can also destroy the balance between the small and large towns.[3]

Then again one wonders whether it is not possible to establish a rational partition of this urban capital. Naturally this partitioning would not be the same everywhere; it depends on the activities of the region and the density of its population. The C.E.R.E.S. (Committee

[1] S. Derruau-Boriol and A. Fel, *Le Massif central français*, Paris, 1963.

[2] R. Dugrand, *Le Réseau urbain du Bas Languedoc méditerranéen*, Paris, 1963.

[3] S. Zajchowska, 'Développement de l'habitat en Posnanie', *Przeglad Geograficzny*, 1960.

of Regional Economic and Social Studies)[1] defines in this way the optimal urban structure of a region:

1. One capital town whose sphere of influence is one of the criteria of the demarcation of the region, with a population of from 200,000 to 400,000 and incorporating all the highly specialised services, the largest industries, and firms supplying specialised and luxury goods 2. A network of towns with a population of 20,000 to 30,000 whose raison d'être is to provide urban services for their own inhabitants as well as those of the neighbouring countryside and to include industries the nature of which permits a certain degree of decentralisation; 3. finally if, and only if, the region has to include certain units of special production of a national interest, specialised industrial towns, each one having a minimal population necessary to house technical industrial units corresponding with their geographical rôle (steel towns, mining towns, chemical towns and large railway centres).

Of course we must also envisage the existence, beneath this fundamental network, of smaller centres to which one can hardly refuse the title of towns. But the scheme suggested seems to correspond with the satisfactory equipment of a region and a rational arrangement should emerge from it.

The problem can also be posed not on a regional scale but on that of a whole country. One wonders what is the optimal proportion of the population living in the capital or the most important town. At the present time, in a third of the large countries, the largest town has more than twice the population of the next largest town.

Such countries may be described as macrocephalous; and the causes are very varied. It is sometimes a case of towns whose communications were orientated towards the exterior during the colonial period: thus Buenos Ayres has 15 per cent of the population of Argentina, Sydney has 22 per cent of that of Australia; whilst Reykjavik with 41 per cent, represents the awakening of Iceland to modern economic life. Elsewhere it is the capital city of an empire that has vanished: Vienna contains 22 per cent of the population of Austria. Denmark is hardly more than the hinterland of Copenhagen, which houses 20 per cent of its population. French centralisation has concentrated 19 per cent of its population within Greater Paris, whilst Greater London contains 18 per cent of the population of England and Wales.

These great metropolitan cities, which have been described as dinosaurs, do, of course, have many advantages. Large centres have

[1] Vol. 11, 'Niveaux optima des villes d'après l'analyse des structures urbaines du Nord et du Pas-de-Calais', Lille, 1959.

thus been constructed which have been in the vanguard of progress: Athens and Rome, with their half million inhabitants are examples from times of antiquity; the incomparable influence of Paris on the whole world would have been impossible if Paris had remained the small capital of the Ile-de-France instead of having, even at the time of Louis XIV, a population of 500,000 in a country where no other towns reached a population figure of 100,000. The energy concentrated on the bend of the Seine, where the labours of ten centuries have accumulated, represents today one of the major achievements of French economic power.

P. Sargant Florence,[1] in weighing up the economic advantages and disadvantages of metropoles, comes down on the side of the advantages. Sven Dahl, on the contrary, advocates medium sized towns (80,000 to 100,000, even up to 140,000 inhabitants) where working conditions are easier, and fares are cheaper, and which offer the same resources in terms of labour force and goods.[2] Above these medium sized towns, it is considered a good thing to have towns with a population of 500,000, equipped in a suitable fashion for administrative or cultural services. Is it necessary to go beyond this? Every day we hear people complaining about the congestion in large cities, housing difficulties, traffic jams and poor hygienic conditions.

It is, of course, hardly possible to make a clean sweep of cities with a population over a million. But it is neither certain nor desirable that cities should continue to prolong indefinitely their curve of growth. Greater London decreased by 0·21 per cent between 1951 and 1961; eighteen new towns have been created in Great Britain, of which eight in the London region, with populations of up to 60,000, have relieved the pressure on the capital.

Plans for decentralisation are the order of the day in the majority of countries, but the problem appears to be particularly serious in France and in England, where it arises both in the capital and in the provinces. In the capital—whether London or Paris—everyone complains of the increasing population influx which makes it impossible to solve the housing problem, of time lost by suburban dwellers in longer and longer commuting distances and of chaotic traffic conditions. It has been estimated that the total cost of living for a provincial family which has moved to Paris is almost double what it would have been in the provinces. Parking problems are becoming excruciating.

The French provinces, on the other hand, are suffering from the draining away of their life force, and they are threatened with paralysis.

[1] 'Economic efficiency in the Metropolis', in *The Metropolis in Modern Life*, 1955.
[2] 'En ny Stadstyp', in *Staden, några Stadbygds-problem*, Göteborg, 1955.

Four-fifths of all theatrical productions are put on in Paris; nearly all books are published there, and 80 per cent of all French cars are made there. From this arises the vigorous and often successful campaign conducted in France since the Second World War against centralisation, the Parisian empire.

This question, however, is no more pressing today than it was twenty years ago. Speed of communication makes possible efficient cooperation between the capital and the provincial towns. Organisations with headquarters in Paris can direct facories and laboratories in several provincial towns, and the same is true for London.

It is important, however, to distinguish between deconcentration and decentralisation.[1] Deconcentration takes place inside an organisation which develops branch factories; decentralisation consists of moving an organisation out into the provinces. The former is already practised on a large scale; the latter is restricted by all the reasons which have sited the firm's headquarters in relation to the great banking and industrial organisations. Success will come rather from the rise of large provincial undertakings which understand how to maintain their liaison with the centre of affairs. The most desirable development is this: as efficient as possible an expansion of the provincial towns, in cooperation with a metroplitan centre which will neither stagnate nor live in isolation from them.

The problem oversteps national frontiers and is found on a world scale. There are in fact towns which can be considered as cities of the world. A characteristic of these is a reputation which extends beyond their own countries. As a rule these towns are 'multi-functional'. But the commercial function is paramount especially the most complex aspects of it: banks, insurance, big business. Some of these towns are also distinguished by their cultural influence; the industrial organisations of others supply the whole world. Politics also play their part in this: the world city is often an imperial one. These towns have been preconditioned by the rôle they played in their own country; their fame is due to their position as centres of communication. In the age of the aeroplane, however, their subtle influence no longer depends on physical conditions; economic and political considerations are more important and herein lies both their stability and their weakness.

[1] Neither must we confuse decentralisation with peripheral growth which is simply one aspect of the growth of an agglomeration; cf. S. Riemer, *The Modern City*, 1955, p. 60.

CHAPTER 30

CONCLUSION

The density of the urban network is increasing; each town within the network is tending to grow larger; the metropolitan cities are also swelling; the limits of these agglomerations can no longer be clearly defined, as they spread over the countryside, which is being drastically changed by this urban invasion. Henri Pirenne has shown that the economic organisation of the countryside was disturbed in the Middle Ages by the formation of towns. In our own time the situation is even more dramatic.

Since the beginning of the industrial era there have been warnings, deploring the exodus from the countryside, condemning the expansion of towns. At the end of the nineteenth century people sighed over the deserted countryside. In this one can perhaps hear the echo of the nature cult made fashionable by Jean-Jacques Rousseau and the Romantics. There was also a little of the egotistical attitude of the town dweller who wished to enjoy urban privileges and to exclude others from them. The expression 'dearth of agricultural workers' was often heard. Cultured townsfolk recalled Virgil: 'O fortunatos nimium . . .' and songs extolling the tillers of the soil were popular.

In our own time this would be more likely to be a cry of distress: the rural areas around large towns are dying. The hinterland of towns on the Côte d'Azur is reduced to a suburban status, and cannot even profit from the advantages of this locality, for water which would enable this countryside to be transformed by irrigation is kept for the tourists in the towns.[1]

The large exodus of agricultural workers has sometimes been catastrophic, certain regions have fallen into decay and wasteland has increased, and a social vacuum is left by the departure of the population. In the underdeveloped countries, the exodus has broken up all the traditional patterns, the people have abandoned their tribes without finding their proper place with another framework; family life has been disrupted: the men go off alone to the towns and often fall a prey to alchoholism. This is not a question of a cycle that all countries, passing through successive changes, have to go through one after another. Urbanisation, hastened by rail travel, automobiles and aeroplanes has been imposed on countries which are very unevenly

[1] B. Kayser, *Cent ans d'expansion cannoise, La campagne devant l'urbanisation*, 1954; id. *Campagne et villes de la Côte d'Azur*, Paris, 1958.

advanced; the normal pattern of evolution and gradual adaptation has been broken up.

It appears, however, that this is an irreversible phenomenon. In order to guide it as well as possible it would seem better to look on the beneficial side rather than to deplore it.

It is in towns that new civilisations have developed, and they have always been the melting-pot where the different races mingle. The mixture of populations in Asiatic towns has often been described. In the U.S.S.R. it is in the towns that the assimilation of different ethnic groups has been achieved.[1] Towns are also the gates through which progress, education and better conditions are able to enter.[2]

On the whole towns fulfil this rôle satisfactorily. In western Europe, where urbanisation is at its most intense, Etienne Juillard[3] distinguishes three types of urbanisation of varying degrees of merit as follows:

1. The town of 'landowners'. We have studied this in the section on town property.[4] The town acts as a parasite vis-à-vis the country—which does not mean that it may not assist the country in certain circumstances. In 1950 the town of Béziers alone (population 65,000) levied on the neighbouring countryside a land tax valued by Raymond Dugrand at 1,500 million francs of that period.

2. The 'insular' town. The town acts as a foreign body in its rural environment. Juillard means by this that it has a sterilising and not an enriching effect. It retains its own wealth and drains the young away from the land. This effect is produced particularly in cases where towns have been recklessly superimposed and thrown into the middle of a backward and defenceless rural environment.

3. The 'urbanising' towns which have exercised this function gradually and have penetrated far into the countryside: we have described the various forms of its influence. It is the only one capable of effecting a rational utilisation of the countryside which will be of benefit to all. The towns of the first and second types also may have, to a lesser degree, an 'urbanising' effect.

Thus the whole group of towns in a region constitutes a strong framework which exercises a powerful influence[5]

[1] C. D. Harris, 'Economic groups in cities of the Soviet Union', *Geog. Rev.*, 1945.
[2] T. Hägerstrand, 'Innovations förloppet ur kronologisk Synpunkt', *Medd. Lunds. Univ. Geog. Inst.*, 1953.
[3] 'L'urbanisation en Europe occidentale', *Études rurales*, 1961.
[4] See above, p. 395.
[5] G. Chabot, 'L'armature urbaine en géographie régionale', in *Urbanisme et Architecture*, 1954.

The trend in favour of towns is inevitable. People no longer lament that agricultural workers are in short supply; today when the combine-harvester accomplishes in two hours what was formerly done in 150 hours with the sickle and the flail, requirements are quite different. Now men are needed to construct the combine-harvester and to manufacture the steel from which it is made and this leads to the drift from the fields to the factory.

Small village grocers are going out of business because of competition from the delivery vans from the large shops in the neighbouring towns which bring fresher and more varied foodstuffs; and even the country people themselves find this an advantage.

The contrast between town and country is no longer quite the same as it was a hundred years ago, for the town and the countryside are getting closer together. The agricultural countryside is becoming urban; we now have (to use American terms) the *non-rural farm* population as well as the *rural non-farm*. In Minnesota, towns with a population of eight to ten thousand are inhabited by farmers who drive to work daily in their own automobiles. In the U.S.S.R. there is a tendency to construct agricultural towns (or *agrogorod*) equipped with modern plant, where the peasants of the district, particularly those of the big *kolkhozes*,[1] converge. In this way near Moscow, on the Sovkhoze Zaria Kommounizna territory, a town has been built where farm workers are housed in blocks of flats several storeys high, with all modern conveniences. In the irrigated regions of California, agriculture is often undertaken by workers from the town, and on the site there is simply an administrative office and a few sheds.[2]

Conversely, the town is beginning to assume a rural character. In Connecticut the suburban parishes are populated by part-time farmers who, in addition to their jobs in the town, also run a small agricultural holding.

And this sort of thing is not confined to the United States. Many a Polish 'peasant' has two sources of income—his small-holding and his factory job; and in France there is now some doubt as to whether to call such people worker-peasants or peasant-workers. Are they still country-dwellers, or townspeople who spend their day in town and their night in the country?

Thus there is a new orientation in the relationship between the town and the country. In the past, the town retained something of its rural character; chickens and pigs ran around in the streets. The medieval peasant could be seen returning to the town in the evening,

[1] P. George, 'L'urbanisation des villages en U.R.S.S.', *Ann. de Géog.*, 1951.
[2] J. Blache, 'Coup d'oeil sur les villes américaines', *Rev. de Géog. de Lyon*, 1955.

his tools on his shoulder, after having tended his crops. Even at the beginning of the nineteenth century, as remarked by Charles H. Pouthas, 'urban life was not much different from country life'.[1] After that time, the town strengthened its position and was clearly distinguishable from the country in its manners and customs, indeed was proud of this difference. But today the contrast is diminishing because, to an increasing degree, the town runs into the country.

The urban manner of living is spreading everywhere. It is without doubt this very type of urban life which best characterises the evolution of a country, which defines its civilisation. It is easy to verify its progressive infiltration into the countryside of western Europe. The peasants' clothes are copies from the fashions of the town; blue dungarees have replaces the smock; radios, washing-machines and refrigerators are found in many homes. Privileges which were formerly jealously guarded by the townspeople are disappearing. The town is no longer the only place where village matters are discussed; theatrical gossip is known at every farm. News of great national events is immediately announced on the radio and received in the most remote hamlets and it is a far cry from the time when a revolution in Paris was not heard of for a week or was never heard about at all.

One might even wonder whether the town is not, to some extent, drained of its substance, to the profit of the countryside which it has conquered. It has lost some of its prestige and its magnetism. In the highly urbanised country districts there is no longer any need to go into town for shopping or even for recreation; the village has its own cinema.

To some extent the town is dissolving into the countryside; there is a risk of having an 'urban *bocage*' where the town will have no more significance than the village has in the rural landscape of fields, hedges and woods.

It is, however, impossible to imagine the disappearance of towns, of their vanishing without anything to replace them. The rôle, the very conception of the town evolves with civilisation. The more efforts there are made to organise each country and to create rational units, the more need there will be for centres where these projects can be worked out and where communal elements can be gathered together. These will be laboratories where equipment and methods will be organised simultaneously. These centres will be the towns of tomorrow. We cannot yet see their outlines clearly, but they will continue and reinforce the urban tradition.

[1] C. H. Pouthas, *La population française pendant la première moitié du XIVe siècle*, Paris, 1956.

BIBLIOGRAPHY

Studies in urban geography have so multiplied in recent years that it is impossible to give anything like a complete bibliography. In the foregoing pages we have cited in the footnotes the local examples on which parts of the text have been based. The following works are of more general import. We wish to draw attention especially, however, to four works in which the sections devoted to towns are themselves miniature manuals of urban geography:

SORRE, M. *Les fondements de la géographie humaine*, Vol. III, pp. 154–546, Paris, 1952.
BEAUJEU-GARNIER, J. *Géographie de la population*, 2 vols., Paris, 1956, 1958.
DERRUAU, M. *Précis de géographie humaine*, Part IV, pp. 463–518, Paris, 1961.
SCHWARZ, G. *Allgemeine Siedlungsgeographie*, pp. 309–501, Berlin, 2nd edn, 1962. (contains an extensive bibliography).

* * *

AAGESEN, A. 'Om isochronkort', *Geog. Tidsskrift.*, 1943.
ABASCAL GARAYOA, A. 'La evolucion de la poblacion urbana española en la primera mitad del siglo XX', *Revista geographica*, Zaragoza, 1956.
ALEXANDERSSON, G. *The industrial structure of American Cities*, Stockholm, 1956.
ALLEFRESDE, M. 'Données nouvelles sur les villes arctiques norvégiennes (Narvik-Bodo-Mo i Rana)', *Rev. Géog. Lyon.*, 1962.
ALLIX, A. 'The geography of fairs, illustrated by old-world examples', *Geog. Rev.*, 1922.
AMEEN, L. 'Stadtsplane typer', *Svensk Geog. Arsbok.*, 1961.
ANDERSON, N. *The urban community*, London, 1959.
ASHWORTH, W. *The genesis of modern British town planning*, London, 1954.
BACHMANN, F. *Die alten Städtebilder*, 2nd edn, Stuttgart, 1965.
BALANDIER, G. *Sociologie des Brazzavilles noires*, Paris, 1955.
BARTELS, D. *Nachbarstädte*, Hamburg, 1960.
BARTHOLOMEW, H. *Land use in American cities*, Cambridge (Mass.), 1955.
BEAUJEU-GARNIER, J. 'Problèmes de l'urbanisation du Royaume-Uni', *Ann. Géog.*, 1958.

BEAUJEU-GARNIER, J. *Geography of population*, London, 1966.
BERGSTEN, K. E. 'Agglomeringstendensen inom svensk bebyggelse', *Svensk geog. Arsbok.*, 1950.
BERGSTEN, K. E. 'Variability in intensity of urban fields as illustrated by birth-places', in *Studies in rural-urban interaction*, Lund, 1951.
BERGSTEN, K. E. 'Staden som region centrum', in *Staden: några stadsbygdproblem*, Göteborg, 1955.
BERNARD, E. M. 'Structures démographiques comparées du département de l'Isère, de l'agglomération grenobloise et de la France', *Rev. Géog. Alpine*, 1957.
BERNOUILLI, H. *Die Stadt und ihr Boden*, Zurich, 1946.
BLACHE, J. 'Sites urbains et rivières françaises', *Rev. Géog., Lyon*, 1959.
BLANCHARD, R. 'Une méthode de géographie urbaine', *La vie urbaine*, 1922
BOBEK, H. 'Grundfragen der Stadtgeographie', *Geog. Anz.*, 1927.
BOBEK, H. 'Die nordamerikanischen Kleinstädte und ihre Entwicklung', *Mitt. der geog. Gesellsch.*, Vienna, 1940.
BOBEK, H. 'Über einige funktionnelle Stadttyper und ihre Beziehungen zum Lande', *C.R. Congrès int. Géog. Amsterdam*, 1938, Sect. III.
BOESLER, K. A. 'Zum Problem der quantitativen Erfassung städtischer Funktionen' in the *IGU Symposium in Urban Geography*, Lund, 1960.
BOESLER, K. A. 'Die Städtischen Funktionen' *Abhandl. des Geographischen Instituts der Freien Universität Berlin*, 1960.
BONETTI, E. *La teoria delle localita centrale*, Trieste, 1964.
BORCHERT, J. R. 'The twin cities', in *Minnesota's changing Geography*, Minneapolis, 1959.
BOUSTEDT, O. 'Die Stadt und ihr Umland', *Raumforschung und Raumordnung*, 1953.
BOUSTEDT, O. 'Die Zentralen Orte und ihre Einflussbereiche', in *The IGU Symposium in Urban Geography*, Lund, 1960.
BOYCE, R. R. and HORWOOD, E. M., *Studies of the CBD and Urban Greenway development*, Seattle, 1959.
BRUNNER, K. H. 'Der Verkehr als Bindeglied zwischen Stadt und Land', *Raumforschung und Raumordnung*, 1956.
BUNGE, W. *Theoretical geography*, Lund Studies in Geography, 1962.
BURGHARDT, A. F. 'The location of river towns in the Central Lowland of the U.S.A.', *Annals of the Assoc. of Amer. Geog.*, 1959.
BURKHARDT, MACKENSEN, R., PAPALEKAS, J., PFEIL, E. and SCHUTTE W., *Daseinsformen des Gross, tadt*, Tübingen, 1959.
BURNET, L. *Tourisme et villégiature sur les côtes de France*, Paris, 1963.
BURGESS, E. W. 'The growth of the city: an introduction to a research project', in *The City*, Chicago, 1925.

CARALP, R. 'Un problème de population dans le Massif Central: Les agglomérations ferroviaires', *Rev. Géog., Lyon*, 1959.
CARRIÈRE, F. and PINCHEMEL, P., *Le fait urbain en France*, Paris, 1963.
CARTER, H. *The Towns of Wales*, Cardiff, 1965.
CASAS TORRES, J. M. *Ciudades, Urbanismo y Geografia*, Madrid, 1957.
CHABOT, G. 'Les zones d'influence d'une ville', *C.R. Congrès int. Géog.*, Paris, 1931.
CHABOT, G. 'La détermination des courbes isochrones en géographie urbaine', *C.R. Congrès int. Géog.*, Amsterdam, 1938.
CHABOT, G. 'Géographie urbaine: la naissance et la localisation des villes industrielles', *Inf. Géog.*, 1947.
CHABOT, G. 'L'armature urbaine en géographie régionale' in *Urbanisme et architecture, Etudes en l'honneur de Pierre Lavedan*, 1951.
CHABOT, G. 'A propos du phénomène urbain', in *Mélanges géographiques offerts au Doyen E. Bénévent*, 1954.
CHABOT, G. 'Faubourgs, banlieues et zones d'influence', *Urbanisme et Habitation*, 1954.
CHABOT, G. L'évasion urbaine', *La Vie urbaine*, 1957.
CHABOT, G. *Les Villes*, 3rd edn, Paris, 1948.
CHAPIN. *Urban growth dynamics in a regional cluster of cities*, London, 1962.
CHARDONNET, J. *Les Métropoles économiques*, Paris, 1959.
CHATELAIN, A. 'Géographie sociale des villes françaises en 1946', *Rev. Géog., Lyon*, 1956.
CHATELAIN, A. 'Démographie urbaine', in *Livre jubilaire M. Zimmermann*, Lyon, 1944.
CHATELAIN, A. 'Les notions démographiques de zones urbaines: de la "cité" à la "banlieue"', *Et. Rhodaniennes*, 1946.
CHATELAIN, A. 'Le journal, facteur de régionalisme', *Et. Rhodaniennes*, 1948.
CHATELAIN, A. 'Les données actuelles de la géographie des journaux', *Et. Rhodaniennes*, 1949.
CHATELAIN, A. 'Les forces démographiques d'expansion urbaine', *Rev. Géog., Lyon*, 1950.
CHATELAIN, A. 'Géographie sociologique de la presse et régions françaises', *Rev. Géog., Lyon*, 1957.
CHRISTALLER, W. *Die Zentralen Orte in Süddeutschland*, Jena, 1933.
CHRISTALLER, W. 'Rapports fonctionnels entre les agglomérations urbaines et les campagnes', *C.R. Congrès int. Géog. Amsterdam*, 1938.
CHRISTALLER, W. 'Das Grundgerüst der räumlichen Ordnung in Europa. Die Systeme der zentralen Orte', *Frankfurter Geog. Hefte*, 1950.

CHRISTALLER, W. 'Die Hierarchie der Städte', *The IGU Symposium in Urban Geography*, Lund, 1960.
CLAVAL, P. *Géographie générale des marchés*, Paris, 1962.
CONANT, J. B. *Slums and suburbs*, New York, 1961.
CONDUCHÉ, R. 'Les commerces non sédentaires dans la distribution commerciale: Marchés et tournées', *La Vie urbaine*, 1960.
CONZEN, G. *Alnwick: a study in town plan analysis*, Institute of British Geographers, Vol. 27, 1960.
COPPOCK, J. T. and PRINCE, H. *Greater London*, London, 1964.
COPPOLANI, J. 'Agglomérations et conurbations dans le Midi aquitain et languedocien', *Revue Géog. Pyrénées et S.-O.*, 1957.
COPPOLANI, J. *Le réseau urbain de la France*, Paris, 1959.
COPPOLANI, J. 'De quelques notions fondamentales et définitions en géographie urbaine', *La Vie urbaine*, 1960.
CORNISH, V. *The great capitals*, London, 1923.
COURTIN, R. and MAILLET, P. *Economie géographique*, Paris, 1962.
COUSSON, A. and BOISSEAU, J. *Les Capitales du monde*, Paris, 1959.
DACEY, M. F. 'The minimum requirements approach to the urban economic base' in *The IGU Symposium in Urban Geography*, Lund, 1960.
DAHL, S. 'En ny stadstyp' in *Staden: några stadsbygd-problem*, Göteborg, 1955.
DAINVILLE, F. DE. 'Grandeur et population des villes au XVIIIe siècle', *Population*, 1958.
DAVIS, K. and HERTZ, H. 'The pattern of world urbanization. The world distribution of urbanization', *Bull. of the Intern. Stat. Institute*, 1954.
DEFFONTAINES, P. 'Du Patrimonio au condominio. Contribution à l'étude de la géographie urbaine du Brésil', *Cahiers Outre-Mer*, 1961.
DENIS, J. *Le Phénomène urbain en Afrique centrale*, Brussels, 1958.
DICKINSON, R. E. 'The morphology of the medieval German town', *Geog. Rev.*, 1945.
DICKINSON, R. E. *The West European city*, London, 1951.
DICKINSON, R. E. *City, region and regionalism*, 2nd edn, London, 1952.
DICKINSON, R. E. 'The geography of commuting: Netherlands and Belgium', *Geog. Rev.*, 1957.
DICKINSON, R. E. *City and region*, London, 1964.
DITTRICH, E. 'Das Stadt-Umland Verhältnis in seiner planerischen Problematik', *Raumforschung und Raumordnung*, 1956.
DRESCH, J. 'Les villes de l'Afrique occidentale', *Cahiers Outre-Mer*, 1950.
DUGRAND, R. *Le Réseau urbain du Bas-Languedoc méditerranéen*, Paris, 1963.

DUMONT, M.-E., 'Les migrations ouvrières du point de vue de la délimitation des zones d'influence urbaine et la notion de zone d'influence prédominante: application à l'agglomération gantoise', *Bulletin Soc. Belge Et. Géog.*, 1950.

DUNCAN, O. D., et al. *Metropolis and region*, Baltimore, 1960.

EGLI, E. *Die neue Stadt in Landschaft und Klima.* Erlenbach-Zurich, 1951.

ENEQUIST, G. 'Vad är en tätort' in *Tätorter och Omland*, Lund, 1951.

ESTIENNE, P. *Villes du Massif Central*, Clermont-Ferrand, 1963.

FEAD, M. I. 'Notes on the development of the cartographic representation of cities', *Geog. Rev.*, 1933.

FEHRE, 'Zur Abgrenzung der Stadtregion', *Raumforschung und Raumordnung*, 1956.

FERRARA, R. *Problemi e prospettive dei transporti urbani in Europa*, Milan, 1959.

FISHER, R. M. *The metropolis in modern life*, New York, 1955.

FORDE, D. (ed.) *Aspects sociaux de l'industrialisation et de l'urbanisation en Afrique au sud du Sahara.* UNESCO, 1956.

FORET, J. 'Urbanisme et vieillesse', *La Vie urbaine*, 1959.

FREEMAN, T. W. *The conurbations of Great Britain*, Manchester, 1959.

FRÖDIN, J. 'Staden som geografisk företeelse', *Ymer*, 1946.

FRÖDIN, J. 'Några dag av modern svensk stadsutveckling', *Ymer*, 1948.

FUCHS, J. R. 'Intraurban variation of residential quality', *Econ. Geography*, 1960.

GACHON, L. 'Les rapports villes-campagnes: le sens prévisible de leurs lignes d'évolution', *Norois*, 1954.

GACHON, L. 'Géographie des rapports villes-campagnes', *Bull. soc. belge Et. Géog.*, 1957.

GACHON, L. 'Les rapports villes-campagnes', *Rev. Géog. alpine*, 1961.

GAIGNARD, R. 'La montée démographique argentine', *Cahiers Outre-Mer*, 1961.

GARRISON, W. L. 'Towards simulation model of urban growth and development', in *The I.G.U. Symposium in Urban Geography*, Lund, 1960.

GEDDES, P. *Cities in evolution*, London, 1915. Republished 1949.

GEISLER. 'Zur Methodik der Stadtgeographie', *Petermanns Mitteilungen*, 1932.

GEORGE, P. 'La banlieue une forme moderne de dévoloppement urbain', in *Etudes sur la banlieue de Paris*, Paris, 1950.

GEORGE, P. *La Ville. Le fait urbain à travers le monde*, Paris, 1952.

GEORGE, P. 'Problèmes posés par l'accroissement urbain spontané dans les pays en cours de développement', *Publications de l'Institut de développement économique et social*, Paris, 1958.

GEORGE, P. 'Questions de morphologie urbaine et d'aménagement des villes'. *Ann. Géog.*, 1958.
GEORGE, P. 'Problèmes géographiques de la reconstruction et de l'aménagement des villes en Europe occidentale depuis 1945', *Ann. Géog.*, 1960.
GEORGE, P. 'Problèmes urbains d'Indonésie', *Ann. Géog.*, 1960.
GEORGE, P. 'Conurbations ou réseaux urbains', *Ann. Géog.*, 1960.
GEORGE, P. *Précis de Géographie urbaine*, Paris, 1961.
GIBBS, J. 'Metropolitan growth: an international study', *American Jour. Sociology*, 1960.
GLASS, D. V. *The town and a changing civilization*, London, 1935.
GLASS, R. *The Newcomers: West Indians in London*, London, 1960.
GODLUND, S. 'Trafik, Omland och Tätorter, in *Tätorter och Omland*, Lund, 1951.
GODLUND, S. *Busstrafikens framväxt och funktion i de urbana influensfälten*, Lund, 1954.
GOKHMAN, V. M. and KOVALEVSKY, P. 'Post war changes in the distribution of population of the U.S. by major regions, and some problems of urban growth', *Soviet Geography*, 1962.
GOTTMANN, J. 'Expansion urbaine et mouvements de population', *Research Group for European Migration problems Bulletin*, 1957.
GOTTMANN, J. 'Revolution in land use, *Landscape*, 1958–59.
GOTTMANN, J. 'L'urbanisation dans le monde contemporain et ses conséquences politiques', *Politique étrangère*, 1960.
GOTTMANN, J. 'The impact of urbanisation' in *The Nation's Children*, New York, 1960.
GOTTMANN, J. *Megalopolis*, New York, 1961.
GREEN, F. H. W. 'Urban hinterlands in England: an analysis of bus services', *Geog. Journ.*, 1950.
GREEN, F. H. W. and EDWARDS, R. F. A. 'A commercial application of urban hinterland studies' in *The IGU Symposium in Urban Geography*, Lund, 1960.
GREGOR, H. F. 'Spatial disharmonies in California population growth', *Geog. Rev.*, 1963.
GRIMAL, P. *Les villes romaines*, Paris, 1954.
GROSS, E. 'The role of density as a factor in metropolitan growth', *Population studies*, 1954.
HAGERSTRAND, T. *The propagation of innovation waves*, Lund, 1952.
HAGERSTRAND, T. 'Städers storlek och läge', *Geog. Notiser*, 1954.
HALL, P. *The world cities*, London, 1966.
HANSEN, V. 'Some characteristics of a growing suburban region' in *Guidebook Denmark. Internat. Geog. Congress*, 1960.
HARRIS, C. D. 'Suburbs', *The American Journal of Sociology*, 1943.

HARRIS, C. D. 'A functional classification of cities in the United States', *Geog. Rev.*, 1943.
HARRIS, C. D. 'The cities of the Soviet Union', *Geog Rev.*, 1945.
HARRIS, C. D. 'The market as factor in the localisation of industry in the United States', *Ann. Ass. American Geographers*, 1954.
HARRIS, C. D. 'The pressure of residential-industrial land use', in *Man's role in changing the face of the earth*, Chicago, 1956.
HARTKE, W. 'Die Zeitung als Funktion Sozial geographischer Verhältnisse im Rhein-Main-Gebiet', *Rhein-Mainische Forschungen*, 1952.
HARTMAN, G. W. and HOOK, I. C. "Substandard urban housing in the U.S.A.: quantitative analysis', *Econ. Geog.*, 1956.
HASSER, K. *Die Städte geographisch betrachtet*, Leipzig, 1907.
HAUSER, P. M. *Le Phénomène de l'urbanisation en Asie et en Extrême-Orient*, Calcutta, 1951.
HENRY, L. 'Villes nouvelles et grandes entreprises. Structure de la population', *Population* 1960.
HIORNS, F. R. *Town-building in history*, London, 1956.
HOFFMANN, F. and KLÖPPER, R. 'Schriftum zur Abgrenzung von Stadt und Umland', *Berichte zur deutschen Landeskunde*, 1955.
HOOVER, E. M. and VERNON, R. *Anatomy of a metropolis*, New York, 1959.
HOSELITZ, B. F. 'The cities of India and their problems', *Annals Assoc. Amer. Geog.*, 1959.
HOYT, H. 'Forces of urban centralization and decentralization', *American Journal of Sociology*, 1941.
HOYT, H. 'World urbanization', *Urban land Institute Technical Bull.*, 1962.
HUGHES, R. H. 'Hong Kong: an urban study', *Geog. Journ.* 1951.
HUNT, A. J. 'Population mapping in urban areas', *Geography*, 1960.
HUNTER, J. M. 'An exercise in applied geography. Geographical planning in urban areas for the 1960 census of Ghana', *Geography*, 1961.
HUTTENLOCHER, F. 'Die städte des Neckarlandes', *Stuttgarter Geographische Studien*, 1957.
IPSEN, G. 'Stadt' in *Handwörterbuch der Sozialwissenschaften*, Stuttgart-Tübingen-Göttingen, 1956.
JACOBSON, 'Tätorter centralitets grad', *Svensk geog. Arsbok*, 1958.
JACQUEMYNS, C. 'Problèmes actuels d'urbanisme', *Cahiers d'urbanisme*, Brussels, 1954.
JEFFERSON, M. 'Distribution of the world's city folk', *Geog. Rev.*, 1931.
JEFFERSON, M. 'The law of the primate city', *Geog. Rev.*, 1939.
JONES, E. 'Sociological aspects of population mapping in urban areas (Belfast)', *Geography* 1961.

JONES, E. *Towns and cities*, London. 1966.
JOURNAUX, A., and TAILLEFER, F. 'Les villes minières du Labrador', *Bull. Ass. Géog. Français*, 1957.
JUILLARD, E. 'L'urbanisation des campagnes en Europe occidentale', *Etudes rurales*, 1961.
KANT, E. 'Umland Studies and Sectors Analysis', in *Tätorter och Omland*, Lund, 1951.
KANT, E. 'Suburbanization, urban sprawl and commutation', in *Migration in Sweden*, Lund, 1957.
KANT, E. *Zur frage der inneren Gliederung der Stadt*, 1960.
KANTOROWICH, R. *Regional shopping centres: a planning report on Northwest England*, Manchester, 1965.
KAYSER, B. 'L'évolution démographique des petites villes', *Rev. Géog. Pyrénées et S.-O.* 1960.
KIUCHI, S. *Urban geography*, Tokyo, 1951.
KLÖPPER, R. 'Methoden zur Bestimmung der Centralität von Siedlungen', *Geogr. Taschenbuch*, 1953.
KLÖPPER, R. 'Die deutsche geographische Stadt-Umland-Forschung', *Raumforschung und Raumordnung*, 1956.
KLÖPPER, R. 'Der geographische Stadtbegriff', *Geog. Taschenbuch*, 1956–57.
KNOBELSDORF, E. V. 'K. Voproci o Tipar sovetskir gorodov', *Geographicheskii Sbornik*, 1957.
KOLONDY, Y. *Géographie urbaine de la Corse*, Paris, 1961.
KONSTANTINOV, O. A. 'Some conclusions about the geography of cities and the urban population of the USSR, based on the result of the 1959 Census', *Soviet Geography*, 1960.
KORCAK, J. 'La comparaison géographique des grandes villes', *Stuttgarter geographische Studien*, 1957.
KORZYBSKI, S. 'Le peuplement des grandes agglomérations urbaines. Londres et Paris aux XIXe et XXe siècles', *Population*, 1952.
KORZYBSKI, S. 'Le profil de densité de population dans l'étude des zones urbaines de Londres et de Paris', *Urbanisme et Habitations*, 1954.
KÖRBER, J. 'Einzugsbereiche zentraler Orte', *Berichte zur deutschen Länderkunde*, 1956.
KRAAL, J. F. and WERTHEIM, W. F. *The Indonesian town*, Amsterdam, 1958.
KRATZER, P. A. *Das Stadtklima*, Brunswick, 1956.
KUHN, A. 'Stadt-Stadtrand-Umland im amerikanischen Schrifttum', *Raumforschung und Raumordnung*, 1965.
LABASSE, J. *Les Capitaux et la région*, Paris, 1955.
LAMBERT, E. 'La permanence des plans urbains et les souterrains voûtés des anciennes villes de France', *Bull. Ass. Géog. Français*, 1946.

LAMBERT, A. M. 'Millionaire cities, 1955', *Econ. Geog.*, 1956.
LASSERRE, G. 'Villes d'A.E.F.', *Cahiers Outre-Mer*, 1955.
LAVEDAN, P. *Histoire de l'Urbanisme:* I. *Antiquité. Moyen Age*, Paris, 1926.—II. *Renaissance et Temps Modernes*, 1941.—III. *Epoque contemporaine*, 1952.
LAVEDAN, P. *Géographie des villes*, Paris, 2nd edn., 1959.
LAVEDAN, P. 'La reconstruction en Allemagne', *Urbanisme et Habitation*, 1954.
LAVEDAN, P. 'La circulation urbaine', *La Vie urbaine*, 1957.
LAVEDAN, P. *Les Villes françaises*, Paris, 1960.
LEE, R. *The city, Urbanism and urbanisation in major world regions.* Chicago, 1955.
LEFÈVRE, M. A. 'Habitat rural et habitat urbain', *Bull. Soc. Belge Etudes Géog.*, 1933.
LE GUEN, G. 'La structure de la population active des agglomérations françaises de plus de 20,000 hab.', *Ann. Géog.*, 1960.
LE GUEN, G. 'Les structures sociales et économiques des villes bretonnes', *Norois*, 1961.
LEIBBRAND, 'Der Verkehr als Städtegründer', *Geog. Rundschau*, 1959.
LENORT, N. J. *Entwicklungs-planung in Stadtregionen*, Cologne, 1961.
LENZ, W. 'Die Millionenstädte der Erde', *Geog. Rundschau*, 1957.
LINDSTAHL, S. 'A plan for investigation of central places in agricultural communities', in *The I.G.U. Symposium in Urban Geography*, Lund, 1960.
LÖSCH, A. *Die räumliche Ordnung der Wirtschaft*, Jena, 1944.
LOWENTHAL, D. 'Tourists and thermalists', *Geog. Rev.*, 1962.
LYUBOVNYY, V. Y. 'Some questions relating to the formation of urban population', *Soviet Geography*, 1961.
MAJORELLE, J. 'Essai sur la localisation des industries en France', *C.R. Congrès int. Géog.*, Paris, 1931.
MARTIN, R. *La Vie urbaine dans la Grèce ancienne*, Paris, 1956.
MAUERSBERG, H. *Wirtschafts und socialgeschichte zentraleuropäischer Städte in neuerer Zeit. Dargestellt an den Beispielen von Basel, Frankfurt, Hamburg, Hannover and München*, 1960.
MAYER, H. M. and WHITE, G. F. *Changes in urban occupance of flood plains in the United States*, Chicago, 1958.
MAYER, H. M. *Industrial cities excursion guidebook*, Washington, 1952.
MAYER, H. M., and KOHN, C. F. *Readings in urban Geography*, Chicago, 1959.
MAYER, R. 'Methoden zur Bestimmung der Stadtgrenzen', *Zeitschrift für Erdkunde*, 1936.
MAYER, R. 'Der Geographische Stadtbegriff', *Zeitschrift für Erdkunde*, 1943.

MAYFIELD, R. 'Conformations of service and retail activities. An example in lower orders of an urban hierarchy in a lesser developed area' in *The I.G.U. Symposium in Urban Geography*, Lund, 1960.

MECKING, L. 'Die Seehäfen in der geographischen Forschung', *Petermanns Mitt.*, 1930.

MEYNEN, E. and HOFFMANN, F. 'Methoden zur abgrenzung von Stadt und Umland', *Geog. Taschenbuch*, 1954–55.

MINGRET, P. 'Quelques problèmes de l'Europe à travers l'exemple de Liège et de sa région', *Rev. Géog., Lyon*, 1962.

MORGAN, F. W. *Ports and harbours*, London, 1952.

MORGAN, W. T. W. *Nairobi, city and region*, Nairobi, 1967.

MORRILL, R. L. *Migration and the spread and growth of urban settlement*, Lund, 1965.

MOSER, C. A., and SCOTT, W. *British Towns. A statistical study of their social and economic difference*, London, 1961.

MUMFORD, L. *The culture of cities*, London, 1938.

MUMFORD, L. *The city in history*, London, 1961.

MUNDY, J. H. and RIESENBERG, P. *The medieval town*, New York, 1958.

MURPHY, R. *The American city*, New York, 1966.

NAGEL, J. 'Die Industrie, wirtschaftliche Funktion der See und Binnenhäfen', *C.R. Congrès int. Géog.*, Amsterdam, 1938.

NASHIMOTO, S. 'The industry in the city of Karbour', *C.R. Congrès int. Géog.*, Amsterdam, 1938.

NEEF, E. 'Das Problem der Zentralen Orte', *Pet. Mitt.*, 1950.

NEEF, E. 'Die Veränderlichkeit der Zentralen örte mittleren Ranges' in *The I.G.U. Symposium in Urban Geography*, Lund, 1960.

NEFT, D. 'Some aspects of rail commuting, New York, London, Paris', *Geog. Rev*, 1959.

NELSON, H. J. 'A service classification of American cities', *Econ. Geog.*, 1955.

NELSON, H. J. 'Some characteristics of the population of cities in similar service classifications', *Econ. Geog.*, 1957.

NICE, B. 'Entwicklung und Probleme der italienischen Grossstädte' in *The I.G.U. Symposium in Urban Geography*, Lund, 1960.

OBERHUMMER, E. 'Der Stadtplan. Seine entwicklung und geographische Bedeutung', *Verhandl des. 16. deutschen geographer tages*, 1907.

OLBRICHT, K. 'Gedanken zur Entwicklungsgeschichte der Gross-stadt', *Geog. Zeitschr.*, 1930.

OLBRICHT, K. 'Die Grossstäde und ihre Entwicklung im letzten Jahrhundert', *Geog., Wochenschrift*, 1936.

OMAN, I. *Great cities and their surroundings. A statistical delimitation problem*, 1957.

OSBORNE, R. H. 'Changes in the urban population of Poland', *Geography*, 1959.
PALOMAKI, M. 'The functional centres and areas of South-Bothnia, Finland', *Fennia*, 1963.
PAPY, L. 'Etablissements portuaires de la Guyane française', *Cahiers Outre-Mer*, 1955.
PARK, R. E. 'Urbanization as measured by newspaper circulation', *American Journ. of Sociology*, 1929.
PASSARGE, S. *Stadtlandschaften der Erde*, Hamburg, 1930.
PEARSON, M. 'Conurbation Canada', *The Canadian Geographer*, 1961.
PERNOUD, R. *Les Villes marchandes au Moyen Age*, Paris, 1958.
PFEIL, E. 'Gross-stadt forschung', *Veröffentl. der Akademie für Raumforschung und Landesplanung*, 1950.
PHLIPPONNEAU, M. 'Les caractères originaux de la vie rurale de banlieue', *Ann. Géog.*, 1952.
PINCHEMEL, G. and P. 'Les villes nouvelles britanniques', *La Vie urbaine*, 1958–1959.
PINCHEMEL, P., VAKILI, A., GOZZI, J. *Niveaux optima des villes*, Lille, 1959.
POKCHICHEVSKII, 'Nekotorie voproci microgeographitcheskogo izoutchenia gorodov CCCR', *Geographicheskii Sbornik*, 1957.
PIRENNE, H. *Les Villes du Moyen Age*, Brussels, 1927; *Mediaeval cities*, New York, 1925.
PROST, M. A. *La hierarchie des villes*, Paris, 1965.
QUEEN, S. A. and CARPENTER, D. B. *The American city*, New York, 1953.
RATZEL, F. 'Die geographische Lage der grossen Städte' in *Die Gross-stadt*, Dresden, 1903.
RICHARD-MOLARD, 'Villes d'Afrique noire', *France d'Outre-mer*, 1950.
RIEMER, S. *The Modern City*, 3rd edn, New York, 1955.
ROBEQUAIN, C. 'Citadins et ruraux du Gabon et du Moyen-Congo', *Ann. Géog.*, 1956.
ROBINSON, G. W. S. 'British conurbations in 1951', *Soc. Rev.* 1953.
ROBINSON, I. M. *New industrial towns on Canada's resource frontier*, Chicago, 1962.
ROBINSON, K. W. 'Processes and patterns of urbanization in Australia and New Zealand, *New Zealand Geographer*, 1962.
ROBSON, W. A. *Great cities of the world*, 2nd. edn, London, 1957.
ROCHEFORT, M. 'Rôle perturbateur des frontières sur le réseau des petites villes en Alsace', *Ann. Géog.*, 1956.
ROCHEFORT, M. 'Méthodes d'étude des réseaux urbains. Intérêt de l'analyse du secteur tertiaire de la population active', *Ann Géog.*, 1957.
ROCHEFORT, M. 'L'Organisation urbaine de l'Amazonie moyenne', *Bull. Ass. des Géog. français*, 1959.

ROCHEFORT, M. *L'Organisation urbaine de l'Alsace*, Paris, 1960.
ROSTAND, G. and WALDMANN, R. *Le Coût des transports urbains dans les agglomérations*. Vol. 2: *Les transports en commun de la Région Parisienne*, Paris, 1962.
ROSTOVTZEFF, M. *Caravan cities*, Oxford, 1932.
ROUGE, M. F. 'Définition des agglomérations', *Urbanisme*, 1958.
RUBIC, J. 'The geographical limitation surrounding a town', *Geogr. Glasnik*. Zagreb, 1949–1950.
RUST, H. *Heilige städten Peking, Benares, Lhasa, Mekka, Medina, Jerusalem, Rom, Moskau*, Leipzig, 1933.
RYDBERG, H. 'Stadsbygd och Landsbygd', *Svensk. Geog. Arsbok.*, 1937.
SALINARI, M. E. *Bibliografia degla scritti di geografia urbana*, Rome, 1948.
SANTOS, M. 'Quelques problèmes des grandes villes dans les pays sous-développés', *Rev. Géog. Lyon*, 1961.
SCHARLAN, K. 'Moderne Umgestaltungen in Gründrisz iranischer Städte', *Erdkunde*, 1961.
SCHOLARS, D. 'The Indonesian town. Studies in urban sociology', *Selected studies on Indonesia published by the Royal Tropical Institute*, Amsterdam, vol. 4, The Hague, 1958.
SCHÖLLER, P. 'Aufgaben und Probleme der Stadtgeographie', *Erdkunde*, 1953.
SCHÖLLER, P. 'Wachstum und Wandlung japanischer Stadtregionen', *Die Erde*, 1962.
SCHREPFER, H. 'Zur Geographie der Gross-stadt und ihrer Bevölkerung', *Zeitsch. für Erdkunde*, 1944.
SCHULTZE, J. *Stadtforschung und Stadtplannung*, Bremen, 1952.
SCHULTZE, J. H. (ed.) *Zum Problem der Weltstadt*, Berlin, 1959.
SECK, A. 'Dakar', *Cahiers Outre-Mer*, 1961.
SEKLANI, M. 'Villes et campagnes en Tunisie. Evaluations et prévisions', *Population*, 1960.
SELF, P. *Cities in flood*, London, 1961.
SENIOR, D. *The regional city*, London, 1966.
SIDDALL, W. R. 'Wholesale retail trade ratios as indices of urban centrality', *Econ. Geog.*, 1961.
SMAILES, A. E. 'The urban mesh of England and Wales', *Trans. Inst. Brit. Geog.* 1946.
SMAILES, A. E. 'The analysis and delimitation of urban fields', *Geography*, 1947.
SMAILES, A. E. *The geography of towns*, London, 1953.
SMAILES, A. E. 'Some reflections on the geographical description and analysis of townscapes', *Trans. Inst. Brit. Geog.*, 1955.
SMITH, T. L. *Fundamentals of population study*, Chicago, 1960.
SMITH, T. L. *The sociology of urban life*, New York, 1951.

SORET, M. *Démographie et problèmes urbaines en A.E.F.* (*Poto-Poto, Bacongo, Dolisie*), Brazzaville, 1954.
SOULAS, J. 'La naissance et le rôle des capitales', *Revue de Synthèse*, 1941.
SOYER, J. 'Les Bastides et leur parcellaire', *La Vie urbaine*, 1960.
SPELT, J. 'Towns and Umlands', *Econ. Geog.*, 1958.
SPORCK, J. A. 'Le rôle croissant des facteurs humains dans la localisation des industries', *Bull. Soc. Belge Et. Geog.*, 1963.
STANISLAWSKI, D. 'The origin and spread of the grid-pattern town', *Geog. Rev.*, 1946.
TAAFEE, E. *Air transportation and U.S. urban distribution*, 1956.
TAUPIN, J.-L. and VIGNAUD, C. 'Italie du Sud: La Pouille (le cadre urbain)', *La Vie urbaine*, 1958.
TAYLOR, G. 'The seven ages of towns', *Econ. Geog.*, 1945.
TAYLOR, G. *Urban geography: A study of site, evolution, pattern and classification*, London, 1949.
THOMPSON, J. H. 'Urban agriculture in southern Japan', *Econ. Geog.*, 1957.
TRICART, J. *L'Habitat urbain*, Paris, (n.d.).
TRICART, J. 'Contribution à l'étude des structures urbaines', *Rev. Géog. Lyon*, 1954.
TROEDSSON, C. B. *Transportation and City Building*, Göteborg, 1954.
ULLMANN, E. L. 'Trade centers and tributary areas of the Philippines', *Geog. Rev.*, 1960.
VANCE, J. E. 'Labor-shed, employment field and dynamic analysis in urban geography', *Econ. Geog.*, 1960.
VEREKER, C., MAYS, J. B. et al., *Urban redevelopement and social change*, Liverpool, 1960.
VERNON, R. *The changing economic function of the central city*, New York, 1959.
VIET, J. *Les villes nouvelles, éléments d'une bibliographie annotée*, Paris, 1960.
WEBB, J. W. 'Basic concepts in the analysis of small urban centers of Minnesota', *Ann. Assoc. Americ. Geog.*, 1959.
WEBER, A. *Theory of the location of industries*, Chicago, 1929.
WEHRWEIN, C. S. 'The rural-urban fringe', *Econ. Geog.*, 1942.
WEIGEND, C. C. 'Some elements in the study of port geography', *Geog. Rev.*, 1958.
WILHELMY, H. 'Probleme der Gross-stadt Entwicklung in Südamerika', *Geog. Rundschau*, 1958.
WIBBERLEY, G. P. *Agriculture and urban growth*, London, 1959.
WILLIAM-OLSSON, W. 'City bildning och Trafic', *Ymer*, 1952.
WINID, W. 'The problem of geographical law governing the distribution of industry', *C.R. Congrès int. Géog.*, Warsaw, 1934.

WISSINK, G. A. *American cities in perspective*, Assen, 1962.
WOOD, H. A. 'The St Lawrence Seaway and urban Geography, Cornwall-Cardinal, Ont.', *Geog. Rev.*, 1955.
ZAREMBA, P. 'Problèmes de reconstruction des villes en Pologne', *La Vie urbaine*, 1962.

* * *

Symposia, official reports, etc.

Geografia gorodov, *Voproci Geografii*, Moscow, 1956.
Geografia nacelenia, *Voproci Geografii*, Moscow, 1962.
Goroda sputniki, Moscow, 1961.
Land use in an urban environment, *Town Planning Review*, Liverpool, 1961.
London: aspects of change, Centre for Urban Studies, London, 1964.
Phénomène urbain (le). *Rev. de l'action populaire*, 1963.
Planification régionale, U.N.O. 1959.
Problèmes fonciers urbains et politiques d'urbanisme, Bulletin No. 7, U.N.O. New York.
Proceedings of the I.G.U. symposium on urban geography, Lund. 1962.
Seminar on urban problems in Latin America, U.N.O. New York, 1959.
Statistiques démographiques des grandes villes (1946–1951). Inst. Internat. de Statistique. The Hague, 1954.
Tätorter ach Omland, Lund, 1951.
Urbanisation en Amérique latine, UNESCO, 1962.

* * *

See also the sections on urban geography in the volumes issued for the annual conferences of the British Association for the Advancement of Science; especially Birmingham 1950, Merseyside 1953, Sheffield 1956, Glasgow 1958, Manchester 1962, Nottingham 1966.

INDEX

No attempt has been made in this Index to duplicate references given in the Table of Contents, pp. vii–xi. The more important references are set in bold type.

Africa, Central, 229, 236, 416
Africa, North, 83–6
Africa, South, 98–9
Africa, West, 95–8, 229
Agglomerations, 18, 20, 27, 52, 243–4, **256–60**
Airports, 146
Aleppo, 338
Algiers, 229, 264, 272, 308
Amsterdam, 318
Ankara, 87–8, 192, 194, 195, 225
Annecy, 162
Asia, 19, 65–8
Atlanta, 76
Atmospheric pollution, 342–3
Auckland, 71
Australia, 69, 70–2

Bahia (Salvador), 17, 81, 224, 308, 363, 378
Bamako, 96–7
Bangkok, 90
Beaune, 215
Berlin, 227, 268
Bidonvilles, see Shanty towns
Birmingham, 297, 308, **348**, 349, 389
Birth-rate, in towns, 372–5
Bombay, 89, 277, 282, 307, 374
Brasilia, 16, **193**, 215, 288, 349, **354**
Brazil, 8, **13**, 80, 362, **364**, 369, 372, 379, 382
Brunswick, 132
Budapest, 62–3
Buenos Aires, 81, 82

Calcutta, 16, 89, 247, 266, 277, **296**, 360, 382
California, 306, 342
Canada, 73, 359
Canberra, 193, 217
Canton, 92
Cape Town, 266, 270

Capital cities, 190–5, 201, 446–8
Caravan routes, 126, 129–30
Casablanca, 85, 121, 270, **271**
Central Business District (C.B.D.), 299–304
Central Place theory, 427–35 *passim*
Chicago, 74, 75, 132, 270, 276, 282, **289, 291,** 316, **330–31, 357,** 415, 431
China, 90–2, 380, 382
Climate, and towns, 42, 57–8, 67, 68, 152, 342; and water supply, 332
Coal, urban use of, 322, 324
Colliery towns, 153, 309
Colonial towns, 229
Commercial towns, 131–3
Commuting, 9, 300, 302, **313–21** *passim*, 401, 414, **417–25**
Conakry, 379
Conurbations, 243–9
Copenhagen, 45, 245, **257,** 446

Dakar, 17, 98, 134, 270, 358, **362, 365,** 366, 367, 387
Death-rate, in towns, 377–82 *passim*
Denmark, 45, 446
Density of population, 263–76
Disease, in towns, 377–81
Donbass, 64, 244
Dormitory towns, 239, 252
Douai, 198, 389

Electricity, 323

Fairs, 130
Faubourg, 228
Fez, 83, 126, 297
Finland, 44, 45, 46, 402, 403
Food supplies, urban, 325–30, 397–399

467

Fortress towns, 119–20, 198
France, 17, 49, 111, 380, **390–1, 430**
Frankfurt/Main, 112, 133, 247, **409**

Garden cities, 344
Gas, 321–3, 324
Gibraltar, 121
Green belts, 237
Gridiron town-plans, 212–14

Hanover, 132
Helsinki, 47, 192, **198**, 217
Hierarchy, urban, 436–42, 445–6
Holiday resorts, 182–7, 411
Hong Kong, 16
Houses, 304–13 *passim*, 321, 335, 344–54 *passim*
Hyderabad, 269, 362, 366, 374

Iceland, 48
Income, in towns, 382–3
India, 5, 11, 12, 20, **89–90**, 360, **361**, **374**, 377, 382, 387
Industrial towns, **156–66**, 325, **388–389**; function of, 164–6; health in, 381; and labour supplies, 159–60; and localisation of industry, 160–2; population of, 360; and raw materials, 157–8; and water supply, 159
Irkutsk, 67
Isochrones, 412–14
Istanbul, 86, 136, 317

Japan, 20, **92–3, 352–4**
Johannesburg, 150

Kampala, 95
Karachi, 15, 88
Kiruna, 152
Kuopio, 211

Land use, urban, 288–97
La Paz, 79, 81
Le Havre, 121, 122
Le Mans, 3, 156, 298, 313
Leningrad, 420

Léopoldville (Kinshasa), 371
Le Puy, 169, 170, 171
Libreville, 379, 386, 387, 391
Liège, 417, 420
Lille, 244, 279, 328, 330, 377, 384, 389, 390, **422–5**
Linear towns, 216
Łódź, 255
London, 49, 306, 340, 343; City of, 276, **298**; death-rate in, 381; food supplies, 398; Greater, 253, **259, 267–8,** 345; growth of, 8, 227; industries of, 295; population of, 268, 272, 299, 447; water supplies, 333, 336; zones of, 270, 298
Los Angeles, 9, 14, 226, 256, 278, 300, 415
Lourdes, 169, 170, 171
Lyons, 132, 166, 223, 406

Manchester, 245, 308, 348
Mannheim, 244
Market gardens, 326, 338, **398–9**
Market towns, 124
Markets, 292, 329, 402–4,
Marseilles, 222
Mecca, 169, 170
Medina, 83, 85, **297**
Megalopolis, 248, 249–51, 323, 433
Melbourne, 69
Metropolitan cities, 200–3, 446–8
Mexico, 81, 82
Milan, 214, 319, 375
'Millionaire' cities, 202
Mining towns, 52, 75, **80, 149–54,** 381, 386, 388, 391
Montreal, 74
Moscow, 11, 65, 192, 194, 221, 288, 316, 347, 384
Motorways, 314
Mourenx, 151

Naval bases, 120–1
Newspapers, 408–10
'New Towns' (in Great Britain), 52, **252–3,** 344, 345, **349–51,** 366–7, 375, 383

INDEX

New York, 16, 196, 224, 307; commuting to, 299, 300, 301, **302,** 318, **418**; food supplies, 329, 330; growth of, 74; land-use, 281; occupations, 389, **419**; population, 266; transport in, 316, 317, 318; water supplies, 331, 334; zones of, 303, 305, 312
New Zealand, 70–1
Nice, 187–8, 328, 333–4, 361
Norway, 44, 45, 46

Oasis towns, 85
Occupations, urban, 108–13, **385–91**
Open space, 341, 383–4
Osaka, 250
Oslo, 47, 192, 235
Oxford, 199, 200, 313

Paris, 8, 226, 227, 235, 259, 263, 306, 381, 384; area of, 268–9, 277–8; atmospheric pollution, 343; as capital, 191, 195, 201, 447–8; commuting, 299, 318, 421; food supplies, **327–8,** 329, 330, 398; fuel supplies, 322, 324; industries, 148, 297; land values, 281–2; open space, 341–2; population, 3, 16, 200, **264, 265,** 271, 272, 285, **295,** 364; rebuilding of, 219–20; sewage and refuse disposal, 337, **339–40;** shops, 294, 304; slums, 307, 348; socio-economic structure, **311–12;** suburbs, 249, 257; town planning, 345, 348, 351; transport, 315, 316, 317; university, 171, 172–3; water supplies, 331–2, 333, 336; zones of, **258,** 310
Pekin, 91
Pilgrimage towns, 168–71
Poland, 375
Ports, 133–46; entrepôt, 147; estuarine, 137–9; fishing, 141; and hinterlands, 137, 140, 142; inland, 139–40; location of, 134–7; and urban development, 142–6

Prague, 62
Pretoria, 286

Quebec, 74

Railways, and urban growth, 3, 66, 74–5, **127–9,** 225, 226, 230, 282, 315–21 *passim*
Recife, 13, 379
Refuse disposal, 338–40
Regions, urban, 435–6
Reykjavik, 48, 190, 327,399
Rio de Janeiro, 190, 192, **223,** 227, 284, 308, 311, 314, 335, 340, 379
Rivers, and towns, 126–7
Rome, 55, 56, 228
Roubaix-Tourcoing, 292–3, 311, 377, 389
Ruhr, The, 244, 250

St Louis (Senegal), 97
Sanatorium towns, 179–80
San Francisco, 213, 224
São Paulo, 16, 82, 269, **270–1, 273,** 283, 287, **301,** 323, 364, 379
Satellite towns, 236, **252–5**
Schefferville, 152
Seaside towns, 182–4
Sewage, 337–8
Sex-ratio, in towns, 358–62, 370–1
Shanghai, 92, 382
Shanty-towns, 82, **227–8,** 271, 272, **308,** 381
Siberia, 66–7
Singapore, 266
Skyscrapers, 73, 77, 287
Spa towns, 180–1
Stockholm, 43, 44, 222, 223, 254
Strasbourg, 266, **275,** 309
Suburbs, 9, 11, 233, **238–42**
Sweden, 42, 45, 46, 403–4
Sydney, 11, 69, 72, 313
Szczecin, 143

Tashkent, 66
Tel Aviv, 15, 88
Texas, 150

Tokyo, 93, 226, 269, **274,** 297, 326, 328, 329, 389
Tourist towns, 51, 58–9
Town-planning, 218–20, 233–7, 343–54; in Great Britain, 344, 345, 349–51
Transport, urban, 313–21, 417–25 *passim*
Turin, 15
Turku, 47, 402, 403

Ukraine, 64
United Kingdom, 20, 49, 50
University towns, 171–5, 408
Uppsala, 171, 172, 173, 199, 200
Urals, towns of, 64
Urban population, percentage of, 5, 19, 77, 84, 92, 94, 95, 356
Urbanisation, cycles of, 50–1

U.S.A. 9, 12, 13, 17, 20, 318, 326, **334,** 336, 344, 356, 359, 363, 364, **365,** 367, 368, **371,** 372, **373, 382,** 389, 392
U.S.S.R. 5, 10, 14, 20, **61–8,** 319, 320, **346,** 383

Vatican City, 196
Venice, 177
Vichy, 181
Vienna, 231–2, 236, 239, 333
Volga R, and towns, 63

Warsaw, 59, **219,** 339, **351–2**
Water supplies, 331–7, 341
Wellington (N.Z.), 71

Yakutsk, 67

Zones, industrial, 292–7; residential, 304–13; urban, 289–97, 302–3
Zones of influence, 425–45 *passim*